SHADOWS
IN THE WOODS

SHADOWS
IN THE WOODS

A Chronicle of Bigfoot in Maine

DANIEL S. GREEN

COACHWHIP PUBLICATIONS

Greenville, Ohio

ISBN 1-61646-284-1
ISBN-13 978-1-61646-284-0

Cover design by Holly Green; Woodland path © Sara Winter

CoachwhipBooks.com

CONTENTS

Introduction	7
Chapter 1	13
Chapter 2	19
Chapter 3	39
Chapter 4	85
Chapter 5	116
Chapter 6	198
Chapter 7	233
Chapter 8	258
Chapter 9	342
Chapter 10	356
Chapter 11	370
Chapter 12	439
Epilogue	446
Bibliography	447
Images	466
Footnotes	470

INTRODUCTION

Let us inquire just what is really going on right
under our noses, and then compare the results first
with the current popular concept of what is what,
and secondly with reality, which we will discover is
something quite else again. (Ivan T. Sanderson)[1]

Reading old Maine newspapers is a hobby that indulges my passion for Maine history. The Pine Tree State enjoys a rich tradition of fine newspapers—*The Bath Independent, Biddeford Journal, Ellsworth American, Bangor Whig & Courier* and *Portland Press Herald* among them. I confess to having a favorite, *The Lewiston Evening Journal*. Writers like Holman Day, L. C. Bateman, Sam Conner, and Edith Labbie brought a wonderful, folksy style to its pages and the Saturday appearance of the *Journal's Magazine* was eagerly read by generations in the south and central parts of the state. Today these old dailies and weeklies are treasure troves of information and replete with insights into times long gone.

It was from old newspapers I learned something I would come to attach importance to, an oft-made observation related to cougars. Not that eastern panthers once lived in Maine, I had already known that. But rather, descriptions of the big cat's vocal repertoire included evocative, anthropomorphic terms—haunting screams, child- or woman-like crying, and desolate wailing.[2]

Every time I would encounter this imagery from a hunting anecdote or a camping narrative it would give me pause. I read

7

these newspaper stories well into adulthood . . . blessed with a wonderful family, a job with Uncle Sam, and as often as possible, indulging my aforementioned interest in Maine's history. But I found myself transported by these stories to a moment in my past when as a gangly teenager I heard something I will never forget.

My home is nestled at the head of a bay on land that has been in my family for seven generations. The property is surrounded by mid-growth woods reclaiming sloping pastures, bordered by two streams and close by several low boggy areas. To say we get blitzed by clouds of black flies and mosquitoes in the warmer months is a decided understatement. Almost 25 years ago as of this writing, I leashed Missy, the family's Lhasa Apso/beagle mix, and headed out for what I thought would be a substantial walk, perhaps to the Maine Central tracks nestled close to the base of Schoodic Mountain. We were only about two hundred yards from the front door when something demonstrated it was sharing the idyllic day with me and my dog. Missy and I had gone most of the way up a gently ascending trail on the western edge of several acres of blueberry fields. We had nearly attained the top of the ridge where the carpet of blueberry bushes ran up to a mixed wood line.

That's when two powerful cries came from the woods about a couple hundred feet away.

Aoorwooooah! Aoorwooooah!

Missy gave a not-too-loud woof as both our heads snapped toward the sounds. The picture that instantly formed in my mind was of something ape or monkey-like—specifically a chimpanzee. There was a crashing through the tangle of brush and then, quickly, just stillness. I saw nothing of what made the calls, which I stress again were loud and powerful. I was fully of the impression that they were made in response to my presence; the cries conveyed *annoyance.* In the moment, as whatever-it-was rapidly retreated into the woods, the process of trying to reconcile the sounds with something I could positively identify had begun.

Foxes, coyotes and dogs do not, to the best of my knowledge, make a cry in any way similar to what I heard that day. Caribou are lamentably gone from Maine environs, but there are plenty of deer, black bears, even the occasional moose traversing our part of the coast . . . but none of these animals seemed obvious candidates. Eastern panthers are not supposed to be in Maine, though the debate over their presence in the Pine Tree State was argued in the 1980s as much as in the present day. As a teenager I wouldn't have known mountain lions made sounds that many had compared to human-like screams. That's why a cougar's screams, so-characterized, became a point of interest to me as an adult. Despite the official word that cougars remain absent from Maine, I was at last in possession of a seemingly plausible (and frankly reassuring) animal that called out from the woods that day. Reassuring in the sense that for a long, long time I had been left with the unpleasant conclusion I may have had an encounter with a Bigfoot.

> We consider it a good principle to explain the phenomena by the simplest hypothesis possible. (Ptolemy, attributed by James Franklin)[3]

Similar to Ockham's razor, Ptolemy's comment on reasoning advocates a simple answer as likely the best for any case. Identifying the agent for an unknown sound in the woods would generate candidates, even if unexpected, mundane. Mundane like you can thumb through a wildlife text and find the appropriate section of interest with associated picture, behavioral characteristics, and general milieu information. A large, hairy hominid, unrecognized by science, is not so accessible or accepted as, say, a black bear or a moose. Large, hairy hominids after all are, in the minds of many, best relegated to ironic or comedic advertising for such fare as beef jerky. In a Bigfoot encounter one is left with the incongruence of what they have subjectively observed juxtaposed with the absence of an elite-opinion and properly sanctioned template.

The idea of Bigfoot encompasses those interested and disinterested, and shared information that is serious and non-serious—a true marriage of science and entertainment. A history and study of the subject reveals an object lesson in the near-biological behavior of information flow; the reflection and continuation (and evolution) of information echoing within the cultural body. With respect to Maine, this series of Bigfoot tales, stories, and reports began with the Native Americans living here prior to European colonization and continues with modern encounters into the present day. It is a subject that appeals to an appreciation of wonder and mystery. Simply put, Bigfoot is fantastic in its permutation through our cultural tapestry and is indeed a phenomenon.

A loose comparison may be drawn between Bigfoot and the New England Sea Serpent, the latter a meme largely, though not exclusively, confined to the 1800s and early twentieth century. Here we find a subject whose treatment ranged from tongue-in-cheek humor to open ridicule by newspapers' editors—a style of approach that sounds similar to Bigfoot's portrayal today. Consider the following serpent snippets from the *Lewiston Evening Journal* covering a period of more than twenty-five years:

> 1879: "The most striking fact about sea-serpents is their tendency to appear when news is dull and the reporters and correspondents of the newspapers are decidedly in need of something startling. The discrimination shown by these animals in this particular is very extraordinary, considering their low type of organization, and the probability that they are totally unacquainted with the usages of modern civilization."[4]

> 1900: The sea serpent has arrived and the summer season may be considered as officially opened. Two South Portland fishermen got the first sight of his snakeship, Saturday, while on the way to Witch Rock.[5]

> 1905: The sea serpent appeared Wednesday afternoon off Willard beach, South Portland, and while the sight was terrifying to the uninitiated, the wiser ones danced up and down gleefully aware of the fact

that henceforth, at least for a time, the eyes of the populace would be directed to the place that was supposed to have received a respectable funeral. Several strollers on the beach caught sight of the serpent at one and the same time and all eyes were riveted on his movements.[6]

The Sea Serpent

What, one may wonder now, would people a hundred years ago be gaining through sharing their accounts of a creature fitting the description of the sea serpent? The list of reputational benefits is apparently a short one, though the last snippet implies there could be a commercial boost enjoyed from serpent sightings.

The claim for Bigfoot's existence requires evidence to support it. Evidence, as distinguished from opinion or belief, is considered to be of factual nature and supportive of a premise or claim. This book is not presenting material as evidence for Bigfoot in Maine and is not intended to convince nor cast doubts in regards to the existence of Bigfoot. Though I would argue that the bar for accepting evidence should be high and the efforts to substantiate a large cryptid should be appropriately scrutinized, these discussions are best left to the auspices of other sources. The present volume is intended to chronicle Bigfoot in Maine in both past and modern contexts. It is hoped the reader will find in this work a useful and informing collection of sightings, stories, and legends related to Maine's Bigfoot as well as the closely associated concept of the Wild Man.

I am sympathetic to the belief held by many in the reality of a large, hairy hominid sharing our world with us. I think that if indeed such creatures exist, they are flesh-and-blood denizens of our wild lands and forests, and not inter-dimensional travelers capable of defying our conceptions of time and space. Bigfoot's supposed reality, though unexpected when encountered and spectacular in its rarity, is nonetheless rooted in mundane corporeality.

For my part, while conducting the research presented herein, I discovered the holistic tableau of Bigfoot in Maine to be richer far beyond my initial expectations. I was surprised at the number of Wild Man sightings reported in newspapers and given pause by the presence of mysterious *somethings*, seemingly like a Bigfoot, in period literature set in Maine environs. To delve further into my own stance, I refer back to the personal anecdote at the start of this introduction—though I do not know what I encountered on that knoll, I am very confident with respect to what it wasn't. But I find myself reluctant to discount possibilities outside the comforting and cherished mundane. This reluctance is as much recognition of other's experiences and convictions as it is my own unwillingness to yield up a certain sense of wonder.

CHRONICLING BIGFOOT IN MAINE—AN APPROACH

Some people romantically perceive New England as a Norman Rockwell painting: a spirited and sentimentalized canvas of folksy charm, dry wit, and self-reliance. But Maine has an undercurrent of things *passing strange*[1]—strange experiences told and retold as long as people have been here.

This book's aim is to chronicle Bigfoot within the state of Maine. Presented herein are Native traditions and legends, modern Bigfoot sightings, along with associated stories and Bigfoot-related folklore. With the intent of not overlooking perhaps distally related topics, the Wild Man in Maine, both as a source of terror haunting the dark backwoods and as entertainment ensconced in fairgrounds and sideshows, is included. Similarly, Bigfoot in Maine's popular culture, important for the transmission of ideas and reinforcement and shaping of concepts, will be examined.

John Green's seminal work, *Sasquatch: The Apes Among Us* (1978) lists over 75 regional names for Bigfoot, including Haskell Rascal, (Texas), Big Mo (Illinois), and the Spottsville Monster (Kentucky). As will be explored in later detail, Maine has its own special colloquialisms for Bigfoot and Bigfoot-like entities, including the Wendigo, Yoho, Maine Ridge Monster, Osgood the Ape, Durham Gorilla, Leach's Point Devil, Leeds Loki, Meddybumps Howler, and the Wild Man of Katahdin.

A succinct depiction of Bigfoot (also widely known as Sasquatch) is given from the following description by researcher Paul Bartholomew: "Red glowing eyes, about 7 feet tall, making a sound like a pig squealing or a woman screaming and walking like a man but looking more like a gorilla."[2] Dr. Jeffrey Meldrum writes, "It appears that the inferred Sasquatch anatomical and behavioral profile is indeed a very real approximation of a North American Great Ape."[3] Physically, Bigfoot resembles man. It is inferred from witnesses' reports that the majority of encounters are with apparent adult specimens. (Sightings of juveniles and young, though relatively uncommon, do occur.) A prevailing belief held by many researchers is that Bigfoot is highly intelligent and capable of reasoning.

The *Sasquatch Field Guide* (2013) by Dr. Meldrum adds distinguishing features of hidden ears, flat face, and common hair colors of black, brown, blond, and white.[4] These creatures are reported to be large—often between 7-10 feet tall (some even taller), possessing a muscular build (even while sometimes sporting a protruding, gorilla-like belly), very long arms, hair over most of its body, and a head set into shoulders with little hint of the neck. Dr. John Bindernagel says: "There is a great deal of consistency" in Bigfoot reports. "Most of which appear to be adult males. The large hair-covered, broad shouldered, massive-chested animal, with a short, thick neck, often described as no neck, long arms reaching below the knee, a face with a flat nose, deep-set eyes, often brow ridges, sometimes a conical head . . . pretty consistent."[5]

Burgeoning interest in Bigfoot during the early 1960s led to the emergence of dedicated, first-generation Bigfoot researchers. Their investigative efforts uncovered a deep historical record of large, hairy, hominid creatures appearing across North America supported by Native American stories and newspaper accounts. John Green, Roger Patterson and others published some of these early accounts, revealing a rich history and continuity of the subject and showing Bigfoot in North America was not a new story. It should be noted that as with the sea serpent, some historical newspaper accounts of Wild Men and hairy *wazzits* may have resulted from a playful editor's penchant for tall tales or indulgence in outright fancies for the sake of entertainment. In such an environment ranging from humor to outright derision, authentic sightings of something unexpected and unknown could result in misidentifications influenced by, and conflated with, the—*wink, wink*—local Wild Man legend.

A very prominent and relatively early depiction of Bigfoot (Sasquatch), "a strange race of hairy giants."

Perhaps the earliest mention made in Maine of a Bigfoot-like creature comes from the *Bangor Daily Whig and Courier*. Though not reporting on an incident occurring within the state, it is relevant for demonstrating the spread of these "news" stories throughout the country. The *Whig and Courier's* position is that the story was "designed to tickle the ears," rather than a serious report.

> The following curious tale, evidently designed to tickle the ears of the marvelous, is copied from the *Montrose Spectator*.
> The Whistling Wild Boy of the Woods—Something like a year ago there was considered talk about

a strange animal, said to have been seen in the south-western part of Bridgewater. Although the individual who described the animal persisted in declaring that he had seen it, and was at first considerably fright-ened at it, the story was heard and looked upon more as food for the marvelous, than as having any foun-dation in fact. He represented the animal, as we have it through a third person, as having the appearance of a child seven or eight years old, though somewhat slimmer, and covered entirely with hair. He saw it, while picking berries, walking toward him erect, and whistling like a person. After recovering from the fright, he is said to have pursued it, but it ran off with such speed, whistling as it went, that he could not catch it.

The same or a similar looking animal was seen in Silver Lake Township, about two weeks since, by a boy some sixteen years old. We had the story from the father of the boy in his absence, and afterward from the boy himself.

The boy was sent to work in the back woods near the New York State line. He took with him a gun and was told by his father to shoot anything he might see except persons or cattle.

After working a while, he heard some person, a little brother he supposed, coming towards him whistling quite merrily. It came within a few rods of him, and stopped. He said it looked like a human being, covered with black hair, about the size of his brother, who was six or seven years old. His gun was some little distance off and he was very much fright-ened.

He, however, got his gun, and shot at the ani-mal, but trembled so that he could not hold it still. The strange animal, just as his gun went off, stepped

behind a tree, and then ran off, whistling as before.
The father said the boy came home very much frightened, and that a number of times during the afternoon, when thinking about the animal he had seen
he would, to use his own words, 'burst out a crying.'

Making due allowance for frights and consequent
exaggeration, an animal of singular appearance had
doubtless been seen. What it is, or whence it came,
is of course yet a mystery. From the description, if
an ourang outang were known to be in the country,
we might think this to be it. As no such animal is
known (without vouching for the correctness of the
story,) we shall leave the reader to conjecture, or
guess for himself, what it is. For the sake of a name,
however, we will call the "strange animal" *The Whistling Wild Boy of the Woods*[6]

A key difference between a Bigfoot sighting today and, say, an
historical encounter in late eighteenth-century rural Maine, is that
in the present moment a dominant popular conception of Bigfoot
permeates our culture. Millions of Americans with no vested interest in the subject have been exposed to segments of the 1967
Patterson-Gimlin film. Bigfoot kindly hawks a variety of products
and merchandise, and even serves as the mascot for sports teams
and Olympic Games. Searching for Bigfoot across America has been
the subject of popular TV programs.[7] Online sites dedicated to the
Bigfoot phenomenon, such as the Bigfoot Field Researchers Organization (BFRO), compile and aggregate sightings by state, including current accounts of Bigfoot encounters in Maine. Those interested may also easily access videos of purported Bigfoot creatures—
many derisively called "blobsquatches." For example, take a certain video, purportedly taken in Maine called the Maine Tree Creature. Hailed at the time of its release as the first blobsquatch of
2010, this particular video admittedly gives the impression of a
hunched over, ape-like creature staring back at the camera. But

the general conclusion was the Maine Tree Creature was actually a porcupine far up in a tree (see chapter 8 on Bigfoot Encounters)— a good example of a blobsquatch.

Dr. Esteban Sarmiento, mammologist and anatomist, says, "To me I see Bigfoot as a problem. There are thousands of people who have claimed to have seen it. And there is no physical evidence for it. So either the animal is elusive or it does not exist."[8] Experts contend it is very unlikely that there are any new large mammals to be discovered—especially in North America.[9] But take Dr. Bindernagel's opinion that a large hairy hominid with appropriate population size to propagate the species has in fact already been discovered, but not accepted by science: "This alternative hypothesis regards the sasquatch as extant primarily on the basis of its tracks, by now well documented in track photographs and as track casts. In addition, however, eyewitness descriptions of sasquatch anatomy and behavior, rather than being ignored or discounted as invalid, are shown to warrant scientific attention, especially since they compare favorably with descriptions of the anatomy and behavior of well-studied mammals known to occur elsewhere in the world—the great apes."[10]

This book includes separate chapters on Wild Man and Bigfoot sightings occurring within Maine. When the classification of an account as a Wild Man or Bigfoot sighting was unclear, or its categorization as folklore or ambiguous tale appeared shadowy, your author's determination and opinion between a hairy hominid, a human returned to a wild state, or something simply inexplicable, was the arbiter on the present volume's arrangement. Between Bigfoot, Wild Man, and other humanoid *wazzits*, there are well over 100 accounts spanning centuries recorded in later chapters demonstrating the breadth and volume of mysterious, human, or human-like creatures encountered within the Pine Tree State.

2

MAINE GEOGRAPHY, CLIMATE, AND LARGE FAUNA

Understanding an animal's biological milieu helps us understand the creature itself. Even legends can come to life if we immerse, even briefly, in their environmental context. Otherwise opaque relationships and associations between the subject of interest and unexpected variables can become transparent, while conclusions never dreamed of at a journey's inception are found welcoming one at trail's end. This chapter presents the briefest introduction to the Pine Tree State's geography, weather, and her larger fauna. Building the landscape thusly will serve as a backdrop for following chapters' Bigfoot sightings and related stories.

GEOGRAPHY

Maine remains a state with a sparse population compared to its overall size. Much of the population is focused in the coastal areas and along major river ways. Take the following quote from Henry David Thoreau's *The Maine Woods* (1864), describing the wildness he discovered for himself:

> The mountainous region of the State of Maine stretches from near the White Mountains, northeasterly one hundred and sixty miles, to the head of the Aroostook River, and is about sixty miles wide. The wild or unsettled portion is far more extensive. So that some hours only of travel in this direction will

carry the curious to the verge of a primitive forest, more interesting, perhaps, on all accounts, than they would reach by going a thousand miles westward.[1]

In the early twenty-first century, the central and northern portions of the state remain, as during Thoreau's 1846 trek through the Maine wilderness, sparsely populated[2] and thickly forested. Maine woodlands have steadily increased during the latter twentieth century as more farmland has returned to a wild state. Drawing again upon *The Maine Woods*, written just prior to the rapid growth of Maine's hunting and sportsmen industry and the popularity of the coastal regions for vacationing, we find another mid-eighteenth century glimpse of Maine portrayed as a place almost exotic:

> There were very few houses along the road, yet they did not altogether fail, as if the law by which men are dispersed over the globe were a very stringent one, and not to be resisted with impunity or for slight reasons. There were even the germs of one or two villages just beginning to expand. The beauty of the road itself was remarkable. The various evergreens, many of which are rare with us,—delicate and beautiful specimens of the larch, arbor-vitae, ball-spruce, and fir-balsam, from a few inches to many feet in height,—lined its sides, in some places like a long, front yard, springing up from the smooth grass-plots which uninterruptedly border it, and are made fertile by its wash; while it was but a step on either hand to the grim, untrodden wilderness, whose tangled labyrinth of living, fallen, and decaying trees only the deer and moose, the bear and wolf, can easily penetrate.[3]

Maine is comprised of over 35,000 square miles, most of which is forest and wilderness.[4] One third of the land in the United States is covered by forests, making forestland the number one type of land in the country. Alaska has the most aggregate forestland but Maine has the highest percentage of land area that is timberland—86%.[5]

CLIMATE

Maine's overall moderate weather conditions have been called invigorating by many and allow the state to enjoy a four-season vacation industry. The warmest month statewide is July and the coldest month is January. Maine's climate is divided into three divisions by the National Weather Service: Coastal, Southern Interior, and Northern Interior.

The Coastal Division extends from the ocean inland about twenty miles the entire length of the coastline. This area is impacted by its proximity to the ocean, resulting in milder winter and cooler summer temperatures. Average annual snowfall is 50-70 inches in the Coastal division.

The Southern Interior Division encompasses about the southern third of the state. In the summer months, the Southern Interior is the warmest division. During very warm summers, temperatures may exceed 90°F for up to 25 days. This southernmost division sees up to 60-90 inches of snowfall per winter.

The Northern Interior Division contains the Maine highlands and northern reaches of the state and contains some of the highest elevations.[6] The northern area of Aroostook County experiences warm summers and some of the coldest and highest snowfall totals in the eastern half of the country. Average annual snowfalls can range between 90-110 inches, the highest average in the state.

Maine's variety of climate means the state enjoys a large range of environments and is a reason why many visitors and residents find the statewide moderate weather to be appealing. "It is also the reason that so many plants and animals reach the northern or southern edge of their range in Maine as well as in Maine's marine waters."[7]

FAUNA

The following account by naturalist and author Tom Seymour of an unusual animal haunting the wilds of Waldo County contrasts the strange qualities of the "whatever it was" with recognizable traits of known Maine fauna.

Of all the animals that go bump in the night, there is nothing that can't be rationally explained. So the books say. Back in 1983, in the part of Waldo that I live in, something went bump in the night, and nobody has the slightest idea what it could have been. Some common animals make very frightening sounds. Porcupines make an extremely odd noise at night. Imaginations can run rampant when a fox catches and kills a snowshoe hare. The hare's screams and the final death rattle are enough to make the uninitiated look to a firearm for protection. An experienced woodsman knows what a dying hare sounds like. He also knows what a porcupine sounds like, and even what a rowdy bear sounds like.

Bobcats are not uncommon in our part of the country, and the high-pitched cry of a hunting bobcat can stir the emotions. The sound of a bobcat is identifiable, as are most animal sounds of the night. Who can mistake the yapping and howling of a family of coyotes?

The animal that haunted Waldo back in 1983 couldn't be confused with bears, porcupines, coyotes or bobcats. A half-dozen people heard the thing, and none of them were willing to go back in the woods at night with a camera to make a documentation. It was a big animal, as witnessed by the snapping and cracklings of large boughs and branches that it would break within human hearing. Its call echoed for a far distance. It had a complex vocabulary, and one of its most bizarre sounds was a loud "whoosing," like a giant exhaling.

One night when whatever it was had been vocalizing in a big way, the County Sheriff was summoned by a local resident to "rescue the woman who was being beaten." The resident had heard the animal and assumed that some foul play had taken place. The sheriff, of course, could find no crime being committed. What kind of animal was it? I longed to know.[8]

Below are condensed descriptions and details on Maine's larger animals, whose appearance and vocalizations could possibly be misidentified as something unusual by observers unfamiliar with them. Since a common form of evidence ascribed to Bigfoot is footprints left in malleable substrate, representative track silhouettes are provided for each animal.

BLACK BEAR

Of the three species of bears inhabiting North America (including brown and polar), the black bear is the smallest and the only one to reside within the state of Maine. According to Maine Inland Fisheries and Wildlife, adult black bears can weigh in the range of 250-600 pounds and measure 5-6 feet from nose to tail. Black bears walk flat-footed resulting in five-toed tracks that may give the impression of human footprints. The state bear population is between 24,000 and 36,000. Full adult size is reached at about 5 years in age. Starvation claims many cubs, but those bears that reach 2 years in age enjoy a survivability rate of 90% with human-related incidents becoming the major cause of death. Black bears lead predominantly solitary lives with occasional groupings at food sources (municipality dumps have been a common area for such gatherings). Bears will remain outside their dens in late fall as long as there are abundant food sources. Maine Inland Fisheries cites years of good beechnut crops as an important determinant for bears entering their dens. Bears may enter dens when snow makes travel difficult and may remain in their dens between 5-6 months. Bears can easily travel several miles in a day. Females tend to stay within a 6-9 mile range of their mother, while males disperse as subadults when 1-4 years old.[9]

Peter Byrne researched Bigfoot during the 1970s and wrote an article on Bigfoot (using the Native American term, *Omah*) for the *Explorers Journal*, in which he relates inferred dietary habits of Bigfoot to being very similar to that of a bear:

> The creatures, presuming they do exist would almost certainly be omnivores. For an omnivore's eating habits one has only to look to the bear. His diet,

American Black Bear

Black Bear Track

which could also be the diet of an *Omah*, consists of berries, shoots, fish and small animals like rodents, birds' eggs, reptiles, frogs, carrion, and even grass. According to witnesses who claim to have watched *Omah*, they eat all of these things.[10]

MOOSE

The eastern or Taiga moose is the subspecies found in the state of Maine. Large bulls can weigh well over 1,000 pounds but typically weigh within the range of 600-800 lbs. An account of a 2,500-pound moose occurred in the November 5, 1899, edition of the *New York Times*. A Gilman Brown of Newbury, Massachusetts, was one of the hunters who encountered the giant moose. He described the beast as gray in color, 15 feet high, and sporting an antler spread of 12 feet.[11] The largest of Maine fauna has not surprisingly entered Maine folklore and legend, the Specter Moose being perhaps the most well-known example. Adult moose can be nine feet in length and stand 6 feet at the shoulder. With an estimated population of 76,000, there are believed to be more moose in the state of Maine than black bear. Large racks of antlers can approach five feet in width and the front hoof width of adult males is about five inches. Moose subsist on browsing leaves and twigs, including ash, maple, birch, and cherry, as well as incorporating aquatic plants into their diet. Bulls will typically be at higher elevations than females, who prefer food-rich lower elevations, e.g. regenerating stands following logging operations.[12]

Moose

White-tailed Deer

WHITE-TAILED DEER

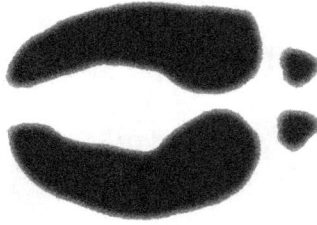

Maine bucks attain weight up to 400 lbs., though the normal range for their weight is between 200 to 300 lbs. Does typically weigh between 120 to 175 lbs. Deer attain a running speed of 40 mph. One of a deer's loudest sounds is a loud whoosh made when startled. Deer produce two coats of hair every year, each adapted to the present climate. A fine pelage is grown in late spring, which is replaced in September by molting to thicker wooly hair under a layer of guard hairs. Deer store fat for the winter months and this reserve can reach up to 25% of a deer's body weight. According to Inland Fisheries and Wildlife, up to 94% of the state is considered deer habitat—from reverting/active farmlands to wetlands and forests. Deer feed on several hundred species of plants, selecting the most nutritious plants at hand. Deer have typical ranges of 500 or so acres. The population of whitetails in Maine is estimated to be over 250,000, for a ratio of 1 deer to every 6 humans.[13]

CARIBOU

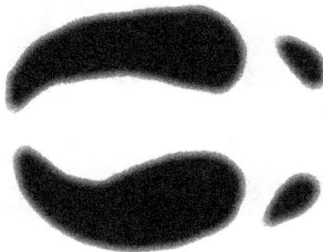

The caribou represent a large species that disappeared relatively recently from the state of Maine. Their mention here is to amplify the critical nature of changing conditions' impact to an animal's milieu and the consequent sensitivity on the part of specific animal populations to that change. It is often given that a host of factors played out in the loss of caribou to Maine. Forestry operations opened up new territory for whitetail deer, which took with them a parasite benign to them but deadly to caribou. Logging also likely had a deleterious impact on the caribou's staple food of lichens and mosses. Hunting, abetted by loggers' opening of new corridors and means to reach caribou, is inferred to have also exerted a significant downward pressure on herds.

At one time the law in Maine read, "three deer, two moose, one caribou," suggesting the relative scarcity of these animals within the state. The following from the *Bangor Commercial* offered a glimmer of hope that herd sizes were stabilized, if not growing toward the end of the nineteenth century.

> Deer and Caribou in Maine (from the Bangor Commercial, Jan 20.) Mr. H. O. Stanley, one of the Fish Commissioners, says that deer and caribou are now so plenty in Maine that if the State should permit sportsmen to kill all they can by still hunting the open months there would be plenty of the game for many years to come. The Commissioner of Fish and Game significantly adds, however, that a sportsman may consider himself lucky if he gets three deer in a season—the limit now fixed by law."[26]

But the aggregate pressures on the caribou resulted in unanimous acceptance of their dwindling numbers. Cornelia "Fly Rod" Crosby, a Maine sportswoman of fame, received the distinction and credit for bagging the last caribou to be shot in Maine in 1896.[27]

Early into the twentieth century sporadic reports were being received of caribou sightings. A *New York Times* article from 1912 reported John S. P. H. Wilson, chairman of the Maine Fish and

Caribou

Florida Panther

Game Commission, had received word from Maine guides and lumbermen of a small herd of caribou in the Woolastook region of northwestern Maine.

The article goes on to say that, "To be sure this big game will never be as plentiful as they once were, even with the best of protection that the Maine Legislature can give and game wardens enforce. The wilderness route is narrowed down much since the exploitation of that part of Canada which was formerly the path of the caribou. Small farms and villages do not appeal to this animal and whatever the reasons the big game quit Maine—whether an epidemic disease, forest fires or, as some woodsmen say, the buck deer, drove the caribou away, or was it lack of good feeding grounds? The animal will never be killed in numbers by sportsmen as in our father's day."[28]

Perhaps the last report of a caribou sighted in Maine was reported in the January 7, 1930, edition of the *Lewiston Daily Sun*.

> Bull Caribou Spied in Moxie Lake Region—Augusta, January 6, (AP)—A report that a bull caribou had been seen by a woodsman in Square Town, near Lake Moxie, Somerset county [sic], today recalled the migration of more than 30 years ago when the plentiful herds left Maine in a single year. No authentic report of a caribou being seen in the State has been recorded since that time.
>
> Today's report came from the commissioner of inland fisheries and game [sic], George J. Stobie, who said an acquaintance of his, Henry Forsythe of Oakland, had seen the animal in the Lake Moxie region. Forsythe is an experienced woodsman and claimed to have had a close view of the caribou, Commissioner Stobie said. Game wardens will investigate.
>
> Prior to the migration, the deer-like creatures roamed the northern woods and were slaughtered in great numbers by hunters. Old timers told of killing as many as 15 in a single herd. The animals were

curious and could be lured easily within range of a gun. When startled in the winter, they sometimes would mill around until the scent of the hunter came to them, thus becoming a target for shots.

Reasons given for the migration, in which herd upon herd crossed the St. John river [sic] into New Brunswick, were many. Some said a shortage of caribou moss was responsible. Kenneth Lee, Maine naturalist and writer, blamed an influx of white tailed deer. He was of the opinion that if one caribou were sighted in Maine, there was a small herd in the vicinity. The animals are great travelers, but they almost never travel alone, he said.[29]

Two failed efforts to bring caribou back to Maine have occurred, one in late 1963 and the second beginning in 1986. Unlike the gray wolf, there are no modern reports of caribou in Maine.

Brainworm, over-hunting, logging, predators, and other environmental conditions are important for understanding and arriving at an explanation for how caribou disappeared and why their reestablishment has proven, so far, impossible. This example of a large fauna's disappearance from the state reveals the importance of the environmental milieu and the myriad ways which natural conditions effect a creature's survival.

EASTERN PANTHER

Eastern panthers are not officially recognized by the State as an extant species of fauna, a conclusion largely derived from lack of physical evidence, including a total lack of tracks/sign found during winter track surveys for wolf and lynx in northern Maine. But sightings continue to persist—some accompanied by photographic evidence, and some reports are made by game wardens and biologists employed by the State. The last cougar confirmed killed in Maine was in 1938. Adult males weigh up to 200 lbs. and up to eight feet in length, including the tail. Track width is 3-4 inches. Cougar vocalizations include what has often been referred to as a human-like scream. Adults require large ranges of undeveloped forest, grassland, and alpine habitats. White-tailed deer count among their more important prey.[14]

Canada Lynx

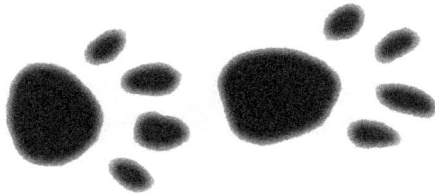

Maine is at the southern range for lynx, which are found primarily in Alaska and Canada. Inland Fisheries and Wildlife estimates the adult population between 600 and 1,200, mostly in northern and western parts of the state where Canada lynx benefit from regenerating spruce stands and fir clear-cuts. These areas are advantageous for snowshoe hares—a key food source for Canada lynx. When walking, the stride (distance between footprints of the opposite foot) may be up to 18 inches. With an abundance of foot hair, lynx tracks are less distinct than a bobcat's, especially in snow, where lynx tracks can be up to 5½ inches long.[15]

Canada Lynx

Bobcat

Bobcat

A medium-sized cat like the lynx, bobcats are often heavier than lynx, though lynx appear larger owing to longer legs and larger feet.[16] Bobcats have an upper weight close to 70 pounds, though less than half of this is closer to an average weight for males. Maine bobcats have similar pelt colorations to lynx. Bobcats tend to have redder pelts and less mottling of their spots than lynx. Favoring habits from dense understory vegetation to more open pastures, bobcats hunt snowshoe hare and other animals, including small rodents, birds, and even deer.[17]

Gray Wolf

According to the Maine Department of Inland Fisheries and Wildlife, the last wolves were hunted down in Maine by the 1890s. Hunting pressure on moose, caribou, and deer reduced these food

Gray Wolf

Coyote

sources for wolves and numerous bounty programs sponsored by
Maine succeeded in their final extirpation.

> The woods above the junction of the two main
> branches of the Penobscot, and all north of Katahdin,
> in Maine, are said to be infested by big gray wolves.
> These marauders from Canada and the Labrador are
> driving the deer, and, the hunters fear, will ere long
> exterminate them. It has been forty or fifty years
> since wolves were seen otherwise than as exceptional
> occurrences in the regions around the north end of
> Moosehead and Chesuncook Lakes.—*Portland*
> *Argus*[18]

Alton Leighton of Cherryfield killed a wolf in Maine in November, 1953.[19] A 67-pound female was shot and killed by a bear hunter near Moosehead Lake in 1993.[20] An 81-pound male wolf was killed near Aurora in Hancock County in 1996.[21]

According to the Maine Department of Inland Fisheries and Wildlife, "Tracks and other evidence suggest there may be additional wolf-like canids in the state, but there is no conclusive evidence of reproduction or establishment of packs."[22] About 75 miles (including the St. Lawrence River) separate Maine from the nearest

This Old Wood Cut Was Used In Coaching Times When Rivalry Was Sharp
Between Two Famous Maine Lines.

A woodcut of a stage on the "Airline" prominently
depicting an attack by marauding wolves.

Quebec wolf packs in the Laurentian Highlands.[23] Gray wolves, also known as timber wolves, would include indigenous Maine animals such as moose, deer, and beaver among their prey.

Large gray wolves can easily attain weights twice that of a coyote. Gray wolf packs normally range from 2-8 individuals. Wolves can live up to 16 years in the wild and are social animals, communicating with their pack mates through howling.[24]

COYOTE

Coyotes are very adaptable and can coexist in close proximity to people. Their howls and yelps can be heard up to three miles away. Adult coyotes can run up to 30 miles per hour. A *Maine Sunday Telegram* story from May 2010 stated there are about 19,000 coyotes in Maine during the fall, though their population decreases to 12,000 during the winter.[25]

BIGFOOT IN NEW ENGLAND

This chapter explores Bigfoot in a broader regional context beyond the boundaries of Maine by presenting a brief look at Bigfoot in the other New England states. New England has certainly had its share of Bigfoot-like creatures in stories and legends, some of which are: the wood devils[1] of New Hampshire; the Winsted Wild Man, Lee's Wild Man,[2] and the Cotton Hollow Giant of Connecticut; the Bennington Monster, Old Slipperyskin, and the Wild Man of Williamstown from Vermont; and Massachusetts' Wild Man of Haverhill.[3] The concurrence of hairy hominid accounts with extant and supporting legends can make the investigation into both the origins and factual substance for these stories problematic. Take for example "Mank" Lederquest, whose role in the Winsted Wild Man story will be talked of below. This gentleman stands out for the legendary proportions he attains *in situ*, brandishing a Winchester whose stock bears notches, each signifying "a wild man gone to his account." In this chapter, the case of the Winsted Wild Man receives the greatest attention due to the breadth of treatment the story originally received—from straightforward and serious, to playful and whimsical.

Since the 1960s, a growing body of investigation into Bigfoot has produced collections of North American sightings and reports. *The Historical Bigfoot*, published in 2006 by Chad Arment provides a large collection of historical sightings and encounters from newspaper accounts that predate the cultural development of the modern Bigfoot phenomenon. A breakdown of New England

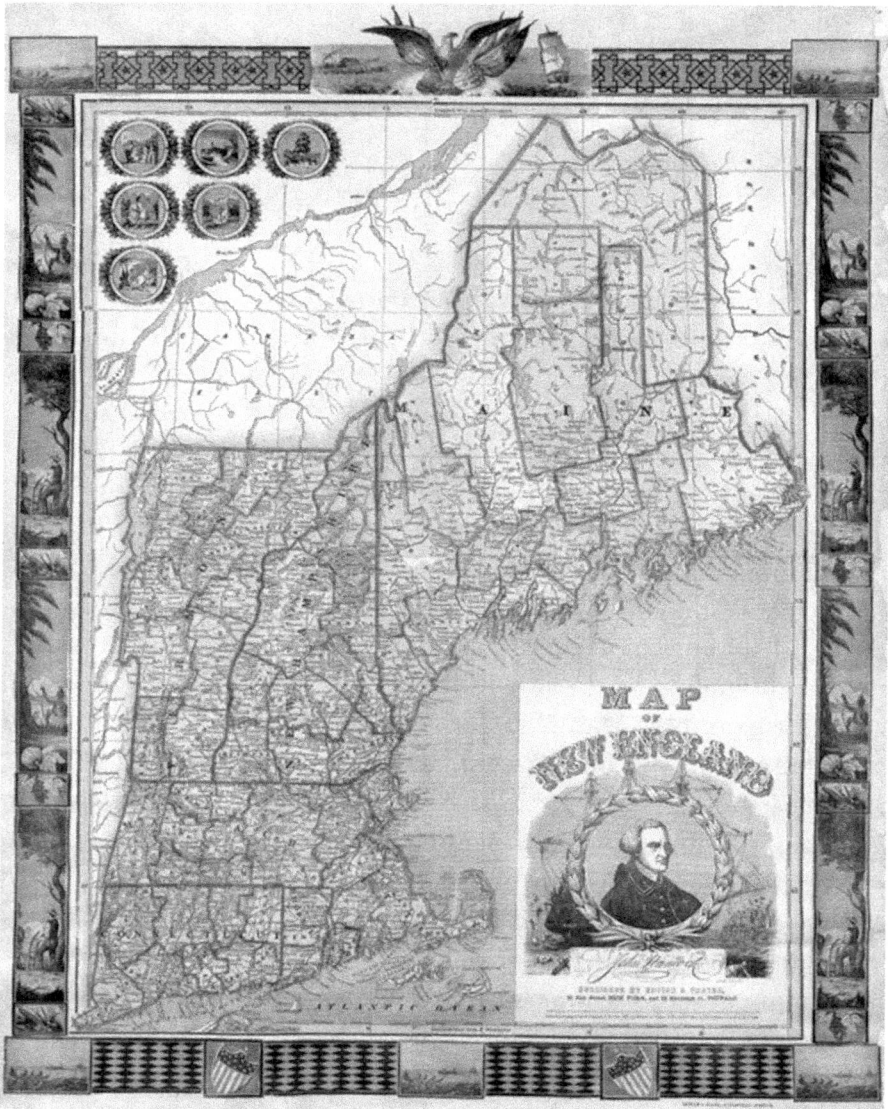

1847 map of New England.

accounts from *The Historical Bigfoot* is: Maine–1, New Hampshire–1, Vermont–1 (Old Slipperyskin mentioned below), Massachusetts–5, Connecticut (including the Winsted Wild Man mentioned below) –10, and Rhode Island–0.

Significant contributions to the investigation into the modern Bigfoot phenomenon (three books and one website) follow, each providing aggregations of regional sightings focusing largely on modern accounts. It is interesting to note the growth and increase in reports of alleged Bigfoot encounters in New England with ensuing publishing dates.

John Green of Harrison Hot Springs, British Columbia, is unquestionably one of the field's preeminent Bigfoot authorities. Green has written extensively on the matter and published his research in several books since 1968. Green began investigating the Bigfoot subject in 1957 while he was the owner and publisher of the *Agassiz-Harrison Advance* newspaper. At that time he learned of several local encounters, including the Bigfoot "classic" at Ruby Creek in 1934, which inspired him to delve into the rich history of the Sasquatch in the local area and, soon after, the rest of North America.[4] John Green's publication of *Sasquatch: The Apes among Us* in 1978 described 14 sightings for New England: Maine–4, New Hampshire–5, Vermont–2, Massachusetts–1, Connecticut–2, and Rhode Island–0.

Janet and Colin Bord's *The Bigfoot Casebook*, published first in 1982, had the aggregate sightings for New England at 15: Maine–5, New Hampshire–3, Vermont–2, Massachusetts–4, Connecticut–1, and Rhode Island–0.

Bigfoot on the East Coast, published by Bigfoot researcher Rick Berry in 1993, provides details on 53 Bigfoot sightings in New England: Maine–10, New Hampshire–8, Vermont–17, Massachusetts–11, Connecticut–7, and Rhode Island–0

A modern repository for Bigfoot reports is the Bigfoot Field Researchers Organization's online Geographic Database of Bigfoot/Sasquatch Sightings and Reports. This resource alone has 61 listings for New England: Maine–13, New Hampshire–10, Vermont–6, Massachusetts–19, Connecticut–8, and Rhode Island–5.

CONNECTICUT

Connecticut's woodlands, once famous for large American chestnuts, are today largely mixed in composition and cover about 60%
of the state.[5] "These forests sweep northward from Long Island
Sound, through the oak- and hickory-dominated woodlands of the
Connecticut River Valley and into the northwestern corner of the
State. Here, the foothills of the Berkshires and New England Highlands begin, along with an increasing predominance of northern
hardwoods."[6]

One of Connecticut's most widely-reported series of Bigfoot
encounters occurred in the late nineteenth century near Winsted:
The Winsted Wild Man. Winsted,[7] settled in 1750, is located in
northeastern Litchfield County, Connecticut. One of Winsted's
famous sons is Louis T. Stone, Winsted-born in 1875. He started
work with *Winsted Evening Citizen* as an office boy in 1889, became a reporter, and later the *Citizen's* general manager until his
death in 1933. It was at this paper that Stone began a reputation
for tall tales and fanciful stories.

An example of his story-telling is that of a wistful Plymouth
Rock Hen that hopped the train at Winsted heading south to Plymouth, leaving an egg on the cow catcher for payment of her 20-
mile fare. Some of Stone's other tall tales included "a farmer who
plucked his hens for market with a vacuum cleaner; a three-legged
bullfrog; a hen that laid a red, white and blue egg on July 4; a bulldog that sat on hen's eggs; a squirrel trained to brush its owner's
shoes with its tail and a man who painted a spider on his bald pate
to shoo away the flies."[8]

Probably his most famous copy appeared in 1895 and described
the sighting of a 'wild man' near Winsted. In 1949, Arthur Peterson's Our Town column for the *Toledo Blade* had this to say about
the Wild Man's genesis:

> Stone was little known outside the borders of
> Winsted when the summer sun of 1895 began ripen-
> ing fruit in Reilly Smith's blueberry patch. Smith and
> Stone met on the street. Smith had been complaining

for days that thieves were stealing his blueberries, and he was at wit's end to control them. Desperation had driven him to the edge of mendacity that day and—apparently hoping that Stone might print a story to scare the thieves away—he told a tale of having glimpsed a kind of a sort of a Something in his blueberry patch that might just possibly have been a . . . WILD MAN![9]

Louis T. Stone.

The chilling story of a monstrous creature haunting the Winsted environs proved so intriguing dailies as far afield as New York sent a battalion of their own men to descend upon Winsted to investigate—some reporters stayed for weeks pursuing leads "over stone walls and across pastures, through woodlots and streams,"[10] but failed to materialize Stone's hirsute protagonist.

As Chad Arment writes, the first stories of the Winsted Wild Man begin at least as early as 1891, making Stone an unlikely candidate (he would have been 16 years old) for conceiving and fabricating a hoax that lasted with reports made well into the twentieth century. In 1973, the managing editor of *The Winsted Citizen*, Robert J. McCarthy, stated that reports of the "Wild Man of Winsted" still crop up every so often.[11]

The New England Munchausen First Made Winsted Famous With His Celebrated "Wild Man" Fable. The Monster Turned Out to Be a Jackass.

A light-hearted depiction of the Winsted Wild Man.

Following are several of the newspaper accounts of the Winsted Wild Man presented in their chronological order.

The Winsted Herald, August 21, 1895: The Wildman—He Appears to Selectman Riley Smith and Scares Riley's Bulldog Out of a Year's Growth—Last Saturday Selectman Riley Smith went up to Colebrook on business. Mr. Smith, while there, went over into the fields and began picking and eating berries from the low bushes in the field. While he was stooping over picking berries, his bulldog, which is noted for its pluck ran with a whine to him and stationed itself between his legs. A second afterward a large man, stark naked, and covered with hair all over his body, ran out of a clump of bushes at lightning speed, where he soon disappeared. Selectman Smith is a powerful, wiry man and has a reputation for having lots of sand, and his bulldog is also noted for his pluck, but Riley admits that he was badly scared and his dog was fairly paralyzed with fear. If any of the readers of the *Herald* have lost a wild, hairy man of the woods, six feet in height, and want to find him, they can go up to Colebrook, and when near the "Lewis Place" wander around in the woods and fields and perhaps recover their lost property."[12]

Built Like a Horse—Wild Man Creates Terror Among Farmers Around Injun Meadow, He Scares Brutes as Well as Human Beings. Posse to Organize and Make a Determined Effort at Capture. Winsted, Conn., Aug 28.—The wild man was seen again yesterday by passengers on Dodd's stage, on route to Winsted from Sandisfield, Mass. He was in the same tract of brush as when seen last Saturday by Selectman Smith, which is five miles from here on the old and lonesome highway leading to Colebrook.

The wild man lives in "Injun Meadow," as it is known to the countrymen. He is thought to be one of a family of three wild men seen two years ago. The man seen by Mr. Smith had no clothes, but was covered with hair. The wild man was seen in Canaan mountain [sic] a few months ago is thought to be the same person.

Farmers in that section are terrorized and afraid to go out of doors after dark, and the robberies of henneries and mysterious disappearances of calves, lambs, and even oxen from Sandisfield and Colebrook farms are blamed upon the wild men.

Five hundred men leave here Sunday morning to hunt for the strange character. They will go out in gangs and surround Injun Meadow, and Cobble mountain farmers have given the use of their teams free, while every men of the posse is warned to go armed.

On Saturday, Riley Smith, while coming over the road, stopped to pick a few berries, but no sooner had he commenced to eat than the wild man emerged from the center of a batch of berry bushes. Smith was about scared to death. His dog commenced to whine, and with its tail between its legs sought refuge in Smith's wagon under a pile of blankets.

Mr. Smith described the man as an awful looking sight. He is large in stature and his head is the most conspicuous part of his body, being nearly the size of a horse's head. His teeth resemble those of a horse in size, but are pointed. His hands are extra large.[13]

The Winsted Evening Citizen, August 27, 1895: Over 100 Able Bodies Search for Riley Smith's "Wildman"—Selectman Riley Smith, who recently had an encounter with a "wild man," offered a reward for the capture of the creature. On Sunday morning over 100 men and boys met at the corner of Lake and Main with all sorts of weapons and ready for the hunt. The areas to be hunted were Injun Meadow, Cobble Hill and Losaw Road. Unfortunately, the hunters returned empty-handed. Nevertheless, there have been other sightings of the creature that include George Hoskins who said he saw the wild man leaving his hen house with two hens under his arms. Furthermore, Jim Maddrah proclaimed he took a Kodak picture of a man with a mass of hair on his head, but none on his body. Jim explained this condition by stating his camera was so frightened it couldn't see straight. However, two ladies from New York say they saw very clearly a large animal cross their path, turn, stand on its hind legs and stare at them. They believe the "wild man" to be an ape or baboon. We also have word from our chief of Police, Steve Wheeler, who stated he tracked a gorilla-man into a swamp where he lost the trail and scent.[14]

Winsted Herald, August 28, 1895: Selectman Smith Stands by His Story—The Story of the Wildman caused quite a little excitement about town. There is little to add to the story, except to say that Mr. Smith states that the man was within a small cleared space, which was surrounded by bushes. Mr. Smith did not see him until the dog whined and ran between his legs. Mr. Smith says the man's hair was black and hung down long on his shoulders, and that his body was thickly covered with black hair. The man was remarkably agile, and to all appearance was a muscular, brawny man, a man against whom any ordinary man would stand little chance.

Mr. Smith is a man who talks but little, he is a man of undoubted pluck and nerve, and his word is first class. When Mr. Smith says he saw the man he did see him, and there can be no question about it.

Quite a number of men in Winsted today stated to us that they were ready to go and hunt for the man.

Well gentlemen, the way is open. If he is still there, he ought, for the sake of the isolated farmers there and women folks, to be captured.[15]

Man Covered with Hair—A resident of Winsted lately became frightened at the appearance of a "naked wild man, covered with hair," in the woods near by [sic] that thrifty place; and about 100 of Winsted's bold and active men went to the woods Sunday to hunt for the wild man. They were armed with bicycles, shotguns, pitchforks, whisky flasks, and other deadly weapons in the presence of which wild men succumb on sight. They didn't find the wild man, but one of the hunters saw a pig—possibly a black pig—covered with hair, and the very fellow that so badly frightened the respectable citizen a few days ago. Another found a hole in the ground, with chicken bones around the opening, which some fox or other hairy creature had left there.

These wild men are frequently seen by some frightened good man or woman, and the strange creatures scream and run off into the thicket. Hunters by the hundred can never find them. They are like the sea-serpent, which is seen only by the captain and crew of some fishing smack, who see their heads, "like the head of a horse," and the foaming water in the snake's track. All who go to the hotel at Nahant, to see the serpent, are disappointed. They only see the roll of a wave or a porpoise. The "wild man," as seen by the frightened good man, is always covered with hair, like a big dog, a pig, or wildcat. He is never seen again till some frightened citizens gets a glimpse of him as he runs into the thicket. A wildcat sits on its haunches, and glares at frightened people. They "scream" when they feel like it, and a hundred volunteers, with terrible instruments, fail to find the thing. Let Winsted be calm. It is too smart a place for worry over wild men, except the wild fellows who carry whisky bottles for defense. [16]

Winsted, Conn., Aug. 29—Passengers on Hall's Stage from Colebrook yesterday saw an animal cross the highway, leap a fence, and then stand on its hind legs. As the stage drew near the animal ran into the woods. The passengers say it was a large gorilla, and it was supposed to be the animal that was heretofore reported as a "wild man," as it was seen in the same locality as that where the "wild man" was said to frequent. The gorilla probably escaped from some circus years ago. During last Winter [sic] a gorilla inhabited the woods in South Norfolk.[17]

Say It Is A Gorilla (Special to the *World*) West Winsted, Conn., Au. 29—The wild man of Winsted is not a man after all, but a gorilla. At least, that is the consensus of opinion of seven people who have seen him since yesterday morning.

But the establishment of that important fact doesn't in the least allay the alarm felt throughout the county of Litchfield since Selectmen Riley W. Smith recounted his adventure in the Injun meadow two weeks ago with a strange creature covered with coarse black hair and possessing a fine set of Roosevelt teeth. John G. Hall, who drives a stage between Sandisfield and Winsted, was the first to report the identity of the strange creature that has so terrified the residents. Hall had three passengers, Wall street [sic] men, it is said, who were on their

SAY IT IS A GORILLA.

The Wild Man of Winsted Is Now Thought to Be a Ferocious Beast.

IT WAS SEEN TWICE YESTERDAY.

Its Savage Yells and Roosevelt Teeth Are the Talk of Litch- field County.

SIGHTED BY A STAGE COACH DRIVER.

Many Say They Have Seen the Animal in the Past—War of Extermination Projected.

A Typical Headline

way to Winsted to catch a train to New York. He had listened with open-mouthed astonishment to their talk of bulls and bears and concluded that they were mighty hunters.

The stage bowled along past Colebrook town and was just nearing Injun Meadows, a wild path of country, with not a dwelling in sight, when suddenly a strange creature bounded into the road from the underbrush and crouched on all fours directly in the path of the horses.

There was no doubt about it being Selectman Smith's old acquaintance. There was the coarse black hair and the glistening white teeth that would have sent shivers down the back of a New York policeman. But at the first glance Hall saw the stranger was not a man but a huge gorilla.

Ready for the Fray—For the last two weeks every man who owned or could borrow a gun or revolver has carried his weapon when called outside of his own home, and Hall, with dreams of glory and fortune, whipped out a heavy colt, determined to bring back the carcass of the animal or die in the attempt. He was backed in this resolution by the knowledge that he had three famous bear-hunters with him, and was determined to show them that all the big hunting was not done outside of Connecticut.

His passengers saw him about to descend from the stage, and in affright asked him what he was going to do. "Oh! Just a little hunting," Hall answered, nonchalantly. "Perhaps you'd like to take a hand?" But the passengers said they wouldn't.

They entreated Hall not to leave them, and convinced him by many arguments that it was his duty as a common carrier to stand by and protect them instead of looking for trouble.

All this time the strange animal stood glaring at them in the roadway, not 150 feet away. Suddenly, with a frightful yell, it raised itself on its hind feet and with one mighty bound disappeared over a stone wall.

Hall valiantly wanted to follow, but his passengers entreated him not to miss the train, as they had important business in town. So he reluctantly started up his frightened horses and scared all Winsted out of a year's growth by telling the story, corroborating it by showing his revolver.

Seen a Second Time—Late this afternoon the animal was encountered near the same place. Miss Sadie Woodhouse and Mrs. Musbone, of New York, who are stopping at Colebrook, were driving towards Winsted when an unearthly shriek, followed by a crunching sound in the underbrush startled the horses into a mad run.

With difficulty the colored coachman reined in the frightened animals just as a hairy, grinning monster made a flying leap across the road and disappeared among the trees. The women displayed remarkable coolness, and the horses were whipped into a smart run that soon took them into a more inhabited part of the road, where it was felt the wild animal would not be likely to follow.

These stories reaching Winsted created the greatest excitement, and nothing else is being spoken of here. Selectman Smith's son, Loren, and Burt Griswold started out to-day with a big bulldog and a wagon loaded down with ropes, chains and weapons, determined not to come back without a satisfactory accounting.

They drove straight to Injun Meadow and carefully scanned the neighborhood, but no wild man or beast did they see. So long did the search continue that the shades of night had fallen, and Winsted began to have grave fears for the fate of the mighty hunters. A relief expedition was talked of, and it was remarkable how many valiant men were unable to lend a hand because their guns were at the locksmith's or had been borrowed by neighbors. But while the talk was still going on the rumble of a wagon and the bark of a dog were heard in the distance, and soon Smith and Griswold made their appearance.

They were tire [sic] and dejected, and Winsted's heart sank as they told of their ill-luck. The young men felt positive, however, that there was convincing proof about Injun Meadow of the recent appearance of the monster, which had probably retired to its hiding place in the depths of the surrounding woods.

It would take a large party to make a thorough search, and Smith and Griswold are now organizing a complete expedition that will either bring back the monster or forever prove it to be the evil one himself, though why the latter individual should trouble quiet, even-tempered Winsted nobody knows.

A Hard Theory to Face—In the mean time [sic] a movement is on foot to offer a substantial reward for the return, dead or alive, of the much-dreaded intruder. There is one thing that throws doubt on the gorilla theory. The gorilla is a native of Africa and could not possibly stand the rigors of a Northern winter. Yet the testimony of many residents about here would go to show that the monster now frightening Winsted has been seen about here for years past.

Joseph Bruey, of South Norfolk, and Charles Benson, of West Norfolk, remember distinctly having been chased for a long distance along the snow-covered road last winter by just such a monster, and finally escaped to a neighboring barn.

Carl Moore, Fred Webster and George Marvin, of South Norwalk, discovered traces of the man or animal near Tobey's Pond, about sixteen miles from here two years ago, and followed them to a cave in the side of a hill. They barred the mouth of the cave with chains and returned next day with reinforcements, only to find the barriers torn down and the occupant gone.

The question of the identity of the intruder has become a serious one, and so has the question of its extermination, for that is the only solution that finds favor here. But before that latter desire is accomplished the citizens of Litchfield County will have to obey an old precept, namely, First catch your gorilla.[18]

The following headline and story comes from the front page of *The Lewiston Evening Journal* of August 30, 1895:

Winsted, Conn., Aug. 30—A Hairy Giant—Yesterday afternoon Hall's stage left Colebrook for Winsted, and the passengers and driver encountered the wild man at close range. Most of them were badly frightened at the apparition, and the broadside of rifle, horse pistol and revolver bullets discharged at him went wide of the hairy target, which only defiantly roared and snapped his long white teeth at them.

Montgomery Shepard was on the box of the stage. He has driven over the route every day for years, and fears not man, beast nor catamount. In deference to the apprehension of his passengers he carried a huge, navy six-shooter between his knees, all ready for a shot. There were five passengers, one of whom was a woman, Mrs. Maysie Hitchcock of Cornwall Hollow.

One of the men was a Philadelphia drummer,[19] who had a loaded double-barreled duck gun, and another was "Mank" Lederquest, who has hunted wild men in Texas and Wyoming, and was armed with a Winchester, which has nine notches upon its stock. Each notch, "Mank" says, represents a wild man gone to his account.

The stage bumped quickly along the road between level fields for a mile or so and then plunged boldly into the Pamunkey Pond woods. The tense look upon the passengers' faces heightened and they all gripped their fire-arms as though afraid that they might take to themselves wings and fly away. Nothing appeared. Over the slope of Burr Mountain and across the log bridge which spans Pickard Brook they went in safety.

But the security was only apparent. At a point about six miles from Colebrook there is rough clearing, with a cabin in the middle, which has long been abandoned. As the stage approached the hut Mark Lederquest exclaimed: "See there! In the yard!" And as the words left his lips the sharp crack of a Winchester was heard.

At the same moment a grotesquely hideous form sprang over the fence and crouched in the road. Everyone recognized it as the wild man. The horses tried to wheel around and run away.

Great and hairy the maniac looked. His mouth was wide open and lined with bristling teeth. His eyes were fierce and bloodshot. A matted growth of hair was on his head and the beard was fully three feet long and blown around by the wind. A rag of cloth was about the waist, and the great claw-like hands were full of what seemed to be rhubarb leaves which the wild man had probably found and was eating in the deserted garden.

Sederquest [sic] fired again and the driver sent a shot from his revolver. If either hit there was no sign. The Philadelphia drummer sat like a carved image, petrified with fear, but plucky little Mrs. Hitchcock grabbed the duck gun, braced the butt against the back of the seat and pulled both triggers. The first charge tore up the highway in a cloud of dust, the second reduced the ears of the nigh horse to ribbons.

"Crack!" "Bang!" "Crack!" went the rifle and pistol, while the horses jumped so that they nearly upset the stage.

Then the wild man, with a growling roar, made a bound which took him to the side of the road. Another carried him over the fence, and in a succession of strange, uncouth leaps he crossed the clearing and was presently lost to sight in the woods.

After a while Driver Shepard succeeded in getting control of the horses and whipped them into a frenzied galop [sic] that brought them within sight of the friendly steeples of Winsted in less than an hour.

It was a pale-faced lot of people who climbed down from the stage at the post-office door. Mrs. Hitchcock was the most self-possessed, and she told the story. The Philadelphia drummer had to be carried to the hotel and put to bed. Mank Sederquest [sic] declared to the assembled townspeople that it was no wild man at all that they had seen, but a gorilla that had years ago escaped from some circus menagerie. The Winsteders laughed him to scorn.

Last night the Winsted Light Guards were under arms at the Town Hall, and all the citizens were ready to turn out when the fire alarm bells should give a certain signal.[20]

Hartford, Conn., Sept. 3—When the Sandisfield stage reached Colebrook, near Winsted, at 10 o'clock this morning, Mrs. Culver ran from her house and stopped the stage driver. She said that the "wild man," or gorilla, whichever it is, spent the night on her doorstep. She was greatly alarmed. Mrs. Culver begged that help be summoned to catch the local terror. Six policemen and a large body of citizens set out from Winsted as soon as the stage reached there, and the country around Mrs. Culver's house will be thoroughly searched for the "wild man."[21]

Winsted, Conn., Sept 15—Town's Bugaboo—Unless several of the most reputable citizens of this town have had optical delusions a wild man is lurking in the Litchfield county woods.

Every day some new uncanny story is heard about him—either a farmer arrives and tells of depredations in his garden or hen roosts, or a bicyclist returning from a country jaunt has come across him.

Much of this information is incredible, and sounds as if it were the offspring of overwrought imaginations or a desire to be mixed up in the sensation of the hour, even at the expense of truth. The story tellers would have Winsted people believe that his unaccountable creature talks with them, barks at them, holds up their horses on the wood roads, and milks their cows in the pasture. One of them brought in a pumpkin with the marks of long teeth in it, and said the wild man bit it.

He is described as having all sorts of shapes, from a baboon to a mild lunatic with scanty clothing and disheveled hair.

So altogether the grocery stores where the farmers trade are the scenes of go-as-you-please story telling matches.

Selectman Riley W. Smith "discovered" the freak. Selectman Smith takes no part in the cracker barrel symposiums, however. He is inclined to be conservative and doesn't take any stock in most of the wild man talk. He is positive about what he saw, and doesn't appear to want to make a sensation out of it.

Selectman Smith is one of the most respected citizens of Winsted, a prominent member in the Congregational church and a prohibitionist. He never was known to tell a story before that was not entirely plausible.

Relating to his recent strange adventure, he has talked but little about town, not keeping it a secret, but merely telling what he saw in a matter-of-fact way when asked about it. The Globe is the first newspaper to get an interview with Selectman Smith.

On the day of his adventure, four weeks ago Saturday, he was chasing a lost hog. The chase led him up a lonely back road between the settlements of Winchester and Colebrook, about two miles north of Winsted. This road is traveled very little. Probably not more than 25 wagons pass over it a year. Although it is three miles long there are only two houses on it, and only one of these is inhabited.

Half through this road or horsepath the selectman saw a clearing and remembered that when a boy he had found blackberries there. The day was very hot, the selectman felt that blackberries would taste about right, and got out and hitched his horse.

But he had best tell in his own words what then happened.

"I stepped inside the clearing and saw at a glance that the berries were there as I expected. But almost immediately with my first glance at the place I saw this strange creature. The instant I saw him his back was toward me.

"He was at the other end of the clearing, 40 feet away. It was only a small place, you see, which had years ago been a garden, when a house stood on the site.

"His hand was reached out as though he was picking the berries. I think I interrupted him in a feast. He stood about 5 feet 10 inches, I should say, and weighed perhaps 180 pounds. He was naked, and his back and legs were covered with hair, apparently two or three inches long. He had very long, black hair from his head also, reaching down below his shoulders.

"Almost the instant I saw this creature he became aware that I was around. I think I may have stepped on a bush and alarmed him. At any rate, he turned his head just so he could see me over his shoulder, and giving out a sound such as I never heard before, gave one bound over the bushes and disappeared into the woods.

"For an instant I was very much frightened, because the sight was so unexpected. I had a big bull dog along, and he crouched at my feet and whined."

When asked if he had formed any theories in regard to the strange creature, Mr. Smith said: [I have] tried to reason the matter out, and I can't come to any conclusions. I can only believe what I saw. I think it was a man of some sort, however, for it ran like a man and had a human voice."

The scene of Mr. Smith's adventure is in the heart of a very thickly wooded tract. Indeed all the country from West Winsted to Sandersfield, 12 miles away, is as wild as could be found in any part of New England. The villages are small and scattering and the woods between are dense, with very thick underbrush. How wild and uncanny is this region is shown by the game caught there. It is no uncommon thing to get foxes, and occasionally a specially daring huntsman will bring in a wildcat. There are miles of hills unvisited by any human being year in and out.

This is not the first rumor of a wild man in this town. Several children a generation ago told stories of an ugly apparition.

Capt Henry Tirrell, a resident, who runs a little steamer on Highland lake [sic], says that 12 years ago when he was a boy he saw just such a creature as selectman Smith describes. It was one day when he was returning from fishing and the hairy man ran away from him. Tirrell told his parents about it at the time and they made sport of the story, crediting it to his "being afraid of the dark."

Another man named Phelps, now dead about four years, once reported in Winsted village that he had seen a hairy man in the woods.

If there are any two men in these parts capable of making a wild man yarn believed they are select-man Smith and Tirrell. Both stand very well in the community and come of the most influential people have been heard to remark, "If it had come from anybody else but Riley Smith I wouldn't have believed it."

Of course there are exceptions. They say that most everybody in Winsted has a morbid love for weird things, and that this will soon be branded the fool town of Connecticut because of the stories of wild men and wild cats and other bugaboo yarns that originate here.

It has been many times shown that some people in the borough are rather gullible. A few years ago Col Horn of this town, who was then practicing law, but is now one of the state labor commissioners, succeeded very easily in fooling his neighbors. Weston, the famous pedestrian, had been scheduled to pass through Winsted on a walk from San Francisco. Col Horn, who delights in practical jokes, went into one of the downtown hotels on the morning Weston was expected, and reported that Weston would be in West Winsted in two hours. The colonel then went home, dressed to impersonate Weston, and walked through the village street with hundreds of his townspeople on either side. He intended it merely as a burlesque, but he found everybody took the thing seriously. They invited him to stop and take dinner in the place, and made him gifts, and the deception was carried on for more than an hour, until the colonel took off his hat, unbuttoned his coat and announced himself.

However, in spite of skeptics, the wild man story seems to have some credence with the police officials.

About 10 days ago Chief of Police Stephen C. Wheeler and deputy sheriff Chester Middlebrooks, with two officers and a party of citizens, spent all day hunting the woods for the man. They went through a very large tract of woods, known as Indian meadow. The day before Edmund Perkins of Norfolk had seen a strange person enter the Indian meadow woods on the Colebrook road. This was Perkins' second sight at the supposed wild man. The day after Selectman Smith saw him, Perkins was visiting his brother in Colebrook. He went out to the barn, he says, and passed within a few feet of a curious looking person.

Perkins' wild man had on a torn coat and a hat without any rim. His hair was very long. He carried a pail, which looked as though it had been over the fire. He pointed to Colebrook Center, and asked: "What's that place over there?"

Perkins went into the house to call his brother out, and when they returned the individual was gone.

The woods around the Perkins farm come very close to the house, and it would have only taken a second for any one [sic] to disappear in the brush.

Previously, however, potatoes had been missed from the place, and several times it was thought the cow which pastures close to the woods had been milked.

Nearly 150 people lined up around Indian meadow on the day of the hunt, but only eight cared to venture into the woods. These made an eight-mile circuit. They found places where the ferns were broken and trampled and the remains of a rude oven.

Many people who believe that there is a wild or demented man in the woods think that the search should be conducted stealthily.

Deacon Manchester, the famous prohibitionist of the town, has suggested to Mr. Smith that they two prepare a tempting layout, which shall include a quantity of whiskey, somewhere in Indian meadow. He hopes to get the wild man intoxicated and capture him alive, and make $10,000 to divide, with Mr. Smith, by selling the animal to a museum.

Deacon Manchester has propounded the scheme with a great deal of serious enthusiasm.

Some of the scoffers suggest it would be far easier to bait the man with pie.

At the present time he is making his haunt in the woods about Pat Danehy's house. Pat lives about three miles north of Winsted, and has one of the most picturesque "ranches" in Litchfield county [sic]. The place is all farm, for even in the front room chickens and turkeys roost.

The hairy man has a special fondness for Pat's onion bed, and up to the time the cold snap came on visited it every night and morning. He has taken chickens from the roost and has carried away armful after armful of sweet corn.

Pat's son Mickey claims to have seen the wild man six times. Mickey is 12 years old, and very bright for his years.

He laid in wait for the creature the other evening. As Mickey tells it, the monster came snapping and growling through the woods that skirt the onion patch. He pulled up some onions and ate them, throwing the tails into the brook at the foot of the patch.

Mickey threw stones at him and he ran away frightened.

The next morning Pat went down to the brook and found the onion tails, which of course proved the boy's story.

The Globe representative visited the onion patch. It was very much trampled, and many of the onions were pulled, some lying loose on the ground. The "tracks" that were pointed out as the wild man's foot prints looked as though they might have been made by any animal with hoofs, possibly by pat's donkey.

Up to date there is only one explanation why anybody should be in the woods about here.

Postmaster Culver of Colebrook has received a letter from a German woman in Meridian, stating that the husband of a friend escaped from home over a month ago with very few clothes on and has not been heard of since. He was in a demented condition.

The latest depredations of this creature, whoever or whatever it is, have been made at places two miles or two miles and a half from the Massachusetts line.

The Litchfield county [sic] hamlets wait in terror for the mystery to be solved. The farmers carry revolvers when they go out to the "chores" and Mrs. Pat Danehy takes a pitchfork when she goes to the spring.[22]

Abominable Half-Human Monster Linked to Barefoot Tracks in Winsted Woods—Winsted, Jan. 1.—
Huge footprints were found Tuesday in the woods
near the east end of Highland Lake. Shaped like bare
human feet but three times their size, the tracks had
a pigeon-toed slant.

Retired broker Argyle Throckmorton, who lives
at the lake, discovered the prints in a muddy stretch.
He stumbled over them about noon. At his home

Cartoonish representation of mid-20th
century Winsted track-maker.

eight hours afterwards he was still visibly shaken by his find.

"Gives a fellow a start, you know," Throckmorton said. His highball glass trembled in his hand.

Feed on Ice Worms, Yaks—The tracks looked like the mysterious footprints found in the snow near Mt. Everest recently. The discovery of those led to the belief that the prints had been made by "abominable snowmen"—legendary half-beast, half-human creatures. "Abominable snowmen" are supposed to look like giant apes, only worse. They feed on ice worms, yaks and Nepalese tribesman.

"But this couldn't be an abominable snowman," Throckmorton said. He explained that Winsted could hardly provide enough Nepalese tribesmen for a subsistence diet, to say nothing of ice worms or yaks.

The retired broker said he followed the tracks to a patch of mud where whatever made the prints apparently lay down to rest. From its impression, he estimated the creature slept on its right side, weighed about 600 pounds and had a tremendously big midsection.

"Perhaps, dear, it was an abominable snowman," suggested his wife. Mrs. Throckmorton is noted locally for her wit.

Glawackus Erectus?—Soon after word spread of the strange discovery several Highland Lake sportsmen gathered at the Throckmorton home. One man argued that the tracks might have been made by a Glawackus walking upright.

The Glawackus was a giant, tawny, cat-like animal that roamed Connecticut 12 years ago. It got its name this way: "Gla" for Glastonbury, where it was first reported; "Wack" for wacky, the way it made most people feel; and "us," a proper Latin ending. Although never captured, it frightened many persons at night with its glowing eyes and piercing scream.

A typical encounter with the Glawackus occurred one November evening when the beast stalked farmer Charles G. Strickland up Bush Hill in Manchester. It made a blowing noise at him, sort of a modulated "whuh" a little over on the bass side. Mr. Strickland said the hair on the back of his head assumed an upright position. Unfortunately, he left that general vicinity too soon afterwards to get a good look at the creature.

"Gentlemen," Throckmorton told the sportsmen dramatically, "what about Lou Stone's wild man? Could this be it?"

The room became so quiet one could have heard a pin on its way down.

Adventures of "Truthful" Stone—The late Louis ("Truthful") Stone brought nationwide fame to Winsted as a source of weird nature stories. His Munchausen-like yarns went to newspapers around the world. Stone tickled America's fancy with: The cat with the hair lip that whistled Yankee Doodle; the highland Lake windstorm that blew a piece of paper into a typewriter and printed the alphabet backward; the Winsted cow so shaken by a garage explosion that she gave butter at the next milking; the bulldog that parked for three weeks on a setting of hens eggs; and the Winsted man who painted a spider on his bald head to keep away flies.

But the story that launched Winsted's reputation as America's looniest town was the tale Stone told of a wild man who stark naked chased a town official and terrorized the community. Every train for two weeks brought more and more reporters to cover the story. As Stone said, "the wild man certainly needed covering."

The Wizard of Winsted would have envied Mt. Everest its "abominable snowmen." And, as Throckmorton recalled, Stone made it a personal tradition on the first of each year to enhance his town's wacky reputation. He sent out one of his weird yarns to newspapers every New Year's Day. Can the Courant do less?[23]

Sightings of a strange creature were not isolated to the late nineteenth century. Two young men, Wayne Hall and David Chapman heard some unidentifiable noises in July 1972. "The two men looked out the window in the direction of Crystal Lake, and in the dim light saw a large, furry, eight-foot tall "man" lope out of the woods, cross the road and head toward a barn."[24] The men were emphatic that what they witnessed was not a bear.

A Wild Man-like creature was spotted again in 1974 at Rugg Brook Reservoir, when a large furry beast interrupted the amorous pursuits of a romantic couple. Patrolman George Corso went to the area to investigate but detected no trace of the creature's presence.[25]

The account of two men witnessing a Bigfoot creature and their encounter's subsequent sensationalized publicity in the tabloid *The Weekly World* appeared in the November 26, 1982, edition of the *Hartford Courant*. John Fuller and David Buckley "had what might be described best as a vision. It was just after midnight and the moon was full. The two were behind the barn of Valley Farms dairy—they work the night shift—when they came face to face with Bigfoot."

> Fuller has a lively recollection of that night and gives a vivid description of the creature that caused the stir. He was 7 feet tall, easy, Fuller said, and all hairy and muscular with a flat nose and big teeth.
>
> "It was like something you'd see in a horror movie," he said. "The way it was looking at us—like a dog, all snarling and showing its teeth. He had his hand in the silage bin. I don't know whether he wanted to eat or whether he was just playing with it."

A subsequent investigation by state police did not determine what the men may have encountered but left investigators with the impression they did chance upon something.

> "We have no doubt that they did see something," said Sgt. Frederick Bird of the Stafford state police barracks, who oversaw the investigation. "But I couldn't tell you what it was. It could've been somebody playing a joke."[26]

MASSACHUSETTS

Forests cover about 3,500,000 acres of the Bay State's land area. Bigfoot sightings have a history in Massachusetts occurring into the present day with recent reports from the October Mountain area in the western part of the state and Cedar Swamp in central Massachusetts.

> Toward the later part of July, 1826, the people of the town were not a little excited by the reports of several well known [sic] persons that a "wild man" had been seen in the woods in the town, who always fled when discovered. Supposing that it must be a man named Andrew Frink, who had, about two weeks previously, suddenly disappeared in a fit of insanity, and for whose recovery a general turn-out and search of the town had been already made, a large hunting party was made up, and after a long search, and great exertions, succeeded in finding and capturing the man. He proved not to be Mr. Frink, but literally a wild man of the woods. It was supposed from his appearance that he was some unfortunate, who, having perhaps met with disappointment in life, had, in a fit of insanity, fled from society.[27]

According to the *New England Folklore* blog, Andrew Frink was found later, his body floating in a stream, the victim of drowning.[28]

A Wild Man of the Mountains. Two Young Vermont Hunters Terribly Scared. Pownal, Vt., Oct. 17.— Much excitement prevails among the sportsmen of this vicinity over the story that a wild man was seen on Friday last by two young men while hunting in the mountains south of Williamstown. The young men describe the creature as being about five feet high, resembling a man in form and movement, but covered all over with bright red hair, and having a long straggling beard, and with very wild eyes. When first seen, the creature sprang from behind a rocky cliff, and started for the woods nearby, when, mistaking it for a bear or other wild animal, one of the men fired, and, it is thought, wounded it, for with fierce cries of pain and rage, it turned on its assailants, driving them before it at high speed. They lost their guns and ammunition in their flight, and dared not return for fear of encountering the strange being.

There is an old story, told many years ago, of a strange animal frequently seen along the range of the Green Mountains resembling a man in appearance, but so wild that no one could approach it near enough to tell what it was or where it dwelt. From time to time hunting parties, in the early days of the town, used to go out in pursuit of it, but of late years no trace of it has been seen, and this story, told by the young men who claim to have seen it, revives again the old story of the wild man of the mountains. There is talk of making up a party to go in search of the creature.[29]

Coca-Cola Ledge (formerly known as Witt's Ledge), a prominent cliff overlooking Route 8 in North Adams, gained its name because the eponymous soft drink's name adorned the rock wall in the 1940s. Subsequent years saw the original logo over-painted by various Greek organizations' letters, but, according to Jeff

Belanger's *Weird Massachusetts,* "The billboard isn't the only monstrosity in the area."[30] The Monster of Coca Cola Ledge, seen atop the cliff and in the surrounding woods, is said to be a Bigfoot-like creature. "The story goes that a monster lurks near the top of the ledge—it's one of those stories parents are happy to propagate. It's the type of tale that keeps kids away from the ledge."[31]

A place known for a sinister history of odd occurrences is the Hockomock Swamp area of Southeastern Massachusetts. The swamp was named Hockomock by the Wampanoag, for "place where spirits dwell." The Hockomock Swamp is part of the so-called Bridgewater Triangle, an area of alleged paranormal activity that includes the towns of Brockton, Easton, the Bridgewaters, Raynham, and Taunton.

According to researcher Christopher Pittman, following reports of a 7-foot tall creature, officials from Bridgewater and state police with dogs hunted and tracked a "giant bear" in 1970.[32]

> "Since the creature was not found, police were never certain what it really was. Although bears have not been seen in the Bridgewater area for many years, they are certain that, whatever the creature was, it was not a hoax. Several very reputable citizens had a good look at the huge animal before it lumbered off into the woods, and definite tracks were found there. In other parts of the country, people trying to make sense of the unexplained have often labeled these large, hairy creatures as bears."[33]

In 1978 Joe DeAndrade was standing on the shore of Clay Banks, a Bridgewater pond close to the swamp. Standing with his back to the water, he was suddenly gripped with a strong urge to turn around.

> "And I turned around, and there, off to the right, maybe 200 yards away, there was this—well, I don't know what it was. It was a creature that was all

brown and hairy, like a big apish-and-man thing. It
was making its way for the woods, but I didn't stick
around to watch where it was going. I ran for the
street."[34]

DeAndrade was not able to reconcile what he saw with anything he could identify. He soon organized Bigfoot-searching expeditions under the moniker of Bridgewater Triangle Expedition
Team. The group went into the swamp to search for the creature
but despite their efforts, no trace of what DeAndrade saw was
found. DeAndrade would later form the Paranormal Investigation
Organization, whose members would receive updates from
DeAndrade on sightings and strange activity in the area.

John Baker from West Bridgewater had his own encounter with
a similar creature in the early 1980s. It was during a pitch black
night that Baker was running muskrat trap lines from his canoe
on the Hockomock River near his home. A feeling developed that
he was not alone and of being watched.

"Something was following me and I knew it was big.
So I took the boat down a small creek to a dry hill
and it kept moving." As he paddled quietly, he could
hear the shambling gate of the man-beast shatter
the thin crust of swamp ice. He stopped and watched
as the shadowy, hair-covered giant strode a few
yards away. "I knew it wasn't a human because when
it passed by me I could smell it," said Baker. "It
smelled like skunk—musty and dirty. Like it lived in
the dirt."[35]

The following story took place in Berkshire County, western
Massachusetts. October Mountain State Forest is located in the
central Berkshires and at 16,500 acres, is the largest state forest
in Massachusetts. Herman Melville is credited with the name of
"October Mountain" being moved by the autumn view from
his home in Pittsfield.[36] The August 23, 1983, headline from *The*

Berkshire Eagle was "Weird Mountain 'Creature' is Reported by Picnickers." Eric Durant and Frederick Parody said while picnicking they encountered something near the former Boy Scout Camp Eagle on Felton Lake in the town of Washington. "It stood on two legs, silhouetted on the trail in the moonlight, and it was huge. I don't scare easily, but it scared me." At the conclusion of a cookout, the two men and two other friends were leaving when "the headlights of their car picked up the creature, lurking behind some bushes, according to Durant and Parody. They said it was erect on two legs and was 6 to 7 feet tall. It was dark brown in color and had strange eyes that glowed, they said. Both were emphatic in saying that it was not a bear. Parody said he has hunted bears in Maine and was quite familiar with how they look."

Parody went back the next day to investigate the area of the sighting. The article states, "Parody said he and one of the others returned to the scene yesterday at midday and caught a fleeting glimpse of the creature again. It moved extremely fast, he said. About all he really saw, he said, were "arms moving" in the woods."[37]

Massachusetts has the distinction of having had a "regionally-produced" Bigfoot film lensed within its borders during the 1970s. Such film fare was usually destined for the drive-in circuit and was typified by low-budgets, questionable or no special effects, and erstwhile if not totally polished acting ability. As is often the case of such pursuits, passion and desire fueled the drive for obtaining financing, enlisting willing actors, and hiring editing room expertise.

FILMED IN LOWELL

BIG FOOT...No longer a Myth, No longer a Legend, A HORRIFYING REALITY, KILLING TO SURVIVE!

SASQUA

THE MISSING LINK ATTACKS!

Chelmsford Cinema-Lowell D.I.
August 13-20th

In August 1975, the Lowell-lensed *Sasqua* enjoyed a brief theatrical release. *Sasqua* was written, produced, and directed by
Channon Scott, who wrote his first theatrical treatment concerning Bigfoot in 1968 while a student in acting class. He recounts
one night he had a dream of returning to Massachusetts, seeking
investors, and developing his story further into an actual script.
The realization of this dream began in 1972.

The film's press book provides a description of the plot:
"SASQUA is the story of a group of motley peace-seekers, led by
Big Jim (James Whitworth), who have deserted the rigors of an
established society to form a commune, and their encounter with
a legendary man-ape known as 'Bigfoot' in the deep forest regions
of New England."

The film is generally considered lost though copies are rumored
to exist. Scott states he still owns a high-8 copy of the film. An
anecdote of *Sasqua's* filming shared by Scott with Polish horror
website *Danse Macabre* highlights certain risks inherent in filming an actor dressed up like Bigfoot. Scott said the production had
moved to a forest setting in Lowell for a scene of the monster running through the woods. "The next thing was we were being visited by the local police dept. Their deputies informed me that an
older lady saw a 'creature' running through her back woods and it
frightened her to no end. We had to stop filming."[38]

NEW HAMPSHIRE

New Hampshire is the only state that Maine borders[39] (Maine's
other boundaries are shared with Canada and the Atlantic Ocean).
The Granite State has warm summers and cold, snowy winters. New
Hampshire has the highest percentage of timberland in the country after Maine.[40] The White Mountains reside in the northern part
of the state dominated by Mount Washington, the tallest mountain (over 6,000 feet) in the northeastern United States.

The anonymous posting below appears on several Bigfoot report aggregation websites and is attributed by the *naturalplane*
blog to a posting originally submitted to a USENET forum.

Many years ago I lived up in Coos County in New Hampshire. Some of the old men would talk about things called "wood devils" that live in the woods. There apparently were a lot more of these creatures back in the 1930s than there are now.

These Wood Devils were tall and very skinny. They are gray colored and very hairy. I guess that people saw them mostly when they didn't expect to. They stay in the deep woods. They can move very fast. When a person walked through the woods he would nearly walk into one before he spotted it. They hide by standing upright and still against a tree. As a person approaches it, the creature will stand against the opposite side of the tree. As the person passes it will move so that the tree is always between the person and the tree. If it cannot hide it will still stay perfectly still until it knows the person sees it. They make awful screams. They have a semi human shape, but their faces don't look at all human. I have never seen one, but the people who said they did were regular churchgoers and would strap their kids for lying. I don't think they would carry on discussions of things they made up."[41]

In the summer of 1942, Preston Elliot, while engaged in cutting wood in Center Sandwich, encountered a 6-7 foot tall creature resembling a gorilla. The bipedal animal followed Elliot for about 20 minutes.[42]

Salisbury in Merrimack County was the location for Walter Bowers Sr.'s encounter while pheasant hunting with a large hair-covered creature. Bowers' sighting was reported by the *Concord Monitor* in November, 1987. The article relates Bowers' exposure to an earlier anecdote of something strange seen by fellow hunters.

About three weeks ago, a hunter from Warner told Bowers he had seen two strange beasts walk across the field early one morning. "I didn't pay much attention," Bowers said. "I just went out bird hunting, two days, three days afterward." As he crossed the field with his 12-gauge shotgun shortly after daybreak, he had a strange feeling he was being watched. He passed between two stands of trees, turned to the right, "and there he was. Standing right out in the middle of the field." Bigfoot, Sasquatch. Whatever you want to call it.

Bowers recalled some of the creature's features and aspects of its appearance:

> "This thing was BIG," he said. "I would say at least 9 feet. Maybe less, maybe more, because I didn't stick around too long to do any measuring. . . . The whole body was covered with hair. . . . I would say it was kind of a grayish color, from where I was standing. Of course the sun was coming up facing me, but it wasn't that bright. . . . The face, I couldn't make that out too good. . . . The hands were like yours or mine, only three times bigger, with pads on the front paws, like a dog. . . . Long legs, long arms. It was just like, I would say, like a gorilla, but this here wasn't a gorilla. . . . I'm tellin' ya, it would make your hair stand up."

After a few moments of mutual observation, the creature ran from the field toward a swamp. Bowers, an experienced outdoorsman with four bears to his credit, believed he had encountered something truly out of the ordinary. The *Concord Monitor* article concluded with a statement of Bowers' firm conviction of what he had witnessed.

> One of these days, he may use his camera and stake out the area using apples for bait. A clear photo of the animal or its tracks might convince the skeptics.

Until then he figures people can have a good laugh
at his expense. "Go ahead," he said shaking his head,
"but it's not gonna change a thing, 'cause I still see
what I see."[43]

Ed Parson writing for the *Conway Daily Sun* contributed an
article in 2006 titled, "Hiking: Myth versus Reality of Sasquatch
in Ossipee Range."

My friend Peter Samuelson, 65, of Fryeburg, Maine
has been prospecting in these mountains for more
than 40 years. Back in the late 1960s, when I was
being introduced to the area by working summers
up at the AMC, he was already prospecting far and
wide in the mountains, carrying heavy packs of supplies in, and heavy packs of stones out.

Parson had already written a series of articles regarding the
Ossipee Range and the "burden of a private race track on its frag-
ile slopes" when he was reminded by Samuelson of a Bigfoot encounter the prospector had experienced in the area.

Samuelson, his dog named Kat, and his girlfriend
Holly Swaffield, then of Wolfeboro, New Hampshire
were out prospecting in the Ossipee Range.
  'We drove in the Gilman Valley Road, parked at
the gate and continued up the old road past the
Tamworth/Ossipee town line,' he said. 'Then we cut
into the woods on the right and headed west up Bald
Mountain.'

The group bushwacked two miles to the ledges of Bald where they
hoped to find pockets of beryl and topaz crystals. Reaching a point
where Connor Pond became visible below them as the trees opened
up, they saw something 100 yards ahead of them on the ledges
which they could not explain.

"It was a small structure, yet made of big stones, stacked on each other," he said. "The roof was flat and made of thatched hemlock bows. There was an opening, like a rustic doorway. We saw a giant man-like creature inside, about seven feet high, and back to us. It was totally covered with tangled gray hair about three inches long."

In the same instant that this all became visible to them, Kat began growling intensely and the creature started to make loud noises indicating it was upset.

"I can't describe the noise," said Samuelson. "Anyway, Holly freaked and we all felt threatened. We hightailed it out of there immediately, in the direction we had come. Only later, part way down the mountain did we pause and ask ourselves, 'What did we see?'"[44]

An unusual report was made at the Hollis police headquarters in May 1977. The town of Hollis resides in Hillsborough County in southern New Hampshire and is listed on the National Register of Historic Places. The Hollis Village Historic District includes several hundred acres in Hollis' historic town center. Paul Bosquet, Hollis' Police Chief at the time, provided details of the report to the *Nashua Telegraph*.

Police here are awaiting the return of a Lowell, Massachusetts man, identified only as Mr. St. Louis, after he reported seeing a 10-foot-tall hairy monster Saturday night at the Hollis Flea Market.

St. Louis had come into police headquarters to report the incident.

St. Louis told the dispatcher that he, his wife and two sons were sleeping in their pickup on the flea market grounds, when they were awakened by the

shaking of their truck. The man said he looked out
and saw what he described as a 10-foot-tall hairy
animal with "human-like" features, shaking the
truck.

Police investigated the scene to discover a covering of pine
needles bore no evidence of footprints. Police concluded a bear
searching nearby trash containers for a meal may have been the
creature that alarmed St. Louis.[45]

RHODE ISLAND

"As it turns out, Rhode Island is covered in thick woods. In fact,
59% of Rhode Island's land is forested, and even though it's a small
state, that adds up to 393,000 acres of woods! Most of this land is
cultivated farms being left fallow, and the surrounding forests reclaiming the pastures and crops."[46] Subsequent to the state's peak
in farming in the early 1800s, forests reclaimed much land abandoned by farmers moving west. This regeneration and increase in
forested land continued to about the 1950s, when development
pressures began to impact Rhode Island's forested land. Rhode
Island is the third overall most densely populated state and ninth
in the nation in percentage of forest cover.[47]

A 1975 encounter that occurred in the village of Wakefield,
Washington County, Rhode Island was submitted to bigfoot-
encounters.com by the witness who at the time was a student at
the University of Rhode Island. He departed his girlfriend's house
and began bicycling to campus about 10:00 pm on a Sunday night.
Sounds of something large in a field next to him focused him that
he was sharing his evening ride with something close by which with
a bipedal-sounding gait, was compressing the ground with each
step. The man's instinct began to warn him that he was in danger.

As I began to ride, out from the shadows steps into
the street light a white hairy looking gorilla directly
across the street from me but on my right side.

Twenty feet at the most, it stood looking at me, eye-ball to eyeball. I noticed that it was standing bent over with its arms straight, knuckles on the ground and knees bent. About six feet tall, big chest and arms, black hair around the upper lip and down along the side of the mouth to the neck, the chest had no hair. The eyes set close together and deep, the nose human looking but very wide. Short neck that blended with the top of the head and back giving the appearance of a gorilla. Shoulders are high, round and plane out to the middle of the head in that position, guessing about 400-450 pounds easy.

Now it decides to chase me and chase it does for about five maybe ten yards, it stood up straight then ran placing one foot in front of other for a few steps, then arms down on its knuckles and back up. I was in first gear my heart was outside my chest somewhere as I pulled away, looking back all the time, it stopped, watched me for a few, turned away, ran and cleared a stone wall about five feet high all in one motion.[48]

A mother and son encountered a large yellowish-white "ape" in 1978 in Charleston, Washington County. The son, twenty-five years after the encounter, provided a narrative of the story to the Bigfoot Field Researchers Organization. His hesitation was explained in his first sentence, and seems to be a sentiment shared by many who wait to tell their story. "I've told very few people this story for fear that they might think I'm crazy."

After a thunder and lightning show earlier in the day, the boy and his mother began driving the back roads near their home looking for deer in an area close to the Indian Cedar Swamp. "The road was very narrow and the brush along the side of the road made it difficult to back the car around to turn around. The rain was falling but not heavy, the wipers were on, the headlights on the car were on as well."

As his mother turned the car around, the headlights captured in their beam something truly unexpected. "Beside the tall fractured stump stood what looked like a large white (yellow white) ape. It was maybe 6 to 7 feet tall, its hair was long, face flat, long massive arms, its head appeared to be without any neck, its chest was broad."

After 5-10 seconds, his mother slammed the car in reverse and the pair sped from the spot, with no further sighting taking place. "We were clear as to what we saw and we have no question in our mind that we saw something or someone in ape-like shape that night."[49]

The *Finding Bigfoot* team came to Rhode Island (third episode of Season 2) to investigate the "Big Rhodey" video. This video was shot by a group called Beyond the Veil Paranormal Investigations Group while shooting "B-roll" footage for a documentary they were producing on local ghosts.

Finding Bigfoot cast member Cliff Barackman posted a behind-the-scenes commentary on the "Big Rhodey" episode to his website. He wrote of his initial reaction to doing an episode in the Ocean State. "When I first heard that we would be traveling to Rhode Island for an episode of Finding Bigfoot, I was surprised. There were very few sighting reports from Rhode Island and I thought that perhaps our time could be better spent elsewhere. I soon saw the video evidence that we were to investigate, and I thought to myself that perhaps there was something going on there that needed looking into. After all, some of my favorite Bigfooting places have no sighting reports from them, and maybe Rhode Island was one of these."[50]

The team was able to use trees and a three-foot-tall stone wall to come to a consensus on the location of the film subject. The re-creation made the *Finding Bigfoot* team confident that the subject appears to be smaller than cast member Bobo (who stands 6 six feet four inches). "It seems unlikely that it was a human, unless the witnesses were involved in a hoax of some sort. However, after spending time with both Dina and Kris, I do not get the impression that they were trying to pull one over on us. I am left a bit baffled."[51]

Further investigation in the Ocean State conducted for this *Finding Bigfoot* episode included analysis of three prints near the northern border of Connecticut which were adjudged to be human in origin.

VERMONT

Vermont's forests are comprised of conifers and northern hardwoods. Mount Mansfield at 4,400 feet is the highest point in Vermont and is part of the north-south Green Mountains. "Just east of Whitehall,[52] sightings have been reported along Route 4 and all across the state of Vermont from Bennington to Colchester."[53]

Vermont was home to an animal called Old Slipperyskin, a cunning, powerful, and sometimes raucous creature of the Northeast Kingdom. An early written account has been found by Vermont's *Northland Journal*.[54] A man named Duluth wrote in his journal in 1759 that upon returning from a raid on the Odanak, the scouts "were ever being annoid [sic], for naught reason, by a large black bear, who would [sic] large pinecones and nuts down upon us from trees and ledges, the Indians being also disgusted, and knowe [sic] him, and call him Wejuk or Wet Skine." Old Slipperyskin persisted into the 1800s. He was named a bear but in unbear-like fashion was not observed on four feet but instead said to move on his hind legs.

A strange creature first sighted in Northfield Vermont in the early 1970s was referred to as the Pig Man. Joseph A. Citro's *Weird New England* includes a farmer's encounter with the creature. "It was definitely naked," the farmer said. "And it ran on two feet."[55] This entity differed from a typical Bigfoot description in that it had the face of a pig. This "neo-savage" was active at night, capitalizing on accessible pets and garbage cans for its subsistence.

Route 7 is a north-south highway which has experienced several Bigfoot reports by drivers, commonly around the Bennington area. Noah Hoffenberg's "Man Spots 'Bigfoot'," was a 2003 article in the *Bennington Banner* describing one of these encounters on Route 7. Ray Dufresne, the article's witness was heading north, "when he spotted a 6-foot-plus tall, 270-pound, "big black thing" walking upright from near the highest point of elevation on Route 7." Dufresne offered he was a hunter since the age of 15 and knew what he was seeing was not any usual fauna, but something quite extraordinary.

Although Dufresne couldn't see the animal's face, he said from his vantage point of about 140 feet away he could see that the creature had very long arms and was covered in long, black hair. He said it walked east into the woods toward Glastenbury Mountain.

Dufresne continued on driving north toward his home, but the experience had certainly astonished him. "Why would someone be walking down the road like that?" Dufresne said he asked himself at the time. "I kicked myself for not turning back."[56]

As Joseph Citro indicates in *Weird New England*, reports of an unidentified creature from the Bennington area have a long record. Along with Bigfoot or Bigfoot-like creatures, the area has a history as a place of high strangeness:

> But it is not a new mystery: It has existed for hundreds of years. The area that includes Glastenbury Mountain has long been known for the odd stuff that goes on there. It was the scene of multiple vanishings during the 1940s and 50s, but long before that, many of the earliest settlers reported spook lights, formless phantoms, unidentifiable sounds, and mysterious odors. This was, and apparently still is, the stomping grounds for the Bennington Monster.[57]

The second episode of *Finding Bigfoot's* third season took place in Vermont. "We've finally come to do some Bigfooting in the center of Vermont," from BFRO founder Matt Moneymaker's opening narration. "I'm pretty excited to be here. I had no idea Vermont really looked like this. There's population down in the valleys, but there's hills and mountains and the whole place looks like the Adirondacks. It's very squatchy in every direction."[58]

The lure for the TV series' visit was a trail camera photo taken on private property September 3, 2010, in western Vermont. The property owners had noticed fruit was missing from an apple tree they usually kept an eye on in order to observe wildlife. The missing apples seemed quite unusual for the height off the ground they were taken. In order to find out what kind of animal was taking the apples, the owners deployed a trail camera which caught a coyote eating fallen apples on two consecutive nights. On September 3rd, the camera caught the image that brought the *Finding Bigfoot* TV show to Vermont. "Around midnight on September 3rd, the camera took a photograph of what appears to be a Sasquatch bending down in a squatting fashion and reaching towards the ground with its left arm. The creature's head is to the left of the photograph, and its butt is sticking out to the right (complete with a hint of the natal cleft) with its legs bending below it, though the knee is obscured by the creature's left arm."[59] Observers of the photo also pointed out objects they interpreted as possibly being a juvenile hanging onto the larger creature. "Interestingly enough, there's a darker-colored figure seemingly grabbing on underneath this thing. You see what appears to be knuckles and darker fur. It's well known that apes carry around their juveniles on their chest, even humans do it, certainly Sasquatches would be expected to do the same behavior."

The image showed white splotches on the creature's back, that could be explained by the creature having vitiligo, a rare condition of skin depigmentation as explained by Cliff Barackman, "As far as these white patches go, there are a couple of very reasonable explanations. First of all, perhaps it's leaf litter, perhaps it's just missing patches of fur from being in the woods. Another possibility is vitiligo. Vitiligo is a skin pigment disease that makes the skin lose their pigment and the hair on top also loses it pigment. Remember, Sasquatches are very genetically close to us, it's very likely they are subject to human diseases as well."

The team split up in pairs and did a night investigation near the location where the photo was taken. One pair heard an owl and saw deer. The other group comprising Moneymaker and Fay, did some wood knocks, and actually heard what they interpreted as return knocks. The men believed there were Bigfoot in the area driving deer down the hill, however their attempt at interception was unsuccessful.

Moneymaker, Holland and Fay conducted a town hall meeting in West Rutland. Moneymaker: "There's never been a Bigfoot conference in Vermont, or any kind of a meeting of Bigfoot researchers, so no one's ever really shook the bushes here to see how many witnesses would come forward. This town hall meeting that we'll do is going to be the first time that's ever happened." The team (minus Barackman who was conducting solo field investigation) investigated several witnesses' claims of encountering Bigfoot.

Barackman summed up his thoughts on the photo which inspired the TV show to come to Vermont in the first place. "The arguments that this [the photo] shows anything but a Bigfoot or a human in a suit seem silly to me. If it is a human in a suit, then this is the only photograph of the suit in existence, and it's most likely that a hoaxer who spent his or her time, money, and effort to make such a suit would eventually want to use it again. The size of the figure is certainly within the human range, but many sasquatches are also in that same size range. If it is a hoax, then the perpetrator not only went out of his or her way to introduce anomalies, such as the possible vitiligo, but also created a juvenile hanging on the front in a decidedly ape-like fashion."[60]

4
WILD MEN AND BIGFOOT IN MAINE
—NATIVE TRADITIONS

Native Americans have been in Maine since about 10,000 B.C. Maine's Native Americans are collectively known as the Wabanaki or People of the Dawn. The Wabanaki are divided into four tribes: The Micmacs and Maliseets, based in Aroostook County, the Penobscots, of Indian Island, on the Penobscot River near Orono, and the Passamaquoddy, based in Washington County.[1] The Wabanaki belong to the Algonquian linguistic family which includes the Native Americans of New England as well as the Canadian Maritimes.

An interest in Native American history, particularly from an archeological perspective, began growing in Maine during the mid-eighteenth century. Hosmer's account below shows this curiosity and includes an interesting reference to a large skeleton.

When the first settlers came here, the island was an unbroken wilderness. No evidences were found indicating that it had ever been occupied by white men, and probably but few had ever landed upon its shores. The Indians had made some parts of it places of occupancy, at certain seasons of the year, for the purpose of obtaining a supply of food from the clam-flats, and evidences of their occupancy were to be found where those shellfish were in great abundance, and the depth of the shells in the ground shows that they must have been centuries in accumulating, and they also cover, in many places, considerable space. Then the land was plowed, the spots occupied by their wigwams were easily discernible, and it is probable that the times of their visitation were at such times when other food was not so easily obtainable. Occasionally skeletons have been found, and at one time two were discovered under the roots of a large hard-wood tree; it had grown to a large size and was in a state of decay, when it was blown over during a storm. One was that of a person of ordinary size, the other of one who was at least eight feet in height, and between the ribs of the larger was found the head of a dart made of copper. They lay nearly side by side, and had been probably engaged in mortal conflict, the larger one mortally wounded by the smaller, and the smaller probably fell by the hands of the larger. This conflict must have happened a long time before discovery, as they must have lain upon the ground, and the tree which grew over them must have been a long time in attaining its growth. They were found nearly sixty years ago, and a medical man then residing in the town gathered up the skeleton of the larger one and preserved it; upon putting the bones together, he stated that the height of the man was what is above stated.[2]

Kathy Strain, a professional anthropologist, conducted research and field work on North American Native beliefs in the Hairy Man for about 20 years. "As an anthropologist in California, I am very familiar with the traditional stories of our local tribes, I noted, as have others, a common motif of a child-stealing giant with a basket on his/her back." Strain wrote in her 2008 book, *Giants, Cannibals & Monsters*. "Intrigued, I began to find other similarities throughout the U.S.—cannibal giants, hairy monsters and human-like beings that lived in the woods. Could these characters be Bigfoot? Of course, none of these stories directly used the term

'Bigfoot', but that would be expected since that term was only invented in the 1950s. But if these stories were about a Bigfoot-like creature, what would it mean?"[3]

The Algonquian-speaking peoples' mythology contains many legends of giants with attendant cannibalistic themes.[4] The Passamaquoddy and Penobscot have the Ki wa''kwe and Potamkin; the Micmac have stories of the Chenoo or Djenu.[5] This latter monster resembles the Stone Giants of the Iroquois and the Windigo of the Central Algonquian peoples. The linguistic term Windigo may be found as Wendigo, Wittiko, etc., and has become a generic name applied to these monsters.[6] Of these mythological beings, the Windigo "survives" into the modern context, being entrenched in popular culture from the titular subject of horror movies to a Marvel Comics supervillain.

Strain offers a caveat in examining Native American myths, encouraging a careful, informed approach to investigating this rich panoply of folklore. "It is very common in Native American traditions to 'alter' or 'hide' the true subject of the story in order to avoid taboos or taboo characters. In order to ease you into acceptance of the taboo creature, elements of a taboo character are often mixed with common animals or humans. You will see this often in this collection of stories where the Bigfoot-like character can talk, has a home, and can do other human activities. While some will take these characteristics literally, one should be cautioned that the true intent of the storyteller may not be available for us to know."[7]

Bruce Bourque, Chief Archaeologist and Curator of Ethnography at the Maine State Museum in Augusta included an enigmatic story in his book, *Twelve Thousand Years: American Indians in Maine*:

> "The First Mainers—Snowy Owl next started out to find the camps of human beings to get him a wife. He traveled far to the south, and on the way noticed how the lakes and rivers were drying up. Desiring to learn the cause of the water shrinkage, he ascended

the valleys and finally reached a place where he saw what he thought were hillocks covered with brown vegetation moving slowly about. Upon closer scrutiny he learned that these masses were really the backs of great animals with long teeth, animals so huge that when they lay down they could not get up. He saw that they drank for half a day, thus taking up all the water in the basins of the land. Snowy Owl decided that someday he would have to kill them.

Snowy Owl proceeded then to find the monsters which he had seen before. He went to where the animals had their "yards." He cut certain trees where he had observed the monsters were accustomed to lean for rest at night, almost through, so that when the monsters would lean on them they would break. When the creatures went to rest at night leaning against the trees, they fell upon the sharpened stumps when the top bent over and broke and could not get up again, and Snowy Owl shot them all."[8]

This story is the Penobscot Indian version of a widespread northeastern Algonquian myth of "great animals with long teeth." Bourque comments, "The creature described in the legend bears a fascinating resemblance to reconstructions of mammoths and mastodons. Could this tale be a dimly remembered and myth-laden account of actual elephant hunts by Indians of the distant past, handed down through generations since the end of the Ice Age?"[9] This fascinating possibility underscores the relevance of studying Native American beliefs and the potential insights that may be gleaned. Myths, after all, may be grounded in fact.

In 1869 John Gyles was captured and held captive by the Penobscot for six years in Maine. He wrote an account of this time including an account of the sweat house in powwows.

There was an old squaw who was kind to captives, and never joined with them in their powwowing, to whom I manifested an earnest desire to see their management. She told me that if they knew of my being there they would kill me, and that when she was a girl she had known young persons to be taken away by a hairy man, and therefore she would not advise me to go, lest the hairy man should carry me away. I told her I was not afraid of the hairy man, nor could he hurt me if she would not discover me at the powwows. At length she promised me she would not, but charged me to be careful of myself.[10]

What then, may be gleaned in terms of Native American perceptions of these creatures? Loren Coleman wrote, "The natives clearly thought of these hairy cannibal giants as non-Indians."[11] Michael Ames and Marjorie Halpin explored the idea that these creatures are explained in terms of human metamorphosis rather than as a separate species.[12] These monsters are non-human in that they are indeed monsters, creatures that have succumbed to a state of change that changed their appearance and their behavior culminating in anthropophagy. "This transformation" write Ames and Halpin in *Manlike Monsters on Trial*, "is often symbolized by prolonged exposure to cold, which eventuates in a frozen heart or a piece of ice in the stomach; the effective cause of the transformation is frequently thought to be possession by an evil spirit or sorcerer and/or the eating of the heart or liver of one of these monsters."[13] Halpin and Ames explore more of these creatures' milieu:

> The giants from these Algonkian myths may be of either sex, sometimes sport stone coverings, and favor the colder climes of the north undertaking migrations in that direction during warmer seasons. Stories of battles between good and evil giants were often depicted replete with uprooted trees swung as weapons. Undoing the altered state or preventing a monster from regenerating after it was killed could

be accomplished by melting the ice inside it. One who took on the form of a monster could return to normalcy by swallowing salt, hot grease, or tallow in order to melt the ice inside. Although monsters, these creatures could be befriended if shown kindness, appearing then in myths as powerful allies for the protagonists.[14]

The apotamkin is a creature from Maliseet-Passamaquoddy folklore and belief-system that acted as a bogeyman. Essentially humanoid in shape, it is covered in long hair and noted for long sharp teeth. As a nursery bogey, parents relate its stories to keep children from wandering away especially onto thin ice.[15]

The Micmacs in Maine told stories of the Gougou,[16] Gugwes, Kookwes, Djenu or Chenoo.[17] The Chenoos were a tribe of cannibals, who lived in the north, and occasionally came among the Indians, seeking their prey. They were so terrible that their yell was fatal. Their weapon was a dragon's horn; and when this was thrust into the ear, it penetrated the head, and, if extended to a tree, wound itself around it in such a manner that it could not be unfastened. This was the only way in which a Chenoo could be subdued; but, even then, it was a very difficult matter to kill him. A fire must be kindled, and every particle of flesh consumed; for, if the smallest part remains, the Chenoo will spring from it again. After the flesh has been consumed, the heart must be melted. This is a block of ice, so hard and cold that it puts out the fire again and again. Cases are believed to have occurred of Indians becoming transformed into Chenoos, and of Chenoos being converted to Christianity. The Kóokwes was a huge giant, covered with hair, a cannibal, and possessed of magical powers.[18] Loren Coleman from his book *Bigfoot! The True Story of Apes in America* writes of them,

> "These cannibals have big hands, and faces like bears," noted folklorist Elsie Clews Parsons in her 1925 paper in the *Journal of American Folklore*. "If one saw a man coming, he would lie down and beat

his chest, producing a sound like a partridge." While the behavior attributed to this creature is puzzling, the theme of the one-tone whistle (i.e., the call of the gray partridge, *Perdix perdix*, of southeastern Canada) appears elsewhere as part of the unknown primate's behavior in America.[19]

The Indians' ruminations of the unknown has taken form in his belief in giants, which everywhere seems to shadow him, and which gives to his stories an air of mystery and tragedy.

> The Chenoo, also, was a cannibal, a giant from the north with heart of ice and stone. He was a monster with extraordinary powers of evil, and the Indians feared these creatures of the imagination all the more because they believed that they were men who had been transformed into giants because of their evil deeds.[20]

The Penobscot told tales of the Kiwakwa, which according to Frank Speck's 1935 article "Penobscot Tales and Religious Beliefs" in the *Journal of American Folklore* means "going about in the woods."[21] They follow the giant, cannibalistic theme and are transformed into their monstrous state by black magic. Usually appearing in winter, insatiable hunger can cause them to eat their own lips.[22] Here is a short tale of the Kiwakwa, or snow-demon: "Two men were hunting on a lake. Suddenly they heard some one whooping along the lake, at the foot of the lake. They went out, and again they heard him whooping. They saw him coming. Right up he comes to where they are. They said to him: 'Won't you eat?' The man said: 'I cannot stop; I must go to where it is cold, to the north.' That man must have been a Kiwa'kw."[23] Also known are the archetypal "Stone giants" of the Iroquois and Algonquians, named for their employment of stones as weapons. These creatures wrestle one another, uproot small trees, and rend the earth when angry.[24]

As mentioned earlier in this chapter, Wendigo became a common term for the giants of Algonquian mythology.[25] The Wendigo was featured in many stories told by Mainers, especially in the northern portion of the State. Loggers, trappers, and even Maine Guides latched onto the Wendigo story and in their retelling added elements destined to become component parts of the legend. A 1905 article from the *Boston Globe* shed light onto how the Wendigo story was becoming an amalgam of Native and European American storytelling.

> The wendigo was a tale saved for the time of falling snow. A wendigo will not hurt you as long as you don't walk across its tracks. One telling of the tale says the wendigo is part Frenchman, part Indian and a good part Old Nick.[26]
>
> A Wendigo will take a shine to some travelers in the woods, shadowing them for days, providing woods for the fire and sitting up among the hemlocks will watch over the camp all night, tending the fire. They will also provide game for those they liked. These are some of the better characteristics of the wendigo, there is a darker side. They never speak. And when you come across their mocassined tracks, it is best to walk around them. For those who dare to walk over the path, the wendigo will know of them at that instant, and will begin a different kind of shadowing, this time not in a helpful vein but seeking prey. Guide 4806 offered this: "You have read in the papers about city folks going into the woods for pleasure or game and never coming back. People think they starve to death and have their bodies eaten by foxes, or else they get stuck in bogs and sink from sight. But those ideas are all bosh.
>
> "The real reason for their disappearance is that they have walked across the track of some wendigo, after which it is all day with them. Nobody knows what becomes of the bodies.

> "The Frenchmen from Canada say the wendigo does not kill them at all, but binds them with vines and swaps them with the Old Nick for new rum or elderberry wine or split. Anyhow, nobody ever sees anything more of them."[27]

A Wendigo's footprints were said to be up to 24 inches long. In the middle of each track is a red stain for the Wendigo is said to sweat blood.[28] Human blood ran cold like water under the icy cap of the Allagash when one realized the Wendigo was in pursuit, it's haunting voice trailing one "like the moaning of the pines."[29] Perhaps the spread of the Wendigo myth had a cultural purpose—one that was transmuted to Europeans, especially woodsmen and trappers—an attempt to prevent cannibalism.

> "All Algonquin traditions associate the Wendigo with winter, the north, coldness, famine, and starvation. If a human ever resorted to eating the body of another human being as a result of famine, he was at risk of becoming a Wendigo—a violent and insatiable monster obsessed with eating human flesh. Perhaps stories about Wendigo served as a cultural deterrent against cannibalism, especially in times of extreme hardship."[30]

THE CHENOO LEGENDS

I. THE CHENOO, OR THE STORY OF A CANNIBAL WITH AN ICY HEART (*Micmac and Passamaquoddy*)—Of the old time. An Indian, with his wife and their little boy, went one autumn far away to hunt in the northwest. And having found a fit place to pass the winter, they built a wigwam. The man brought home the game, the woman dressed and dried the meat, the small boy played about shooting birds with bow and arrow; in Indian-wise all went well.

One afternoon, when the man was away and the wife gathering wood, she heard a rustling in the bushes, as though some beast were brushing through them, and, looking up, she saw with horror something worse than the worst she had feared. It was an awful face glaring at her,—a something made of devil, man, and beast in their most dreadful forms. It was like a haggard old man, with wolfish eyes; he was stark naked; his shoulders and lips were gnawed away, as if, when mad with hunger, he had eaten his own flesh. He carried a bundle on his back. The woman had heard of the terrible Chenoo, the being who comes from the far, icy north, a creature who is a man grown to be both devil and cannibal, and saw at once that this was one of them.

Truly she was in trouble; but dire need gives quick wit, as it was with this woman, who, instead of showing fear, ran up and addressed him with fair words, as "My dear father," pretending surprise and joy, and, telling him how glad her heart was, asked where he had been so long. The Chenoo was amazed beyond measure at such a greeting where he expected yells and prayers, and in mute wonder let himself be led into the wigwam.

She was a wise and good woman. She took him in; she said she was sorry to see him so woe-begone; she pitied his sad state; she brought a suit of her husband's clothes; she told him to dress himself and be cleaned. He did as she bade. He sat by the side of the wigwam, and looked surly and sad, but kept quiet. It was all a new thing to him.

She arose and went out. She kept gathering sticks. The Chenoo rose and followed her. She was in great fear. "Now," she thought, "my death is near; now he will kill and devour me."

The Chenoo came to her. He said, "Give me the axe!" She gave it, and he began to cut down the trees. Man never saw such chopping! The great pines fell right and left, like summer saplings; the boughs were hewed and split as if by a tempest. She cried out, "Noo, tabeagul boohsoogul!" "My father, there is enough!" He laid down the axe; he walked into the wigwam and sat down, always in grim silence. The woman gathered her wood, and remained as silent on the opposite side.

She heard her husband coming. She ran out and told him all. She asked him to do as she was doing. He thought it well. He went in and spoke kindly. He said, N'chilch," "My father-in-law," and asked where he had been so long. The Chenoo stared in amazement, but when he heard the man talk of all that had happened for years his fierce face grew gentler.

They had their meal; they offered him food, but he hardly touched it. He lay down to sleep. The man and his wife kept awake in terror. When the fire burned up, and it became warm, the Chenoo asked that a screen should be placed before him. He was from the ice; he could not endure heat.

For three days he stayed in the wigwam; for three days he was sullen and grim; he hardly ate. Then he seemed to change. He spoke to the woman; he asked her if she had any tallow. She told him they had much. He filled a large kettle; there was a gallon of it. He put it on the fire. When it was scalding hot he drank it all off at a draught.

He became sick; he grew pale. He cast up all the horrors and abominations of earth, things appalling to every sense. When all was over he seemed changed.[31] He lay down and slept. When he awoke he asked for food, and ate much. From that time he was king and good. They feared him no more.

The Chenoo and the Lizard.

They lived on meat such as Indians prepare. The Chenoo was tired of it. One day he said, "N'toos" (my daughter), "have you no pela weoos?" (fresh meat) She said, "No." When her husband returned the Chenoo saw that there was black mud on his snow-shoes. He asked him if there was a spring of water near. The friend said there was one half a day's journey distant. "We must go there tomorrow," said the Chenoo.

And they went together, very early. The Indian was fleet in such running. But the old man, who seemed so wasted and worn, went on his snow-shoes like the wind. They came to the spring. It was large and beautiful; the snow was all melted away around it; the border was flat and green.

Then the Chenoo stripped himself, and danced around the spring his magic dance; and soon the water began to foam, and anon to rise and fall, as if some monster below were heaving in accord with the steps and the song. The Chenoo danced faster and wilder; then the head of an immense Taktalok, or lizard, rose above the surface. The old man killed it with a blow of his hatchet. Dragging it out he began again to dance. He brought out another, the female, not so large, but still heavy as an elk. They were small spring lizards, but the Chenook [sic] had conjured them; by his magic they were made into monsters.

He dressed the game; he cut it up. He took the heads and feet and tails and all that he did not want, and cast them back into the spring. "They will grow again into many lizards," he said. When the meat was trimmed it looked like that of the bear. He bound it together with withes; he took it on his shoulders; he ran like the wind; his load was nothing.

The Indian was a great runner; in all the land was not his like; but now he lagged far behind. "Can you go no faster than that?" asked the Chenoo. "The sun is setting; the red will be black anon. At this rate it will be dark ere we get home. Get on my shoulders."

The Indian mounted on the load. The Chenoo bade him hold his head low, so that he could not be knocked off by the branches. "Brace your feet," he said, "so as to be steady." Then the old man flew like the wind,—*nebe sokano'v'jal samastukteskugul chel wegwasumug wegul*; the bushes whistled as they flew past them. They got home before sunset.

The wife was afraid to touch such meat. But her husband was persuaded to eat of it. It was like bear's meat. The Chenoo fed on it. So they all lived as friends.

Then the spring was at hand. One day the Chenoo told them that something terrible would soon come to pass. An enemy, a Chenoo, a woman, was coming like wind, yes—on the wind—from the north to kill him. There could be no escape from the battle. She would be far more furious, mad, and cruel than any male, even one of his own cruel race, could be. He knew not how the battle would end; but the man and his wife must be put in a place of safety. To keep from hearing the terrible war-whoops of the Chenoo, which is death to mortals, their ears must be closed. They must hide themselves in a cave.

Then he sent the woman for the bundle which he had brought with him, and which had hung untouched on a branch of a tree since he had been with them. And he said if she found aught in it offensive to her to throw it away, but to certainly bring him a smaller bundle which was within the other. So she went and opened it, and that which she found therein was a pair of human legs and feet, the remains of some earlier horrid meal. She threw them far away. The small bundle she brought to him.

The Chenoo opened it and took from it a pair of horns,—horns of the *chepitchcalm*, or dragon. One of them has two branches; the other is straight and smooth. They were golden-bright. He gave the straight horn to the Indian; he kept the other. He said that there were magical weapons, and the only ones of any use in the coming fight. So they waited for the foe.

And the third day came. The Chenoo was fierce and bold; he listened; he had no fear. He heard the long and awful scream—like nothing of earth—of the enemy, as she sped through the air far away in the icy north, long ere the others could hear it. And the manner of it was this: that if they without harm should live after hearing the first deadly yell of the enemy they could take no harm, and if they did but hear the answering shout of their friend all would be well with them. But he said, "Should you hear me call for help, then hasten with the horn, and you may save my life."

They did as he bade: they stopped their ears; they hid in a deep hole dug in the ground. All at once the cry of the foe burst on them like screaming thunder; their ears rang with pain: they were well-nigh killed, for all the care they had taken. But then they heard the answering cry of their friend, and were no longer in danger from mere noise.

The battle begun, the fight was fearful. The monsters, by their magic with their rage, rose to the size of mountains. The tall pines were torn up, the ground trembled as in an earthquake, rocks crashed upon rocks, the conflict deepened and darkened; no tempest was ever so terrible. Then the male Chenoo was heard crying: "*N'loosook! choogooye! Abog unumooe!*" "My son-in-law, come and help me!"

He ran to the fight. What he saw was terrible! The Chenoos, who upright would have risen far above the clouds as giants of hideous form, were struggling on the ground. The female seemed to be the conqueror. She was holding her foe down, she knelt on him, she was doing all she could to thrust her dragon's horn into his ear. And he, to avoid death, was moving his head rapidly from side to side, while she, mocking his cries, said, "You have no son-in-law to help you." *Neen nabujjeole*, "I'll take your cursed life, and eat your liver."

The Indian was so small by these giants that the stranger did not notice him. "Now," said his friend, "thrust the horn into her ear!" He did this with a well-directed blow; he struck hard; the point entered her head. At the touch it sprouted quick as a flash of lightning, it darted through the head, it came out of the other ear, it had become like a long pole. It touched the ground, it struck downward, it took deep and firm root.

The male Chenoo bade him raise the other end of the horn and place it against a large tree. He did so. It coiled itself round the tree like a snake, it grew rapidly; the enemy was held hard and fast. Then the two began to dispatch her. It was a long and weary work. Such a being, to be killed at all, must be hewed into small pieces; flesh and bones all be utterly consumed by fire. Should the least fragment remain unburnt, from it would spring a grown Chenoo, with all the force and fire of the first.

The fury of battle past, the Chenoos had become of their usual size. The victor hewed the enemy to small pieces, to be revenged for the insult and threat as to eating his liver. He, having roasted that part of his captive, ate it before her; while she was yet alive he did this. He told her she was served as she would have served him.

But the hardest task of all was to come. It was to burn or melt the heart. It was of ice, and more than ice: as much colder as ice is colder than fire, as much harder as ice is harder than water. When placed in the fire it put out the flame, yet by long burning it melted slowly, until they at last broke it to fragments with a hatchet, and then melted these. So they returned to the camp.

Spring came. The snows of winter, as water, ran down the rivers to the sea; the ice and snow which had encamped on the inland hills sought the shore. So did the Indian and his wife; the Chenoo, with softened soul, went with them. Now he was becoming a man like other men. Before going they built a canoe for the old man: they did not cover it with birch bark; they made it of moose-skin. In it they placed a part of their venison and skins. The Chenoo took his place in it; they took the lead, he followed.

And after winding on with the river, down rapids and under forest-boughs, they came out into the sunshine, on a broad, beautiful lake. But suddenly, when midway in the water, the Chenoo laid flat in the canoe, as if to hide himself. And to explain this he said that he had just then been discovered by another Chenoo, who was standing on the top of a mountain, whose dim blue outline could just be seen stretching far away to the north. "He has seen me," he said, "but he cannot see you. Nor can he behold me now; but should he discover me again, his wrath will be roused. Then he will attack me; I know not who might conquer. I prefer peace."

So he lay hidden, and they took his canoe in tow. But when they had crossed the lake and come to the river again, the Chenoo said that he could not travel further by water. He would walk the woods, but sail on streams no more. So they told him where they meant to camp that night. He started over mountains and through woods and up rocks, a far, round-about journey. And the man and his wife went down the river in a spring freshet, headlong with the rapids. But when they had paddled round the point where they meant to pass the night, they saw smoke rising among the trees, and on landing they found the Chenoo sleeping soundly by the fire which had been built for them.

This he repeated for several days. But as they went south a great change came over him. He was a being of the north. Ice and snow had no effect on him, but he could not endure the soft airs of summer. He grew weaker and weaker; when they had reached their village he had to be carried like a little

child. He had grown gentle. His fierce and formi-
dable face was now like that of a man. His wounds
had healed; his teeth no longer grinned wildly all the
time. The people gathered round him in wonder.

He was dying. This was after the white men had
come. They sent for a priest. He found the Chenoo
as ignorant of all religion as a wild beast. At first he
would repel the father in anger. Then he listened and
learned the truth. So the old heathen's heart
changed; he was deeply moved. He asked to be bap-
tized, and as the first tear which he had ever shed in
all his life came to his eyes he died.[32]

II. Story of a Kookwes—Some little boys were out
hunting. A *kookwes* (giant) was prowling round,
watching for his prey, hunting for people. In order
to attract boys, he imitated the noise of the cock-
partridge, the drummer; this he did by slapping the
palms of his hands upon his breast. The little boys
heard the noise, were deceived by it, and fell into
the trap. The huge giant (the giants are amazingly
strong) was a cannibal, and covered with hair like a
regular gorilla; he seized the boys, and intended to
dash their heads against a stone; but he mistook an
ant-hill for a stone, and so merely stunned them all,
except one, who was killed. The giant then placed
them all in a huge *boochkajoo* (birchen vessel),
strapped them on his back and started for home. The
boys soon recovered, and began to speculate upon
their chances for escape; it certainly must have
seemed rather a hopeless undertaking, but we never
know what we can do until we try. One of the boys
had a knife with him, and it was agreed that he
should cut a hole through the *boockajoo*, and that
they should jump out one after another, and scud
for home. In order not awaken suspicion, they waited
until they heard the limbs rattling on the bark, as

the giant passed under the trees, before the process of cutting commenced. As soon as the hole was large enough, one slipped out, and another and another, until all were gone but the dead one; the giant was so strong that he never perceived the difference in the weight of his load.

When he arrived home, he left his load outside and went into his wigwam, where he had a comrade waiting for him, to whom he communicated his good success. On opening the cage, the birds had flown, all but one (*tokoo sogoobahaijik*). They proceeded to roast the prey by impaling him on a stick and placing him before a hot fire; then they sat down by the fire to watch and wait till he was cooked.

The children soon reached their home and spread the alarm. A number of the men armed in hot haste, and pursued the giant; before the meal was cooked, they reached the place. Whiz! came an arrow, and struck in the side the giant who had carried off the children; he made a slight movement, and complained of a stitch in the side. Soon another arrow followed, and another, but so silently and so swiftly that neither perceived what they were. The fellow fell slowly over, as though falling asleep; and his companion rallied him on being so sleepy and going to sleep before his tender morsel had been tasted. Soon he also began to be troubled; sharp pains began to shoot through him, and as the arrows pierced him he also fell dead.

(The above story was related to me by Peter Toney, as an illustration of the stupidity as well as the physical strength of the giants. It will be observed how in this they resemble their brethren of European fiction; those that "our renowned Jack" slew were some of them remarkably stupid,—the Welsh giant, for instance.)[33]

III. THE GIRL CHENOO (*Micmac*)—Of the old time. Far up the Saguenay River a branch turns off to the north, running back into the land of ice and snow. Ten families went up this stream one autumn in their canoes, to be gone all winter on a hunt. Among them was a beautiful girl, twenty years of age. A young man in the band wished her to become his wife, but she flatly refused him. Perhaps she did it in such a way as to wound his pride; certainly she roused all that was savage in him, and he gave up all his mind to revenge.

He was skilled in medicine, or in magic, so he went into the woods and gathered an herb which makes people insensible. Then stealing into the lodge when all were asleep, he held it to the girl's face, until she had inhaled the odor and could not be easily awakened. Going out he made a ball of snow, and returning placed it in the hollow of her neck, in front, just below the throat. Then he retired without being discovered. So she could not awake, while the chill went to her heart.

When she awoke she was chilly, shivering, and sick. She refused to eat. This lasted long, and her parents became alarmed. They inquired what ailed her. She was ill-tempered; she said that nothing was the matter. One day, having been sent to the spring for water, she remained absent so long that her mother went to seek her. Approaching unseen, she observed her greedily eating snow. And asking her what it meant, the daughter explained that she felt within a burning sensation, which the snow relieved. More than that, she craved the snow; the taste of it was pleasant to her.

After a few days she began to grow fierce, as though she wished to kill someone. At last she begged her parents to kill her. Hitherto she had

loved them very much. Now she told them that unless they killed her she would certainly be their
death. Her whole nature was being changed.

"How can we kill you?" her mother asked.

"You must shoot at me," she replied, "with seven
arrows. And if you cannot, I shall kill you."

Seven men shot at her, as she sat in the wigwam.
She was not bound. Every arrow struck her in the
breast, but she sat firm and unmoved. Forty-nine
times they pierced her; from time to time she looked
up with an encouraging smile. When the last arrow
struck she fell dead.

Then they burned the body, as she had directed.
It was soon reduced to ashes, with the exception of
the heart, which was of the hardest ice. This required
much time to melt and break. At last all was over.

She had been brought under the power of an evil
spirit; she was rapidly being changed into a Chenoo,
a wild, fierce, unconquerable being. But she knew it
all the while, and it was against her will. So she
begged that she might be killed.

The Indians left the place; since that day none
have ever returned to it. They feared lest some small
part of the body might have remained unconsumed,
and that from it another Chenoo would rise, capable
of killing all whom she met.[34]

IV. THE STORY OF THE GREAT CHENOO, *as told by the
Passamaquoddies.* (Passamaquoddy) What the
Micmacs call a Chenoo is known to the Passama-
quoddies as a Kewahqu' or Kewoqu'. And this is their
origin. When the k'tchi m'teoulin, or Great Big
Witch, is conquered by the smaller witches, or
M'teoulinssik, they can kill him or turn him into a
Kewahqu'. He still fights, however, with the other
Kewaquiyck. When they get ready to fight, they

suddenly become as tall as the highest trees; their weapons are the trees themselves, which they up-root with great strength. And this strength depends upon the quantity or size of the piece of ice which makes the heart of the Kewahqu'. This piece of ice is like a little human figure, with hands, feet, head, and every member perfect.

The female Kewahqu' is more powerful than the male. They make a noise like a roaring lion (pee'h-tahlo), but sharper (shriller) and more frightful. Their abode is somewhere in Kas mu das doosek, in some cold region in far Northern Canada.

In summer time they rub themselves all over with poo-pooka-wigu, or fir balsam, and then roll them-selves on the ground, so that everything adheres to the body,—moss, leaves, and even sticks. This was often seen of old by Indian hunters.

Once a newly married Indian couple had, accord-ing to Indian custom, gone on the long fall and win-ter hunt. One day when the man was away an old Kewahqu' came and looked into the wigwam. The wife was frightened, but she made up her mind a once: she called him Mittunksl, or "my father." The old Kewahqu' was very proud to be called father. When she heard her husband returning she ran out and told him that a great Kewahqu' was in the camp, and that he must call him M'silhose, or "father-in-law." Sog goingin he did this, and the Kewahqu' was still more pleased. So they lived with him, and hunted with him. He was very skillful in the chase. When they came to broad and deep waters the Kewahqu' would swim them with his son-in-law on his back. He could run faster than any wild animal.

One day he told his children to go away to a great distance. "There is a great female Kewahqu' coming to fight me. In the struggle I may not know you, and may hurt you." So they went away as fast and as far as they could, but they heard the fighting, the most frightful noises, howls, yells, thundering and crashing of wood and rocks. After a time the man determined to see the fight. When he got to the place he saw a horrible sight: big trees uprooted, the giants in a deadly struggle. Then the Indian, who was very brave, and who was afraid that his father-in-law would be killed, came up and helped as much as he could, and in fact so much that between them they killed the enemy. The old Kewahqu' was badly but not fatally hurt, and the woman was very glad her father came off victorious. She had always heard that a Kewahqu' had a piece of ice for a heart. If this can be taken out, the Kewahqu' can be tamed and cured. So she made a preparation or medicine, and offered it to him. He did not know what it was, nor its strength, so he swallowed it, and it gave him a vomit. She saw something drop, so quietly picked it up: it was the figure of a man of ice; it was the Kewahqu's heart. She, not being seen or noticed, put it in the fire, when he cried, "Daughter, you are killing me now; you destroy my strength." Yet she made him take more of the medicine, and a second heart came out. This she also put on the fire. But when a third came he grabbed it from her hand, and swallowed it. However, he was almost entirely cured.

Another time an Indian village was visited by a Kewahqu', but he was driven away by magic. The people marked crosses on the trees where they expected the Kewahqu' to come. There was a great excitement among the Indians, expecting to hear their strange visitor with his frightful noises. It was the old people who gave the advice to mark crosses on the trees.

Another time an Indian of either the Passamaquoddy or Mareschite tribe was turned to a Kewahqu'.
The last time he was seen was by a party of Indian
hunters, who recognized him. He had only small
strips of clothing. "This country," he said, "is too
warm for me. I am going to a colder one."

This story from the Passamaquoddy Anglo-
Indian manuscript of Mitchell supplies some very
important deficiencies in the preceding Micmac version. We are told that the heart of the Chenoo is of ice
in human figure. This human figure is that of the
Kewahqu' himself, or rather his very self, or microcosm.
It is this, and not the liver, which is swallowed by the
victor, who thus adds another frozen "soul" to his own.
Of the three vomited by the Kewahqu', two were the
hearts of enemies whom he had conquered. He could
not give up his own, however. It is much more according to common sense that the woman should have
given the cannibal the magic medicine which made
him yield his heart that that he should voluntarily
have purged himself. In the Micmac tale he merely
relieves his stomach; in the Passama-quoddy version
he, by woman's influence, loses his icy heart.

It is interesting to observe that the use of the
Christian cross is in the additional anecdote described as *magic*.

It is the main point in the Chenoo stories that
his horrible being, this most devilish of devils, is at
first human; perhaps an unusually good girl, or
youth. From having the heart once chilled, she or he
goes on in cruelty, until at last the sufferer eats the
heart of another Chenoo, especially a female's. Then
utter wickedness ensues. It is more than probable
that this leads us back to some dark and terrible
Shaman superstition, older than we can now fathom.
There is a passage in the Edda which its translator,

Thorpe, thinks can never be explained. "I believe," he writes, "the difficulty is beyond help." The lines are as follows:

> "Loki scorched up
> In his heart's affections,
> Had found a half-burnt
> Woman's heart.
> Loki became guileful
> from that wicked woman:
> thence in the world
> are all giantesses come."

Of which Thorpe writes, "The sense of this and the following line is not apparent." Whatever obscurity exists here, it is evident that it means that Loki, having become bad, grew worse after having got the half-burnt stone of a woman's soul. That is, his own heart, half ruined, became utterly so after he had added to it the demoralized *hugstein*, soul-stone, thought-stone, or *heart* of a woman. If we assume that stone and heart are the same, the difficulty vanishes. And they are one in the Chenoo, who, like Loki, illustrates or symbolizes the passage from good to evil, which a German writer declares is quicker than thought, or that very same *Hugi* which the Norse myth puts forwards as swiftest of all runners. Loki, not as yet lost, gets the stone heart of a giantess, and becomes an utter devil at once. The Chenoo becomes an utter devil when he has swallowed the thought-stone of a giantess, and so does Loki.[35]

LEGEND OF GOUGOU

The author of the following piece, James Mac-Pherson Le Moine, was a member of the Literary and Historical Society of Quebec and

author of several books on Eastern Canadian history. Miscou
Island is a Canadian Island off the northeastern shores of New
Brunswick in the Gulf of St. Lawrence.

> Miscou of old, we think ourselves safe in consider-
> ing anything but a genial place of abode; not even to
> the most sanguine fisher, was it an earthly paradise.
> In addition to its traditions of sickness, desolation,
> death, war, and piracy, Champlain, the great histo-
> riographer, peoples it with forms uncanny and un-
> lovely, calculated, if possible, to enhance the weird
> interest the spot already possesses.
>
> In sketching it, he winds up rather jocosely, we
> are inclined to think, by marking it out as the head-
> quarters of a Satanic fiend—a female devil, who de-
> lighted in torturing the sons of men.
>
> What was the female devil like?
>
> "Old Harry" has ever, from our tenderest years,
> to our susceptible mind, typified a male devil; that
> is admitted on all hands to be bad enough, but what
> his lady, or any female member of the brood might
> be, this we unhesitatingly admit to be beyond our ken.
> According to the text of the illustrious discoverer, a
> fearful monster, in shape and size like a female giant,
> without, seemingly, the least affinity to fish, flesh,
> or fowl, haunted the humid margin of Miscou. The
> terror-stricken Indians knew it as the "Gougou." Of
> its sex, in their minds, no uncertainty existed—it
> ranked under the feminine gender. Had it the trade-
> mark of a Syren? Nothing indicates it had a tail, with
> those womanly attractions sung by poets:
>
> "Desinit in piscem, mulier Formosa superne."[36]
>
> It was certainly amphibious; sometimes, like that
> famed Syren, the Goddess Calypso, it inhabited an
> island. Like Ulysses's charmer, it was keen after
> men, red Indians especially; not to enlist them,

however, as lovers, but merely as tid-bits for its morning meal—a *bonne bouche* previous, probably, to retiring to the "Orphans' Bank," where a few porpoises, or an adult whale, would constitute its dinner. From Champlain's testimony, plainly it was an uncomely, nay, a repulsive monster—*un monstre effroyable*—and the founder of Quebec, the happy spouse of the blooming Hélène Boulé, the prettiest woman in New France, was of an appreciative turn of mind. The "Gougou," for all that, in shape resembled a woman—"*un monstre qui avait la forme d'une femme, mais fort effroyable.*" Had any one except those devoured ever been close enough to the giantess to form a correct opinion? We are again left in the dark. A St. Malo miner, it is true, le Sieur Prevert, while "prospecting for a pocket," had passed so close to the abode of the *monstre effroyable* that he heard the extraordinary hissing, *siffements étranges*, of the fiend. However, whilst thus in quest of a "Big Bonanza," whether a pocket or a vein, le Sieur Prevert, together with his ship's crew and some Indians, was fortunate enough to escape a pocket he was not looking for, the *grande poche*, great pocket, described by Champlain as the receptacle of Madame Gougou's booty. Sieur Prevert, be it remembered, was a miner, and unless his story had been corroborated to Champlain previously by Indians, we confess we would be inclined, like the stories of other miners, to accept it, *cum grano*. There is a fishy flavor about it, requiring many "grains of salt" to render it palatable.

But again this Gougou haunts us. Where, then, was the alleged resemblance to one of the softer sex? The Gougou, we are told, when seen by men, uttered "extraordinary hissings," *sifflements éstranges*. Will any one dare pretend it might not have been a fashionable Syren—Syrens, it is well known, are most common on the sea shore—showing off, before so many Ulysses, her powerful *staccato trills*, like a Calypso, a fast girl of that period, might be expected to do? What, in verity, constitutes a female "*monstre effroyable*"? Did Madame Gougou, out of her teens, sport high-heel shoes, a Grecian bend, a crinoline like Mont Blanc, a chignon Alpine in its dimensions? Here again Plutonian darkness awaits us.

Still, in this age of inquiry and intellectual development, shall we throw up the sponge and proclaim our inability to explain what sort of creature might be the Miscou Giantess, who could swallow red Indians like shrimps or doughnuts? Which "missing link" would the venerable Darwin assign to it? If it was not a "mermaid fair," could it be

That great sea-snake under the sea,

who,

From his coiled sleeps in the central deeps,
Would slowly trail himself sevenfold?

Or else, would it be a gigantic specimen of Victor Hugo's Devil Fish (like the one recently found at Newfoundland) who still lived in the popular mind, from the terror he had caused by having drawn beneath the seething sea, to his slimy and deadly embrace, some noted Indian warrior, whilst bathing, etc.[37]

PUKWUDGIE

The Pukwudgie is a small being from Wampanoag legend. A Pukwudgie's features are similar to humans, with notable differences being larger noses, fingers and ears. Pukwudgie's are grey in color with periodic references to their skin glowing.[38] These creatures would not seem to be cut from the cloth of Bigfoot, but their human-like appearance is noteworthy when one considers the possible representation of a juvenile Bigfoot.[39] A Pukwudgie is said to have the following traits:

> 2-3 feet in height
> Able to appear and disappear at will
> Ability to transform into what appears as a walking
> porcupine
> Employ magic, use poison arrows, and can create fire
> Control Tei-Pai-Wankas, which are believed to be the
> souls of Native Americans they have killed[40]

How exactly to spell and pronounce the word "Pukwudgie" seemed to be as much of a debate in the mid-1800s as it is today. Dr. Henry R. Schoolcraft used the words" "Pukwudjinees" and "Pukwudjees" in his works without differentiating between the two. Longfellow wrote about "Pukwudgininees" in his *Song of Hiawatha*; they slew the mighty Kwasind with pinecones and fir cones. In *Western Woods and Waters*, grammar expert John H. Abrahall determines that Longfellow's term is the correct one.

> According to Mr. Abrahall, the word means "little wild men of the woods (or rather mountains)," "wild" being a translation of the word "puk." "Wudjoo" or "Wudgin" meant "mountains" as demonstrated in Eliot's translation of The Book of Job "into the tongue of the Massachusetts Natives, 'Ininees' means 'little men.'"[41]

Newbery Medal-winning author Elizabeth Coatsworth, while living in Nobelboro, Maine included the following Pukwudgie anecdote in her book, *Maine Memories*. John Snow whom she mentions as having experience with the Pukwidgie was a Penobscot who lived in the vicinity of Coatsworth's Chimney Farm and was known locally for his basket weaving.

John's father was white, his mother was Indian. In his mind the old stories and traditions lay in dimmed and broken images; yet, as a sort of gypsy dweller beside the white men's houses, he still remembered something of the tales which his mother had told him as a little boy. He remembered about the great sea monster, and about the small people known as Pukwudgies. They stood no higher than a man's waist, and their heads were long, or so said the few who had seen them, for they were shy, and hard to get a glimpse of. They would visit a hunter's camp while he was away, and if they were in a good humor they might cook him a dinner and leave it for him to find simmering in his pot over the coals of a fire; or they might scrape a hide for him, or do some other kindness. But if something had happened to put them in a bad humor, they would upset all his things, throw his supplies into the bushes, hide the blankets—there was no end to their mischievous tricks.

The Pukwudgies were childish, but they were very wise. They knew the future, and if one approached them properly they would answer questions. As a man John Snow remembered that his mother had once taken him as a little boy to a rock near the mouth of the Machias River. There were old marks on it. She might have brought a present with her—he could not remember; but he did remember that he and she stood by the rock, and that she called out a question and that the Pukwudgies answered it. Was it Echo which spoke? If so, surely an Indian Echo.[42]

Like the Puckwudgie, the Nagumwasuck and Mekumwasuck of the Passamaquoddy were also little people. Nagumwasuck features include:

A perpetually sad countenance
Only 7 inches tall
Incredibly ugly, and very thin
Shy and evasive

The Nagumwasuck serve as guardian spirits for the Passama-quoddy Indians giving every appearance of loyalty and attachment to their human neighbors. However, they are self-conscious of their ugliness, and will take retributive measures against any who would dare laugh at them.

The Mekumwasuck are slightly larger than the Nagumwasuck and apparently not as attached to humans as is possible with the Nagumwasuck. Attributes of the Mekumwasuck include:[43]

3-4 feet in height
Completely hairy with long beards
Able to swim underwater
Favors mountains, grottos, rocky terrain, and river-
    banks
Wont to play pranks on humans

Joseph Citro in his book, *Passing Strange*, recounts two incidents in the early 1970s of sightings and interactions with Mekumwasuck, both of which, interestingly, involved the Catholic Church. The latter sighting of one of the diminutive creatures caused a church-sponsored dance to end early and abruptly.

5

WILD MAN ENCOUNTERS

Including Maine Wild Men in the present work highlights the subject's oft-reported similarities with our modern conception of Bigfoot. The bulk of this chapter collects in chronological order Wild Man stories, reports, and rumors, all occurring within Maine. Understanding and exploring the treatment of the Wild Man can provide insights into the often fuzzy delineation between stories of wild, feral men, and inferred great apes unsanctioned by science and popular opinion. These historical accounts occur well before the modern cultural development of Bigfoot. The Wild Man of the nineteenth and early twentieth centuries was nationally a cultural phenomenon, but with consequences—several accounts depicted real people, many of whom undoubtedly suffered from some mental illness. The aggregate of reports is interesting when taken as both a corollary to and level of support for what in the modern context is called Bigfoot.

Hairy hominoid anecdotes in the eighteenth century, similar to Sea Serpent stories, permeated society like a Victorian meme; their passage through the cultural fabric abetted by an extensive body of reports in popular magazines and newspapers. Early American newspaper accounts of these wild beings often employed terms like *"wild man"* or *"whazzit"* when referring to their protagonist. Often these Wild Man descriptions averred that the creatures were unquestionably humans—hermits who had adopted a lifestyle immersed in the wilderness and living essentially like animals. They are then, in effect, feral humans.[1] Sometimes a wild man story would appear to bleed between two high-elevation categories—at once depicting a hairy, feral, uncivilized creature of superhuman strength and stature intersecting with the hirsute subject's use of English speech, wearing of various articles of clothing, and living in crude shacks or abandoned camps. An example newspaper account covering Wild Men writ large follows:

> "What is It? As most of our readers are probably aware, there is at present roaming over the United States, and, for aught we know, making occasional excursions into British America and Mexico, a singular creature known as the 'Wild Man.' We have commented on his eccentricities, which are many and various."

This 1871 story from the *New York Times* leverages the "Wild Man" term, and its descriptions of behaviors and characteristics create a modern Bigfoot-like image.

> "He is preternaturally hirsute and ferocious, swift, and strong. He has flaming eyes, and, generally speaking, a horrent and uninviting aspect. The few rash mortals who have approached him near enough for conversation aver that he speaks an uncouth and unintelligible gibberish. So far the received descriptions of him are unanimous. The apparent discrepancy with regard to his stature, which varies in different narratives from about four feet to ten, is easily accounted for by the mental agitation under which these observations of his person must necessarily be made."[2]

The description, "preternaturally hirsute" is evocative of our modern conception of Bigfoot. And *ten* feet tall,—exaggerations only of editorial liberty for entertainment value?

The term Bigfoot, as given to a man-like forest giant, did not exist in the popular mindset prior to the *Humboldt Times'* report detailing Jerry Crew's discovery of huge footprints in the fall of 1958. Earlier than this, the impact of writer J. W. Burns' term *Sasquatch*, in terms of popularizing hairy hominids, likely had more resonance in western Canada than in the eastern United States.[3] Prior to having a relatively standardized description and a name, an encounter with a hairy hominid left observers to employ the expressive and evocative terms at hand. These terms are informed by social values and institutions (which certainly includes the press). If Bigfoot is indeed an extant creature, then it may well have been reported as a Wild Man generations ago.

Tales of Wild Men represent a universal archetype exemplified by a long tradition of the Green Man in Europe. Anthropologist Myra Shackley recounted in her book *Wildmen* (1983) European Wild Man folklore which accompanied immigrants to the New World. "The Wildman, in various manifestations, forms part of the culture and mythology of almost every society since records began, but the problem we face is—how to distinguish myth from reality?"[4]

Researchers Bartholomew and Regal indicated that humans becoming feral and undergoing a reversion to a primitive state created the foundation and perception of Wild Men in the Old World. "During the later Middle Ages, "wild people" were thought to be ordinary humans, such as hermits and eccentrics: the socially marginal and mentally deranged, who had turned to life in the wilderness and began to regress, growing a thick coat of hair and foraging for food like a wild animal."[5]

As Bluhs indicates, a change in the Wild Man mythos occurred in America as the country moved through the decades of the nineteenth century, from the early republic, to the post bellum period and into the age of industrialization. "Stories about wildmen after the middle of the nineteenth century not only troubled the line between humanity and animality, but that between authenticity and fakery as well. Wildmen were symbols of a premodern world— a world more authentic than the plastic one built by industrialization."[6]

Example of Medieval Wild Man depiction,
Pen-and-ink illustration from "Ballade d'une
home sauvage" France, circa 1500.

There may be a cautionary note regarding the development of wild man explications: too-abrupt consignment of an unknown hominoid to socio-psychological grounds rather than an anthropological cause may be a myopic mistake. To avoid the insufficient horse blanket for an elephant, careful qualification is recommended for studying Wild Men. If one Wild Man story, by hitting all the right notes, underpins one's faith in the existence of Bigfoot, then there will be at least one if not several other Wild Man accounts depicting humans living extraordinarily. The aggregate examination of Wild Man provides an exegesis reflecting in situ cultural norms, journalistic styles, and descriptive terms beholden to their time and place.

> In order to gain a clearer perspective on the fictive Wild Man in late nineteenth- and early twentieth-century American culture, it's necessary to look beyond both fiction and ahistorical theories of mind and spirit that explain the figure as a psychological projection or a necessary myth. Part of the creatures' enduring allure to both physical scientists and the general public is the distinct possibility, however unlikely, that they actually exist somewhere in earth.[7]

And here we return to the American popular image of the Wild Man depicted in the eighteenth century's print media. No less than Mark Twain evoked the Wild Man in a "light-hearted and humorous response to fantastical yet increasingly frequent newspaper reports of hairy wild men skulking in the woods at the borders of American towns and villages."[8]

> He was represented as being hairy, long-armed, and of great strength and stature; ugly and cumbrous; avoiding men, but appearing suddenly and unexpectedly to women and children; going armed with a club, but never molesting any creature, except sheep or other prey; fond of eating and drinking, and not

particular about the quality, quantity or character of the beverages and edibles; living in the woods like a wild beast; seeming oppressed and melancholy, but never angry; moaning, and sometimes howling, but never uttering articulate sounds. Such was "Old Shep" as the papers painted him. I felt the story of his life must be a sad one—a story of suffering and disappointment, and exile—a story of man's inhumanity to man in some shape or other—and I longed to persuade the secret from him.[9]

Some Wild Man accounts may likely be attributable to mental illness. An historical approach to the study of mental illness in the United States often reveals the lack of treatment and appropriate care afforded to the afflicted. Among the earliest colonists, some regarded the mentally ill as possessed by demons or victims of witchcraft. Treatment techniques included delivery to alms houses, imprisonment, or simply left untreated at home under the care of family. Lacking modern advocacy and treatment, people with mental illness often suffered as much from society's dignity-destroying treatment as from the illness itself.

New England had more than a few Wild Men of record—for example, The Otis Cannibal of Berkshire County, Massachusetts. This character was undoubtedly a human living in the Otis Hills and participating in local agricultural shows by setting up a little tent on the side with a sign out front, "Only 5¢ to see the oldest cannibal in Berkshire County."[10] With a foundation in P. T. Barnum's sometimes lurid carnival exhibitions and the growth of freak attractions, Wild Men became a common and expected exhibition in fairs. "Attending Wild Man displays—or reading about them—was simultaneously a way of resisting the cultural changes and easing the transition into the new world."[11] For understanding the popular context, the Wild Man was important for confirming man's wild origins and served as either a warning against how close we are to slipping back to a savage role, or paradoxically, as a comforting reminder of our connection to the wilds.

This chapter rounds out with a mention of the Wild Man in the setting of Maine fairs and concludes with a brief comment on hermits. How both of these groups—sideshow attractions and recluses—were regarded and perceived reveals something of people's past attitudes and hints at our modern preconceptions with such subjects.

MAINE'S WILD MAN ACCOUNTS AND SIGHTINGS

The graph below depicts the Wild Man sightings contained in this chapter by the month of occurrence. The spike in August is caused by the Durham Wild Man sightings in late summer, 1913. The two months with the highest number of Wild Man sightings correlate to berry harvests, notably blueberries. The three-month transition from the end of winter into the first two months of spring does not contain any sightings from the sources this author was able to examine.

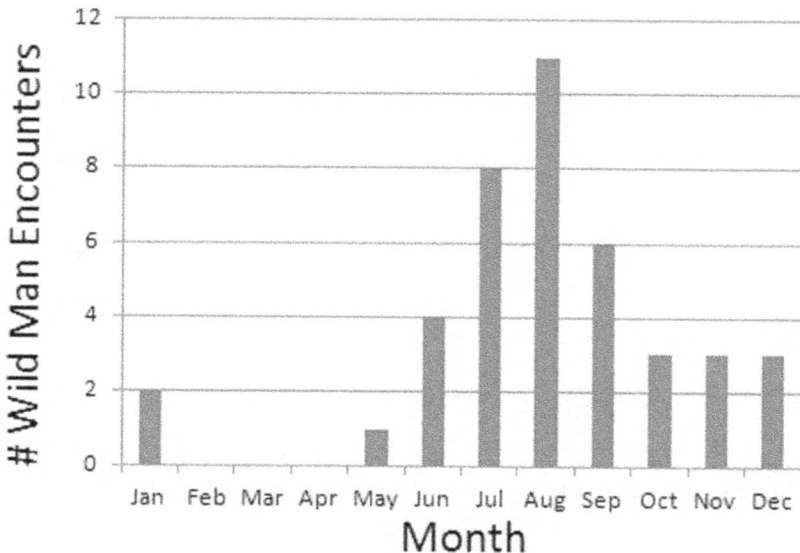

Aggregate number of Wild Man sightings by month.

WILD MAN SIGHTINGS BY YEAR

[~1788; reported in 1862]

Michele Souliere, proprietor of The Green Hand bookstore in Portland and force behind the *Strange Maine* blog, tracked down an anecdote from Josiah Pierce's *A History of the Town of Gorham, Maine*, published in 1862, which she subsequently published on her blog. An interesting feature of one of the encounters related in this story tells of a Wild Man speaking English and wielding a jackknife.

[1] Wild Men—About 1788, there was a general belief in Gorham, of certain strange men were wandering about this town, Scarborough and Westbrook. They were called "Wild men." Between the months of July and October, it is asserted, there were seen in the fields and in the woods, human beings ragged, and having long shaggy hair and beards, picking berries, green corn and peas. Upon discovering any other person, they would run away. Sometimes they were seen going out of barns early in the morning. Cows were frequently found to have been milked during the night in yards. A Miss Webb, rising very early one morning, said she saw one of the wild men going out of her father's yard, and one of the cows had been milked. Mr. Barnabas Bangs was looking for his oxen in a pasture where there were many trees and bushes, and he came suddenly upon one of these men sitting upon a log, eating a dead robin. Mr. Bangs asked him why he did not go to some house and cook his bird? The fellow rose, and brandishing a large jack knife, replied, "I will let you know the reason." Mr. Bangs, being unarmed, speedily left the place. Two boys, Ebenezer Hall and Israel Hall, were one day picking blackberries, and saw two of these wild persons coming towards them; the boys being frightened concealed themselves in the bushes. The boys

said one of them was a woman, and that they were
white people. It was said that a man in the vicinity
of Brandon's Mills, near the line of Scarborough,
being one day out in the woods with his gun, came
upon one of these men, who was eating a young pi-
geon. The Scarborough man pointed his gun at him,
and told him he would shoot him if he did not tell
him who he was, and from whence he came. The
strange man said he was one of twenty-five sailors,
the crew of a large vessel that was cast away on the
coast. No such shipwreck was known by our citizens
to have happened. Two brothers, Abraham and Eli
Webb, were one night driving a team with a load of
boards from Saccarappa to Stroudwater, and they
said they had a fair view of five of the Wild men in a
field by the side of the road; they were picking green
peas. It is said that the last time these wild men were
seen was in Scarboroguh, near Gorham and Buxton
lines, when a Mr. Lubby is said to have counted fourteen of them, in a grove of young pine trees. Not
much importance is to be attached, I suppose, to the
foregoing relation, yet there is no doubt that such
men were seen; that they were foreigners, mysterious persons. Some supposed them pirates, others,
that they were a company of the Acadians, or neutral French, who had been expatriated from Nova
Scotia. But who they were, where from, or what became of them, seems never to have been ascertained.

This account was given me in writing, some years
ago, by an aged and intelligent gentleman of Gorham, who was a boy of ten years of age when these
strangers were said to have been seen. My informant
fully believed in the truth of the story.[12]

[1871]

[1] The mysterious wild man who has been the subject of a sensation for some years was shot dead by a

citizen residing near Livermore pass, who he attempted to kill. The man was simply a dangerous
lunatic, but his name and nativity have not been ascertained.[13]

[1893]

The *Mansfield Daily Shield* story below is interesting for no less
reason than the *Daily Shield* published another Maine Wild Man
story only a few months later in September, 1893. *A Wild Man's
Prisoner* also appears in the Fulton, NY, *Weekly News and Democrat* of February 1, 1893.[14] Research has produced no corroborating
articles from Maine papers, indicating the next two stories as likely
works of fiction.

[1] 'A Wild Man's Prisoner'—The Unpleasant Adventure of a Hunter in the Maine Woods—Nick Carthen
is an old guide and hunter in the Maine wilderness
and knows the whole wild region like a book, in fact,
he has known some parts of it for forty years. Since
tourists and hunters have invaded all that region in
such numbers he hunts and traps no more on his own
account, but serves others as a guide. Like most old
hunters, he likes to talk of the good old times and
relates all his early adventures with a gusto—save one.
    The incident he does not like to think of and has
told only on occasions of extreme sociability was
when he was held prisoner by a wild man and saved
by the sagacity of his dog—a rare Bengal hound. He
was in his camp in "Township 14, in the bend of the
Allagash," as he describes it, and was getting his
evening meal when the door was suddenly pushed
open, and there stood a man over six feet high and
of most ferocious appearance. It was winter, but his
brawny arms were bare and his finger nails had
grown to regular talons. He was dressed in a motley
garb of animal skins and had an old rifle and knife.
He ravenously devoured all the food on the table,

not speaking a word, and then took his stand at the door after getting the guide's rifle.

"The next thing I knew," says Nick, "I was looking into the barrel of my gun that he was pointing at me. Then his lips seemed to move, and he uttered a sort of noise like an Injun'd make when his tongue's split, drawing his knife as he did so and laying it on the floor. At first I didn't tumble to what he meant. Then his noise got louder, and again he replaced his knife in his belt and drew it out again, motioning me to do the same. Yes, that were it; wanted me to give up the only weapon I had, and when I did not see the force 'o this and told him so he just grew terrible, shut his teeth together with a noise like a steel trap and fingered that there firearm in a way that soon made me comply. I was just in the tightest hold I'd ever been in. 'There,' I said, 'take it,' and I pitched my pelt skinner on the ground in front o' me. P'raps it was well I did, for it sort o' pacofied [sic] him, as after he'd reached and got it he lowered his gun, though he still held on to it as he leant with his back to the doorway.

THE WILD MAN SECURES THE RIFLE,

"Hours passed like this, and nighttime came. So long as I didn't move he was all right, but if I stirred the least bit that loonat brought the musket to his shoulder and hut looked h—l at me. 'Long about 1 o'clock in the morning he 'lowed me to throw more wood on the fire. Then I took the chance to stretch myself out on the b'arskin, for I felt so tired I could hardly keep awake. This, after muttering a piece, he didn't object to, and when I opened my eyes it was daylight, and there sat that cuss just the same.

"Just then my dawg, who had been coming in and out through a hole in the front timbers cut a purpose for him, brought me in a rabbit and laid it at my feet, as I'd trained him to do. This attracted the madman's 'tention, and, cocking his gun again, he made a motion which said plain enough, 'Hand that over here.' With a healthy curse, said quiet, though, to myself, I pitched it to him, and shoot me if he didn't start to tear the skin off with his teeth and eat it raw as it was, using one hand, though, and hanging onto his weapon with the other. Ugh! It nearly made me sick to see him, for he didn't look anymore like a man, but sort o' turned into a wild animal."

At this point in his story Nick usually digresses, dwelling at some length on the sagacity of dogs in general and that Bengal hound in particular, and relates how he had taught him to carry messages to Jake Hood, his nearest neighbor, and how he managed to fix a note to the dog's collar and send him off. Two hours later Jake arrived, and after a desperate struggle the madman was secured.

"It appears," says Nick in conclusion, "that he was a schoolmaster down at Portland, who went sort o' crazy 'cause his sweetheart died, and who just went way off into the woods by himself nearly twenty years afore he tumbled onto me. Folks all thought he was dead, but you see he weren't, and there he'd been living the life o' a wild animal like all alone. That's what made him look so fierce. We notified the authorities, and they came and fetched him. I never saw him again, but heerd he pegged out 'bout six months afterward, but not before he'd murdered one keeper and nearly strangled another."[15]

[2] Some Odd Stories—Interesting Tales of Adventure on Sea and Land—A Wild Man of the Maine Woods, Said to be a Cannibal—In the Clutch of the Monster—He is Smoked Out—Knocked Senseless and Captured—(Copyright, 1893, by American Press Association)—One fall, several seasons ago, accompanied by Roy Lambert, I made my way into the wooded region of northern Maine, intending to spend several weeks amid the pines, the balsamic odor of which I hoped would prove beneficial to my weak lungs. Lambert is a big, healthy fellow and an enthusiastic sportsman, so he went along for the shooting he expected to find.

We selected the Moosehead region, and at Kineo we secured a guide, a wiry, dark faced half breed, his blood being a mixture of French and Indian. Winego Joe had the reputation of knowing the northwestern portion of Maine so he could draw an accurate chart of it from memory, and he was said to be trusty and reliable. When we decided to ascend Moose river [sic] to Lake Brassau, Joe tried to dissuade us.

"Wild man he be zere," declared the half breed, shaking his head warningly. "He keel mana ze man! He do eat ze humane flesh! He ees one cannibal! No go zere—no, no!" "What's that?" cried Lambert, with sudden interest. "A cannibal? Oh, come now, Joe, that's too steep!"

The guide protested it was true, and then he took from his pocket a printed poster that offered a reward of $200 for the capture of the wild man, who was fully described, identification of the creature being easy on account of a livid scar in the shape of a horseshoe on his left cheek just below the eye. He was "wanted" on account of two murders it was supposed he had committed.

"Well, this will add spice to the trip," laughed Lambert. "if we can capture this cannibalistic wild man, the $200 will come pretty near defraying the expenses of this outing." So up Moose river [sic] we went, for all of Winego Joe's warning. We found a deserted wood hut on the northern arm of the Lake Brassau and took possession of it. The place was pretty and convenient to the water, and we thought ourselves well fixed for a week or ten days. There was plenty of small game about, and that satisfied us for the time, although we hoped to get a pop at a bear, an Indian devil or something of the kind before the trip was ended.

I am naturally a tired man, but Lambert is indefatigable, and it was his pleasure to tramp, tramp, tramp, through the woods, while I often preferred to lie on my back beneath the trees and let the mysterious whispering of the breezes lull me to sleep. The third day after we reached Brassau proved too chilly to sleep under the trees, so I took my afternoon nap in the cabin while Lambert and Joe went for a tramp through the woods. I must have been asleep two hours, when I was suddenly awakened by feeling a clutch of iron at my throat. I opened my eyes to see a hideous face close to my own, the expression upon it being more fiendish than anything I had ever imagined. The eyes of the creature seemed to fairly burn, and he licked his lips with his tongue as some wild animal might. His teeth were yellow and snaggy, while beneath his left eye was the crimson horseshoe scar. I knew in an instant that I was in the power of the wild man, and I made a desperate attempt to sit up and break away. Then I found I was like a babe in his hands. He growled and snarled like an animal all the time I was trying to throw him off, but I could see he did not have the least trouble in handling me.

The creature was so dirty that an unpleasant odor came from his body, which was covered by hair. All the clothes he wore were composed of the untanned skins of animals, and they but partially concealed his frame. Never before in all my life had I been so frightened, for the strength of the creature showed me how helpless I really was. I ceased to struggle only when I was completely exhausted, and then he swiftly bound my hands and feet, making me utterly helpless. After this was done he went over to the fireplace and scraped some coals from the ashes. In a few minutes he had a fire started.

A HIDEOUS FACE.

I watched him, wondering what he meant to do, knowing well enough that it concerned me in some way. Every now and then he would glance at me, and ran his tongue over his lips each time he did so. There was something so greedy and savage in those glances that my blood was frozen. However, after a time I managed to ask: "What do you mean to do with me?" He produced a long butcher knife, and after feeling of the edge grunted: "Hungry—fire—roast—eat—um!—much good!" Then he approached me with the evident intention of cutting my throat with that knife. I uttered a cry of terror and kicked out with my bound feet. He was struck in the stomach and sent flat. Before he could rise I heard a thud on the roof, and then the hut began to fill with smoke from the fire. By the time the wild man got upon his feet the smoke was so dense we could barely breathe. With a snarl the creature rushed for the door, which he had fastened after entering ripping it open, he leaped outside. Then I heard a heavy blow and a fall, followed by Lambert's voice crying exultingly: "I've got him, Joe! Quick—we'll tie him!"

"Make him fast!" I yelled. "He has the strength of a Sampson!" Make him fast they did, and then my friend came in and set me free. He had knocked the wild man senseless with a terrible blow that would have fractured the skull of an ordinary person. Lambert soon explained how they happened to appear at such an opportune moment. They were returning to the cabin, when my friend decided to try the fishing in a little cove of the lake, and he sent Joe for his tackle. The guide saw the wild man when he entered the hut, and he ran back to tell Lambert. They both came up and found the door fastened. Then Joe with greatest craft, climbed a tree and dropped on the roof, where he immediately spread himself over the top of the chimney, causing the room to fill with smoke and driving the wild man out. Lambert crouched by the door and knocked the creature over. We delivered him up to the authorities and received the offered reward. He afterward died in a mad-house, raving for a taste of human flesh.[16]

[3] Hunting a Wild Man—A Strange Creature Terrifies the Town of Grafton, Me.—Stage Road Blockaded and All Doors Barred—Bethel Citizens Called Upon Today for Re-Enforcements—Bethel, Me., Nov. 21 (Special).—Grafton, in Oxford county, [sic] is situated on the county road leading from Bethel to the Lakes. A stage passes over the route daily from Bethel. For more than 80 years this town has been noted for deer, bears, moose and foxes, which are very numerous among the mountains. Speckled mountain, in Grafton notch, is 4500 feet in height, while just east of the town there looms up Saddle Back, where many bears are trapped every fall. The farmers along the route are generally well-to-do and get a good living.

Everything for many years past has been very quiet until about two weeks ago a human being, supposed to have come from the island of Cuba, made his appearance, lurking about the town, disturbing the inhabitants—by day, somewhat, but chiefly in the night. Three men came down from there this morning and from them I learned the following: They were passing up a bypath one day when they noticed him sitting down under a boulder. He came up within a few rods of them and finally he spied them. Quick as a flash he darted across the road into the bushes, running with all his might and main. While getting across Bear river [sic] they pounced upon him. He turned and faced them with gleaming eyes, as if to say, "Why are you meddling with me?" They asked him what he ran for and he exclaimed, "I was running after the devil."

The other night, this same being at about 7 o'clock called at the house of Mrs. Brown in the Notch. Mrs. Brown is a widow living alone. She heard some one at the back door and taking a lamp in her hand she went and opened the door. He stood so near that her face came close up to him. She came very near fainting but she slammed the door in his face, passed in and locked the door. Then she went into the sitting-room and pulled down the curtains. At length, gaining strength, she peeped under one of the curtains where she saw the creature peeking under the same curtain. She asked him what he wanted. "Oh, nothing," he bellowed out at the top of his voice.

During the past week travel over the road has been practically suspended. The road is blockaded. Few living in the town dare to travel abroad in the night.

Strange to say not one in the vicinity know how the creature gets his living as he never calls for food.

The excitement is up to fever heat. Some conjecture that he is either a wild man or has committed an awful deed and secretes himself in his back county to keep from being caught.

The mystery is still growing. Doors are barred day and night and the whole town is besieged as was Petersburg when General Grant was in command of the Union Army.

This afternoon a dispatch was received from Grafton for reinforcements from Bethel and at 3 P.M. arrangements were being made in Bethel to send up aid.[17]

[1894]

[1] There is more or less excitement about Milo concerning an alleged wild man who is supposed to be putting in his time in the woods along the Bangor Aroostook railway between Milo and Sebec.[18]

[1895]

[1] Maine once boasted of a small tribe of wild folks, including women. They lived twenty miles north of Skowhegan, near the village of West Athens. Thirteen years ago it was said that in all they numbered forty. Rude huts and caves were their habitations, and they were a terror to the country for miles around. Sheep and cattle stolen from farmers and wild animals caught in traps furnished them with subsistence.

It is said that many of them could not speak any language and held communications only by odd noises and grunts. The men wore practically no clothing. On one occasion four of them were enticed to Norridgewock and locked up.[19]

[2] The game season developed a wild man up in South Norridgewock, which turned out to be demented old Abe Brown, an inmate of the jail, who escaped and became lost in the woods. Abe asked a man the way home, but thinking that Abe was a tramp, the stranger passed on. When Abe asked Dr. Brown and Herbert Frederick how to get back to jail he was knocked down and choked, but managed to get in a whack with a club on Brown's head. Then he was glad to get back to jail.[20]

[3] A man thought to be the wild man of Scarboro Marsh, near Portland, was found Friday in a hovel eating a broiled cat. A man of such epicurean tastes

should live in Bangor, where he might be able to introduce a devilled cat as a side dish to Bangor's favorite lunch—Frankfort sausage.[21]

[1896]

[1] There had been circuses in the area leading to some thought could the wild man be an escaped cast of an entertainment menagerie. The Walter L. Main Circus traveled through Maine in June, 1896, known for its "3-Ring Circus, 5-Continent Menagerie."[22]

[2] Residents out in the Grove Street area were much upset over what they insisted was a "wild man" living in the woods between there and Sabattus. He was first seen in the neighborhood of the late James Carville house but disappeared. Then he showed up over near the Washington Phillips home on the Webster road [sic]. Chickens in that section have been stolen recently and the blame was laid on the mysterious man. He is reported as having a heavy, ragged beard, to be bald headed and wearing no cap or hat. His coat and pants were torn and dirty and he wore no shirt. Blood curdling yells seemed to issue from the woods in that East section of Lewiston and he had been seen as far out as the Granite Mineral Spring on the old Lisbon road [sic]. Sabourin on several occasions found his cows had been milked. Strange lights and the playing of a violin had been reported as issuing from the depths of the peat bog on Sabbatus Street near the Garcelon farm. Take it all in all quite a bit of excitement had been aroused.[23]

[3] Wild Man at Large in Minot—Several Women have Seen him Rush Through the Brush Along the Highways to Minot Corner—Lewiston, Me., July 20—

The people of the neighboring town of Minot are very much wrought up over the appearance at several points in that town within the last few days of a strange man who acts like a madman and seems to avoid the sight of the general public as much as he is able.

The man was first seen on Friday morning by Miss Maud Knowlton. She works in the corn factory at Minot Corner, and was driving from Minot to the factory to work at the time. She was driving through a piece of woods about half a mile from the corn factory when suddenly an old man with long beard and hair jumped out of the bushes directly in front of her horse and stood there for a moment and then jumped into the bushes again, yelling all the while. Her horse was frightened and began to back, but she whipped him up and drove rapidly on, not stopping until she reached the factory, where she told the story. She says he was an old man, tall and heavily built, with very long hair and beard and wore no clothing.

Saturday morning he was seen again. A party consisting of Mrs. Cecil True and daughter, Inez, and Mrs. John Massie and two other women were blueberrying in Caleb Butler's pasture, on the road from Mino Corner to Perkins ridge, when they heard a noise like the breaking of limbs. On looking in the direction of the noise they saw a naked man climbing over a brush fence. He disappeared without making any outcry. Their description of him agrees with that given by Miss Knowlton.

[Reminiscent of a "veritable wild man, devoid of covering," he scared a little girl heading to the town center of Harvard, Mass., church bells were rung and all the people turned out to search for him.[24]] On Sunday a searching party of about 100 was formed and the woods and fields about Minot Corner were searched all day, but without success.

There is much speculation as to who the man can be. The only men known to be missing hereabouts are Foster Lee Randall of Lewiston and Alfred Walker of Auburn. Mr. Randall has been missing for some weeks, but though the general public know nothing of his whereabouts it is understood that his attorneys know where he is, so it cannot be him. Mr. Walker of South Auburn bought a ticket to Portland at the lower Maine Central station in this city last fall and has not been heard of since, but it is considered very doubtful if he is the wild man of Minot.

The selectmen are expected to take steps at once for the capture of the man.[25]

[4] Wild Man Seen at South Auburn—John T. W. Stinchfield Gives a Good Description of Him—Probably the Same Man Seen by Minot People—Auburn, ME., July 31—The wild man who has been frightening the people of Minot, recently, has been seen again, this time by John T. W. Stinchfield, a farmer of South Auburn, and a member of the Auburn city council.

He tells the following story of his experience: "I live about four miles south of the business section of Auburn, in the Rowes corner [sic] neighborhood. Monday forenoon I was out building fences on my farm, about half a mile from my house. I had put a top rail on a length of fence, and, on glancing up, I saw a man standing not over a rod away.

"It was no person that I have ever seen before. He was an old man, about 6 feet tall, and he stood up pretty straight. He had a full gray beard, reaching to his waist. He wore an old slouch hat, tattered coat and trousers, and an old pair of shoes, which were tied on with strings. His trousers hung in shreds and appeared to give but very little protection. His coat was in no better condition, being so ragged that he had to keep the collar turned up and fastened in order to make it stay on. He had long hair, as well as long whiskers, and both were gray. Around his waist was a rope belt with a big knife, slid in between the rope and the man's coat. I should judge that the blade of the knife was from 10 to 12 inches long.

"I had lain down my ax to put up the fence rail, and the first thing I did when I looked up and saw that I had company, was to pick it up. After I had the ax I faced the man and tried to talk with him. I asked him all the questions I could think of, but he didn't utter a word. All he did was to stand there and stare at me. He had a wild look in his eyes, and I set him down as an insane man.

"After looking at me for some minutes the man turned and ran away into the woods."

Mr. Stinchfield says that he has no idea who the man is. There is no one missing from his neighborhood, except Mr. Walker who disappeared from Auburn last fall, but he has known Mr. Walker many years, and he knows it was not he. He does no know Foster Lee Randall, but the description he gives is not that of Mr. Randall. It is probably the same man who was seen by the Minot people.[26]

[5] Aroostook—Star Herald—A wild man has been for some time past frequenting the woods near Minot and South Auburn, being seen first by Miss Knowlton, of Minot. The latter's recital of her experience has been widely published, and the Mechanic Falls Ledger states that as a result she received a letter a few days ago from Mrs. J. W. Wilder, of Washburn, Aroostook county, saying that the account of the appearance of the insane man corresponded with that of her brother Aaron [surname illegible] who

was lost last fall. If the man is caught Mrs. Wilder wishes to be notified, and she has sent a photograph to Miss Knowlton, by which he may be identified.[27]

[6] A Wild Man—The Bona Fide Jibenainosay Captured at Last—He was Caught by Deputy Sheriff Huskins at Lisbon Falls—Has Hair Down on his Shoulders and Beard a Foot Long—Officer Huskins' Interesting Story of the Man's Capture and Conversation—There was a rumor Monday morning that the wild man—the man about whom so much has been written, had been captured and was in close confinement at the county jail in Auburn.

A Journal man ran down the rumor and found there was something in it, but unfortunately the story didn't have the kind of an ending that he had hoped for. To have chronicled the capture of the wild man—the wild man who appeared to several people in the Minot woods and again to John T. W. Stinchfield, for Mr. Stinchfield, who went up to the jail to see him, says so, and what he says may be set down for the whole truth.

However, the man in Auburn jail looks wild enough and acts strangely enough to be set down as a wild man, and his story goes to show that he is more than an ordinary tramp. This man, who has been given the name of wild man was arrested near Lisbon Centre, Nov. 23, by Deputy Sheriff Huskins. Mr. Huskins was notified by telephone that there was a wild or crazy man in the highway who needed looking after. Mr. Huskins took a team and driving out toward Lisbon Centre he ran across the man. He was tall and thin. His clothes were very ragged and he had on no hat and but a poor apologies for a pair of shoes—being, in fact, an old pair of overshoes filled with holes. He wore a full beard nearly covering his face, and his hair, while not hanging over his shoulders, had grown well down in the neck. His hair was matted and his whiskers were sadly twisted. He stood all of 6 feet 6 inches and was straight as an arrow. He appeared to be about 40 years of age. His hair was light, but had no gray in it. Mr. Huskins stated his business and the strange man seemed to be willing to get into the carriage and ride off with him, for the air was raw and the poor fellow was shivering with cold. Mr. Huskins tried to talk with him on the way to Lisbon Falls, but he found his companion non-communicative. He replied several times that he came from Real Estate and that he was going to Real Estate. Mr. Huskins set the man as a decidedly uninteresting conversationalist. He took him before Trial Justice Coolidge and Mr. Coolidge sent him to Auburn jail for 30 days for vagrancy. He gave Mr. Coolidge his name as George Treeves, and he is shown on the commitment papers as hailing from "parts unknown."

Deputy Sheriff Huskins was in Auburn Monday morning, and about every other man he met wanted to know about the wild man he had captured. To all of them he gave the simple story of the arrest, and when pressed for his opinion he said that he believed that the man was nothing more nor less than a tramp, who, in addition, was "a little out in the upper story."

The Wild Man as the man who gives his name as George Treeves is called over at the county jail, tells conflicting stories. He talks very little, and has made few acquaintances among the prisoners. He has, however, unburdened himself to one or two. He tells Foreman Knowles of the jail workshop that he has been round these parts since early in the fall, that he has carried a gun and has lived in the woods a great deal. He says, too, that women have been afraid of him at different times when has asked for food. Mr. Knowles concludes from his story that he has been tramping it for several months, choosing to live for the most part in the woods and depends on what game he could shoot for food. He thinks the man intelligent and that he is sane. He is a good worker and learns quickly whatever they set him about.

The writer called on the aforesaid wild man at 11 A. M., Monday, in company with Professor Bateman, who has had wide experience as a phrenologist. Mr. Bateman examined the man's head (at long range) and gave it as his candid opinion that it was the head of a wild man. "I should swear," remarked Mr. Bateman, "that this is none other than a wild man of the most profound type. Any one with half an eye can readily perceive it. I am glad the fellow is safe behind bars. From all I have read in the Journal of the cases where people have seen strange men in the woods and along seldom frequented highways I should say that this man answers the descriptions. I am very glad to know that you have at last nailed your man."

Not the Auburn Wild Man—But the story that this was the Minot-South Auburn wild man was upset when at 11:30 A.M., Monday, John True W. Stinchfield called at the jail workshop. A wild man or a crazy man or a tramp, as will be remembered, appeared to Mr. Stinchfield while he was building a fence in the woods, last fall. Mr. Stinchfield looked up and there stood the man barely a rod away. He was tall and straight and his clothes were in tatters. Around his waist was a belt and in the belt was a big butcher knife. He stood there and stared at Mr. Stinchfield about ten minutes and then he moved off, leaving Mr. Stinchfield to make his escape, which he lost no time in doing.

Mr. Stinchfield always said he could identify the man at sight. When shown the man at Auburn jail Monday, he shook his head and said, "No, that's not the man. He's tall enough, but he isn't so stout of build and his whiskers and hair are shorter, and besides, the man I saw was all of twenty-five years older, and his whiskers and hair were as white as snow."

Treeves, the alleged wild man, has served seven days of a 60 days' sentence. Before his time is up possibly someone may be able to identify him as the person who has been in the pastures and woods of this section. He says that he is a farmer by occupation and then in the next breath he says he works in a chair shop. He says, too, that he was in the Dover, N.H. jail a year ago for vagrancy. He says that he lives on Grant Street, in Newton, Long Island.[28]

[1897]

[1] Again the Wild Man—This time he is apprehended in Auburn and sent to Auburn Jail—His feet and ears frozen and his feet to be amputated—A strange example of human drift whom no one knows—A problem as to his disposal—The wild man—with all his hair and with his burning eyes and fluttering hands, came into court in Auburn on Wednesday, again the object of the law's decrees, which are alike the law's pity and mercy.

You remember that this strange waif was once before the authorities in Auburn several weeks ago. At that time he was the burden of no end of charges of outlawry and the subject of diverse fearful tales of wild men and demons. He proved to be innocent of all these and to be nothing worse than a poor helpless giant, who was so hirsute that the only purpose for which he could be of use was to frighten children and to advertise the effectual advantages of the razor.

Officer Jenkins of Auburn found the man this time and it is about two weeks since he was released from Auburn jail. The cold weather has intervened and where the man has been and he has suffered no one can tell. He says nothing and, in all his long term in the jail, he hardly spoke more than four or five times. He gave his name this time as George F. Teves, which is not the same as when last apprehended in Lisbon and he was sent up Wednesday from the Auburn municipal court to Auburn jail for 60 days. Once in jail, and even before, it was evident that he needed the assistance of a physician for when found he had only a pair of slippers on his feet and his heels were bare of stockings to the very snows. Examination show that the man's feet and ears were frozen and that his feet will probably have to be amputated at the instep.

"If this is the case," said Sheriff Hill on Wednesday morning, "I can hardly see how we can care for the man here. In addition to this, he is insane and a great care and he ought to be at the insane hospital at Augusta where the proper care and attention can be given him. We cannot do it. The doctor says that without a doubt the poor man will lose his feet. They will have to wait a few days to see what turn they will take but there is no hope to the contrary. This is no place for anything like that and I am going today to Augusta to see Gov Powers and arrange to have a commission appointed to look into the man's sanity and have him, if insane, removed to the asylum."

The "wild man" is six feet and a half in height, is clad in tattered garments and when first caught had a rope about his waist and was a thing to strike terror to the strongest heart. He seems to suffer but

little and says nothing. No one knows his anteced-
ents, where is his home, who he is or whence he
came. He can tell nothing. He is about 45 years old,
powerful in physique but the light of his countenance
has gone out. He is a strange study—a bit of human
drift on a winter sea.[29]

[2] Sheriff Hill saw Gov. Powers at Augusta, Friday,
relative to the "wild man" at the county jail. A hearing will be given by the Governor and council next
Thursday.[30]

[3] The November 17, 1947 edition of the *Portland
Press Herald* ran its 'History from our Files' on page
4. Included under the '50 Years Ago Today' was the
following:
　　Peaks Island has a wild man that is said to con-
ceal himself in the woods by day. A searching party
has been formed to scour the woods to find him. He
broke into an unoccupied cottage and made the place
look as though a cyclone had hit it. He smashed fur-
niture, tore down partitions, and made a general
wreck of the interior.[31]

[1898]

[1] "You know it was Huskins that captured the wild
man who was terrorizing the whole southern part of
the county," said another man. "Every one was afraid
of the wild man. He had been seen crossing roads in
a nude condition."
　　"Had been reported to be armed with a long knife
and to have a beard full of burdock burrs and grass
seed. He had killed sheep and milked cattle in the
pastures. He had suddenly appeared and disappeared to lonely women left alone at distant farms.
Everybody feared him. He was seen at night, moving

about the Lewiston peat bog with a will-o-the-wisp for a lantern, and had been chased by locomotives along the M. C. R. R. track at Sebatis. He had slept in Papa Joe Bradbury's sheep pen and had stolen Tom Barnard's chickens. Everything that happened in the county was laid to him. Was a white face seen staring in at the windows of a farm house at night? Then it was the wild man.

"Huskins was travelling along the Ridge road [sic] one day when he saw something dodging out beside the road far ahead. He didn't think much of it, but when he came up to it and found that it was a very odd looking man, he suddenly remembered the wild man and leaped out and clapped his hands on his shoulders. 'I want you.' He said. The tramp looked at the officer, at the deep woods on all sides and down at the gnarled stick he held in his hands. Huskins pushed the man toward the team and he climbed in. He was only half-dressed, and the only question that he could answer was where he lived. He said he had been in the woods all summer.

"He was evidently not a fool, but was insane. He looked and acted wild, and minus his clothing would have been the wild man all right. But not till the man reached Auburn jail was he recognized as the fiend who had been torturing the imaginations of country people. And not till he was recognized up here did it occur to Huskins that he had captured the famous wild man."[32]

[2] Bad Man With Red Hair—Or the Wizzard [sic] of Witch Spring—For the last two weeks complaints have been numerous regarding the bad behavior of a man who has haunted the woods and roads near Witch Spring. He is described as being apparently a sailor, wears a blue cloth suit, has a freckled face

and red hair, his whiskers, also red, luxuriantly growing almost to his waist. He has chased children and insulted women and girls who have passed the spring. Look out for him.[33]

[3] Speaking of that Witch Spring wild man, when you consider the tax bills sent Bathites year after year it's rather remarkable that the suburban forests aren't full of wild men![34]

[4] Wild Man?—A Man Acts Queerly in the Woods about Witch Spring.—Complaints were made to the police Monday that two ladies, driving past Witch Spring, Sunday, had been chased by a man who, clad only in an undershirt, ran out of the woods.

Monday a West Bath farmer reported that he had heard yells in those woods. Two officers took a team and drove out there Sunday but failed then and have since failed to make any capture.[35]

[5] Aroostook County Woman Believes Danville Wild Man is her Better Half—The Mechanic Falls Ledger prints a story which furnishes a possible clue to the identity of the wild or insane man who has been seen several times in Minot and South Auburn within a few weeks. This is the ledger's narrative:

Since the wild man appeared to Miss Knowlton in the Minot woods, he has been seen by several people living within a few miles of the place where first seen. He was seen a few days ago by a man named Stinchfield living in the edge of Auburn. This time he had on clothing, and carried a long, dangerous knife.

Miss Knowlton's experience has been widely published in the newspapers and as a result of which she received a letter from a Mrs. J. W. Wilder of

Washburn, Aroostook County, Maine a few days ago, saying that her account of the personal appearance of the wild man corresponded with that of her brother who disappeared from his home in Washburn last fall.

This man went out hunting with a friend in September 1895 and has never been seen or heard from since.

The men separated after entering the woods agreeing to meet at a certain place at nightfall but this man did not return. The woods in the vicinity were searched for him in vain. A clairvoyant was consulted who said he was dead and that his body would be found under a pile of brush. There were many piles of boughs in the forest about Washburn and they were overturned but no body was found beneath them. What became of him is a mystery.[36]

[1899]

[1] Sabatis' Annual Wild Man—he makes his debut in a wild chase after two raspberry pickers on Hedgehog. Sabatis, Me., July 20 (Special)—It is the idea of some people in this town that the missing Orin Swett of the Portland, Freeport and Brunswick Steamship Company is prowling around the outskirts of the town, seeking whom he may devour. That is about the only solution that has been given to the apparition that scared two young raspberry pickers nearly out their wits on Wednesday morning. Gus Meister and Charlie Ewing, lads of 13 and 15 years, were picking berries in a thicket over beyond "Hedgehog"—anyone in Sabatis knows where Hedgehog is—when terrific yells saluted their young ears, and a hairy monster, or so he seemed, although to be sure Mr. Swett was not a man of unusual build, began a wild chase after the boys. Indeed, I fancy

that after his first appearance he had to do some lively sprinting tricks to keep the flying heels of the agonized youngsters in sight. It is commonly reported about town that this fearful being chased the two boys a distance of a mile or more at a breakneck speed, ready to tear their hearts out of their breasts if they had tripped or by any other means have fallen into his clutches. Messrs. Meister and Ewing could not be found on Wednesday afternoon to interview them as to the veracity of this tale. Conservative people in Sabatis are inclined to think, however, that the wild man may possibly be the irate owner of the raspberry patch, who was looking for trespassers.

Berries are usually thick in this region, and a party that started on Tuesday from Dr. F. E. Sleeper's house, found twenty-two quarts or more up in East Wales somewhere. The blueberries are ripening along the Pond road between Sabatis and Greene,

A blueberry-picking foray from early 1900s.

and on Tuesday Mr. and Mrs. Charles Dinsmore went on a carriage drive to the Brunswick blueberry plains and brought home about two bushels of the berries.[37]

[2] Peaks Island Wild Man—Strange Acting Man is Frightening Cottagers, Campers and Lovers—Portland, Me., July 19—There is a wild man sensation on Peak's Island in Portland harbor. Campers report having seen a strangely acting person for the past day or two, who has disappeared in the woods when approached. This evening, just before dark, a young couple were badly frightened by evidently the same individual who suddenly sprang before them and then made off into the underbrush with sharp cries like those of a frightened dog. A hat was found in the locality, bearing on the inside rim the mark of D. P. Haskins, 237 Sixth Street, Indianapolis.[38]

[3] Weird and Uncanny—A Wild Man Reported to Be Inhabiting a Cave in the Woods About South Gardiner—Has Been Seen by Several and Has Caused Even the Bravest to Quake—Is Described as Old with a Flowing Grey Beard with Every Appearance of Being Crazy—South Gardiner, Me., Aug. 19—"The wild man of Borneo, who has just came to town," is the talk of South Gardiner just now.

This naturally quiet town is thoroughly stirred up over a well-founded report that a crazy man is prowling around the town nights and hiding in the woods days.

When the rumor was first started about four days ago, it was thought by many to be the creation of some highly imaginative person, and that the yarn had grown some at each repetition, until some people really believed it.

But the minds of the majority have changed ere this, and to say that the nightly visits of this "wild man" to various houses in town have caused great excitement mildly expresses it. It is a subject with which every South Gardinerite is familiar and in most cases is talked over in a matter of fact way.

"It is no laughing matter, but on the other hand a serious affair," said a well-known citizen to the Journal man Thursday.

"Do you really believe that there is a crazy man haunting this quiet village?" asked the reporter.

"Do I really believe there is a crazy man about the town," repeated the person addressed. "I don't believe nothin' about it, I know it to be a fact."

"Have you seen the maniac?" continued the reporter.

"Seen him? Well, I guess I have, and in mighty close quarters, too."

"Tell us the circumstances."

"Well, I will, if you promise not to use my name in the paper."

Being assured that his name should be withheld from the public eye, he continued:

"I had heard something about a wild man in town, but did not believe a durned word of it, and was very free to express my opinion that it was nuthin' more than a yarn. I had read in the papers that a crazy man had been seen in Orono, but didn't 'low there was much truth in it, and thought that probably that story had suggested the idea to somebody who was quick to start a wild man story in South Gardiner.

Well, getting down to facts, I started after my cows last Monday night about 6:30. When I reached the pasture they were not at the bars as usual, and I started into the thick undergrowth to see if I could not hear them browsing. I had not taken half a dozen steps when I heard a human voice ring out. I turned quickly to one side to locate the sound, and there not three rods away was the worst looking specimen of humanity that I had ever seen."

"How did he look," interrupted the scribe.

"Well, as near as I can tell, he was a man sixty years old, his face was covered with gray, streaming whiskers, and his hair hung in mats and reached almost to his shoulders. He was very thinly clad, wore no hat, but carried an overcoat on his left arm and held a large club in his right hand. Evidently he saw me first and was coming toward me at a lively step, shouting something as he ran. I turned and made for the main road, and although he gained on me, was so near the highway that he did not overtake me and gave up the chase and returned to the woods.

"I did not go into the woods again, but waited at the bars until the cows came. I did not say much about my experience that night, and, in fact, this is the first time that I have given the whole account, as I thought I would be the laughing stock of the whole town, but now that others have been chased by this man, I am willing to tell my story."

The Journal man talked with several of the townspeople, and although their opinions differed somewhat, yet the majority are convinced that a strange human being is lodging in the woods. One night recently Mr. Guy Lawrence was returning from Gardiner quite late in the evening, when he heard a queer noise in the woods, some rods south of the Meader place. He stopped his team to make an investigation but, the language, used by the unknown person was too threatening to encourage one to enter the woods unarmed, and Mr. Lawrence did not linger long in that vicinity.

Since then a Gardiner boy who was riding his bicycle on the river road, was frightened almost into hysterics by some unknown person who is said to have dashed out of the bushes which line the road, and tried to club the boy. However, the boy managed to get away unhurt, but terribly scared.

Others who have seen or heard the wild man are Ernest Hall, Bert Mansfield, Fred Sperin and Brian Demmons.

Tuesday morning a wild-looking man and probably the one that is causing so such excitement in South Gardiner, called at Mr. Sperin's and asked for something to eat. Mrs. Sperin was alone and closed and locked the door. The visitor disappeared quickly and has not been seen in the village since.

The children dare not venture outside of their parents' yards after dark and many of them cannot be induced to leave the house. Several small boys were playing near Johnson street [sic], a few evenings ago, and were chased to their homes and badly frightened.

It is said that there is a cave in the woods, not far from the village and some think that this strange human being has taken up his abode there.

There is talk of organizing a posse of armed men and carefully search the woods, and try and capture the man.

There are a few, however, who are firm in the belief that it's all a "yarn," and started just to scare people. If that's the case it is safe to say that it has had the desired effect, because it has caused one of the greatest sensations in the history of the town. The fact that the "yarns" came from some of the most highly respected citizens, makes them more creditable.

There is never a great smoke without at least a small first and in this case there is practically no doubt, but that a "wild man" really exists in the vicinity of South Gardiner.[39]

[1900]

Maine author T. M. Gray is a member of the Horror Writers Association who has written, along with many horror short stories and novels, several nonfiction works on Maine ghosts and New England paranormal oddities. In her book, *New England Graveside Tales*, she recounts the following story she heard in her youth:

> [1] As a teen, I spoke with a woman whose father saw one while logging in Maine in the early 1900s. He was returning from camp when "a large, naked, hairy man" suddenly dashed out of the woods and ran across the old woods road in front of the man's truck, causing him to slam on the brakes. He and his fellow lumberjacks tried to locate the creature, but it had run off into the dense forest. They never saw it again.[40]

In periods past, people with mental or behavioral disorders suffered further from incorrect or absence of diagnosis and received treatment charitably called lacking. 'Ran Amuck!' was the headline of the article concerning "Uncle" William Morse. With a by-line of South Paris, Maine, the July 29 *Lewiston Evening Journal* story began,

> [2] "The inhabitants of this place were thrown into a wild state of excitement Friday morning when Mr. Wm. Morse, one of the oldest residents of the town, appeared on the hotel steps of the Andrews House with a revolver in his hand and discharged it freely.
>
> Fortunately the shots did not hit anyone. It seems rather remarkable as there were many persons on the street, and not far distant at the time. After discharging the weapon he rushed away down the mill road evidently making for the woods. It is believed that he has gone crazy and intends to commit suicide.

Nothing has been seen of him up to the time of this dispatch. Sheriff Tucker has been notified and he with a posse of armed officers are now in search of the demented man. Should they discover his whereabouts he will undoubtedly put up a strong fight.

"Where is Uncle William?" that is what everybody is asking on the streets."[41]

[3] The Turner Wild Man was called by writer Holman Day "a sort of sea serpent of the interior."[42] He was seen many times by farmers in their fields and by teams when he leapt out onto lonely roads, making wild motions with his arms.

On the road in Turner he confronted a woman driving along while on her way to Mechanic Falls. He gesticulated wildly before running rapidly across a field. This woman was as close as 10 feet from him. She reported the person or creature was above average in stature and was entirely covered with hair. She said the face, though wild, was not fierce.

Others having had encounters with the Turner Wild Man corroborated the hair-covered body, but the above woman was one of the very few to be close enough to observe the face. Day reported that this wild man had been seen in a dozen counties of Maine and is apparently the same. Posses had been formed to go into the woods to capture the creature but it had been able to escape the detection of pursuers and searchers every time, leaving the scene of its encounters with startled people without a trace.[43]

[1905]

[1] Poland's Wild Man. The Town Hunting Him. He Carries a War Club Decorated with Flowers—He Appears to be Vicious—Poland, Me., July 22 (special)—Poland has a wild man—at least, he looks wild. He made his appearance, Friday, and terrified a number of people by his mysterious actions. He tried to kill one man by throwing a railroad iron at him

and startled two families by throwing stones through the windows.

The town was in arms Saturday hunting down the wild man. The search was in charge of men from Poland Spring, two of whom have seen and talked with him. Practically everybody in town has seen him, but he did not arouse suspicion with all of them as he appeared rational enough when conversing. He carries a railroad iron, such as is used in coupling cars. On the end of it is a bouquet of flowers. This he brandishes as his war club. If he should hit a person with it, they would not live to swear out a complaint against him.

The wild man is described as being about 30 years old. His face is covered with a growth of beard, not very long. He wears a derby hat and dark suit of clothes. His face is very red as if burned from long exposure to the sun. He weighs about 150 pounds, and is about five feet nine inches tall.

No one knows who this mysterious man is. He made his appearance early Friday, but did not attract so much attention then as later in the day. By dark the whole town had heard of his attempts to do damage. By means of the telephone system the farmers throughout Poland and New Gloucester were notified to be on the look out for him. Mr. A. B. Ricker of the Poland Spring House has made every plan to capture the man and reports of his capture are expected every minute. Some of the young men and boys at Poland Spring have joined in the hunt. They like no better sport than hunting a wild man in the Maine woods.

The wild man has come in the height of the summer season to furnish sensations for the sojourners at our summer resorts. Since the sea serpent washed ashore at Old Orchard and the Sturgis commission got to work there has been nothing out of the ordinary to interest them.

In all probability Poland's wild man is someone whose brain has been affected by the excessive heat of the past two weeks. He may be a hobo that has wandered in from no-one-knows-where, or he may reside in some of the near-by towns. He's a real wild man, though, and dangerous, and until he is caught Poland people will not sleep well nights. All berrying parties in the town have been postponed and the women who are left alone by their husbands when hunting for the man of mystery, keep the doors and windows securely locked and a weapon handy.

It is believed that Mrs. A. L. Durgin of Harris Hill is the first person who saw Poland's wild man, Friday. He first appeared at the kitchen door at about 6 A. M. and asked for a drink of water. He then said he wanted to come in and rest. Mrs. Durgin objected. She didn't like to have a stranger in her house and besides she didn't like the appearance of the caller. She felt that something was not right. The wild man insisted on entering, but was stopped by a big St. Bernard dog which Mrs. Durgin set on her unwelcome visitor.

Mr. Durgin appeared on the scene at this time and his presence cooled the courage of the wild man. "Are you looking for work?" asked Mr. Durgin. The wild man replied that he was and asked the way to Auburn and the Rickers, both of which were pointed out to him. He disappeared down the road.

The man appeared to have slept in a hay stack as his clothes were covered with hay. Mr. Durgin went to his work soon after breakfast and left Mrs. Durgin alone at home with the dog. Nothing more was thought about the incident until 10:30 A.M., when her visitor of the morning appeared again. He did not come up to the house this time, but lay down under the shade of a maple tree in a corner of the yard.

He had his ugly looking war club with him this time. Mrs. Durgin saw him, but felt secure with the protection afforded by the dog. She believes that he came back with the railroad iron prepared to kill the dog. "I had no fears of his doing it," laughed Mrs. Durgin in reciting her experiences to the Lewiston Journal. "I should have feared for his safety if our St. Bernard had gotten after him, however."

The wild man sat under the tree for about an hour and then went down the road again. Ralph Clark, an employee of the Rickers, saw Poland's wild man, Friday afternoon. "I was on my way to Poland station about 1.30 P.M., when I came across a mysterious acting and wild looking man," says Mr. Clark. "He was walking in the road, carrying a branch over one shoulder and a railroad iron in his other hand. The iron was about a foot long with a nut on the end of it. As I passed him he raised this as If he intended striking me, but I looked at him and made a motion to defend myself and he stood still while I drove past. I kept a lookout that he didn't creep up and hit me from behind. When I returned from the station about 5 o'clock the wild man was in about the same place as when I first saw him. He was walking this time. He had exchanged the branch for a dead limb of a pine tree. I watched him closely, fearing that he would strike me."

"Did you ever see him before?" was asked.

"Not that I know of." Replied Mr. Clark.

"Do you think he was a hobo?"

"No. He was too well dressed to belong to that class."

Poland's wild man appeared at the home of John Estes on the Danville Junction road late Friday afternoon, and broke two windows with his war club. The glass cut off some of his posies which were fastened to the iron.

Many other people than those whose names have been given have seen and talked with the man. Although 50 or men have been hunting for Poland's wild man since daylight, Saturday, they have learned no trace of him. He disappeared in the night as mysteriously as he appeared, Friday morning.

Whether he has killed himself in the woods or has escaped to worry the inhabitants in some other town, no one knows. If he has escaped and is still wild, he is liable to appear at somebody's door at any time. The people in New Gloucester, Poland, Auburn, and Minot have been notified to keep a lookout for the man as described early in this story. Anyone seeing him is requested to make him a prisoner and report the same to the proper authorities. Nothing had been heard from him up to noon, Saturday.

Sheriff Cummings of Auburn and Deputy Sheriff Cain of Lewiston were at this place, during the night, Friday, looking for the wild man, the sheriff having been notified by the people. The officers returned home before daylight, Saturday, without locating him. Sheriff Cummings and Deputy Sheriff Cain searched several barns and one old deserted camp. Two men were guarding the camp to keep the wild man from escaping for it was believed that he was hiding there. But when they saw the officers they went to meet them. Both men appeared frightened. The sheriff and his deputy entered the camp and searched it thoroughly up stairs in the cellar, but could find no evidence that the wild man had been there.[44]

[2] Wild Man Caught—Identified as Harding of Lewiston—Boarded a Pullman and Rode on Top of Car to Portland. Lewiston, Me., July 23—The wild man who created such a sensation at Poland Spring has been captured and identified as Peter Harding of this city, who is now believed to be violently insane.

After evading Deputy Sheriffs Smith and Luce of Auburn, and a large posse of men who searched for him in Poland all day Saturday and during last night, Harding boarded the Pullman this morning, and, unnoticed by the train officials, climbed to the top of the car, where he rode to Portland and was captured by the officials in that city.

He was bareheaded and his clothes were badly torn. He will be taken in charge by Lewiston authorities.

Peter Harding, who has a sister here, is a mule-spinner,[45] and until recently had been working at Hartland woolen mills. He returned home about two weeks ago. At times since then he has shown signs of insanity.

At 9 o'clock last Thursday evening he got up and dressed, and told his sister that he was "going to heaven." She tried to persuade him to remain at home, but he left the house and was not seen again to be recognized until captured in Portland.

He is about 30 years of age and unmarried. The people at Poland and vicinity feel greatly relieved at his capture, as his actions made him an exceedingly dangerous person to be at large.[46]

[3] The insane man who came to town riding on the top of a Pullman car, Saturday, and who was lodged by the police at the station, had proved to be one

Peter E. Harding of Lewiston, and, Monday, officers from Lewiston came to Portland and took him
back to his home. It has proved also that Harding is
the wild man who terrorized the citizens of Poland,
on Friday, by his strange wanderings and unaccountable actions.—Portland Express.[47]

[4] Gorham is the latest town to get excited over a
wild man scare.[48]

[1907]

[1] Wild Man in Riggsville—Strange Stories of presence of an Unknown Frequenter of Woods—That
Riggsville has a wild man is the rumor circulated
thereabouts, and while the descriptions of the
stranger are rather varied, his presence has caused
considerable comment and some of the ladies in that
vicinity feel timid about driving or going out alone.
The stranger has never been seen on the highways,
but has been heard to creep through the underbrush
and thickets and seems to have a natural fondness
for gulleys [sic] and swamps, and some people tell
of seeing him from a distance as he has jumped and
leaped to get out of sight of human visage. Where he
came form or what his object is in remaining so concealed is a hard question to answer, but people will
not begin to breathe easy until he is captured and a
statement wrung from him.[49]

[2] Is Not Insane—Nothing to Denote it Found in
Auburn Wild Man—Some Trouble, Tho. Jean Marie
Lebere, the Auburn wild man is still languishing in
Auburn jail—as much as the mysterious stranger
would languish anywhere, for he seems a cheerful
sort of fellow—but as yet he has shown absolutely

no signs of actual insanity. It is probable that when his 90 days' sentence for vagrancy expires he will be turned loose to go his way. There is nothing shown as yet to warrant holding him further.

The Frenchman is as docile and quiet as a lamb in jail, and makes no more trouble now than the most servile prisoner, not because there is any servility about him, but because apparently he wishes to make the best of a bad situation, and gain the good feeling of his jailers.

No one would believe to look at him that he was the wild and hairy man who fought desperately in the woods to prevent arrest by Deputy Lyford. He is good looking and clean featured. He is apparently about thirty-five years of age;—before he was shaved he might be any age between thirty and sixty.

Only his eyes remain unchanged and they are as bright and clear and intelligent as ever. "I should know those eyes among a thousand," says Deputy Lyford who went to visit his former prisoner Tuesday morning. The deputy caught a glimpse of the eyes and know his man in an instant. He knew the deputy too, shook hands with him, and was apparently more than glad to see him.

He still talks in broken English of "my head, she broke," and appears much more like a man who has lost his memory thru an accident than one actually insane. There is something the matter with him certainly, but in the opinion of his jailors it is not insanity. There is much interested supposition as to how and what he might have been before he was hurt and many believe that he was a man of education and means.

In one of the Boston Sunday papers was a big picture of Lebere with a story to the effect that he might be a Frenchman of high blood who had lost his memory, wandered to America in search of fortune and finally became a wild man. Lebere's friends in jail have not reached the stage yet where they believe him a lost count, but that he was more than a mere laborer they do believe.[50]

[1912]

[1] Rushing hatless, coatless, from his home, Sheldon Greenlaw ran into the woods near his house in Limington Tuesday afternoon, and in spite of the tearful entreaties of his wife, who saw him when he left the house like a madman, he disappeared from sight among the trees.

Mrs. Greenlaw at once called the neighbors, and searching parties were organized. All night long, armed with torches, and calling the missing man by name, bands of men scoured the woods in the vicinity of Limington, and although several traces were found, the man himself was not seen.

A vest, which Mrs. Greenlaw identified as belonging to her husband, was found in an old woods road, while early Wednesday morning, the ashes of a fire, still smouldering, were discovered in a little open spot. Pine needles were heaped beside the fire, and it is thought by the searchers that he passed the night there.

The only reason that can be found for Mr. Greenlaw's insane actions is that the extreme heat of the past few days has affected him. He is not known for a drinking man, nor has he ever been subject to fits.

Great anxiety for his safety is felt by the members of his family, to whom this is a great blow. Mrs. Greenlaw says that she followed him right up to the edge of the woods, but he did not look around, and only ran the faster when she called.

Men are still searching for him in the woods, and it is hoped that he may recover his senses and return to his home.[51]

[1913]
Durham Wild Man

The Durham Wild Man of 1913 was championed by both of Lewiston's papers at the time, *The Lewiston Evening Journal* and *The Lewiston Daily Sun*. There are a dozen articles presented below making it singularly Maine's most reported Wild Man story.

[1] Durham—Rumor of a Wild Man Seen—Durham's latest sensation is a wild man. He was seen by several but they were frightened so they could not give a very good description. He runs for the bushes when seen.[52]

[2] Durham Men on Wild Man Hunt—Wm. Welch is Threatened by Almost Nude Stranger—Durham, Me., Aug. 20 (Special)—Durham is thrown into wild excitement by the report of a man running about in the wooded areas with little to wear and whose only appearance in the eyes of the natives has been threatening in character. Today every available man in West Durham is on this man-hunt and the women and children in town are all tightly locked into their houses, fearful to look out of the windows or go to the springs for water or the woodshed for fuel.

The "Wild man" was seen but once, when on Tuesday morning, as William Welch was driving on the Plummer road the stranger stopped his team and told him to turn about, that he "could not pass here." The man, according to Mr. Welch, was about 35, with long beard and hair, his body lank and almost nude. His eyes gleamed as if he were insane. He brandished a club as he ordered Mr. Welch to turn back.

Mr. Welch obeyed orders but on meeting Mr. Carter of Bethel, who is employed with a portable sawmill in town, both set to the spot where the strange man had appeared. The prints of his bare feet were visible in the road, but the man was not seen again.[53]

[3] Durham Wild Man Not Yet Captured—The wild man of Durham has not yet been captured but searchers are confident of locating him soon. So far as reported the strange acting man was not seen in his usual haunts yesterday and there are some who believe that he was aware of the search being made for him, and remained in his den. At different times, the wild man has been seen by several different persons, but not until Tuesday when he held up William Welch, had he shown signs of violence.[54]

[4] All-Day Hunt for "Wild Man"—Reveals No Clue to Durham Mystery—Other Glimpses of Stranger—Durham, Me., Aug 21. (Special)—An all-day search by the officials and people of Durham has failed to throw any light upon the mysterious "wild man" whose appearance in town has driven the women and children into retreat and put the men on guard.

The search was taken up in the forenoon, Wednesday, by Deputy Sheriff Eugene Harding of Lisbon Falls, with about 20 assistances from among the men here and in the afternoon another group of ten local citizens who spent several hours in trying to find the stranger.

According to fuller reports made Wednesday by the people congregated for the search it seems that this "wild man" has been seen several times. His first appearance was on Saturday, Aug. 9th, when Guy Towns was out with his dog and gun hunting in the woods which practically surround the old farmhouse of Seward Merrill, which has been closed up and deserted for a long time. This farm of about 40 acres is almost wholly in woodland. It was while in this tract that young Towns says he suddenly came upon a naked man who was standing near the deserted house. The moment the lad saw him he became so frightened that he ran for home as fast as he could go. There he told his story, but it received little credence, his folks thinking he had been deceived by something which might have looked like a man.

A few days later Mrs. Carrie Sawyer saw a man, who had no clothing on, working among the cows in Edwin Crockett's pasture, evidently trying to keep one of the animals still long enough to milk her.

Still later while a party of berry pickers were in the woods, one of the little girls left the rest of the people to go to a spring, not far away, for water. She came running back saying that someone had thrown a stick at her, tho she could not see anyone around. Her father at once went to the spring but saw no one, although he discovered the tracks of a man's bare foot near the spring.

Tuesday of this week William Welch was met by the mysterious stranger although his verbatim story differs a little from the first reports of the encounter. Mr. Welch tells the Lewiston Journal representative that he was not driving but was walking to Lisbon Falls where he wished to obtain some medicine. He suddenly noticed a man, wearing only a brown shirt, dodge back into the bushes by the road, Mr. Welch proceeded along his way, but when about two rods from the spot where the man had hid, the latter leaped out and saluted Mr. Welch.

"Where are you going?" he asked. "To Lisbon Falls." Replied Mr. Welch. "You can't go by me!"

yelled the stranger, brandishing a club which he held in his hands, "If you do you're a dead man!"

Mr. Welch, having no weapon at hand and feeling ill anyway, had no alternative but to retreat until he found a neighbor to whom he told his story. Then both returned to the spot by the road but could not find the man.[55]

[5] Durham Wild Man on Rampage—Rattles Windows, Steals Cukes and Milks Cows at Night—Footprints Friday Morning Clearly Trailed to Houses—Man Toes in Badly with Right Foot—Prints measure Eleven Inches—Excitement Keen—Durham, Me., Aug. 22 (Special)—Durham's "Wild Man" made the night lively for residents near the Bend, Thursday evening, throwing several households into terror by his peculiar actions, tho but one person saw his shadow, so elusive was he at his mischief. His bare footprints, however, have been found around several buildings this morning, where his presence had been known the night before, and this has disposed of any supposition that the imagination of the people visited was too highly wrought up or responsible for the excitement.

The "wild man" was first heard about eight o'clock by the folks in the house of M. W. Eveleth. Only the women were there then and at first they thought someone knocked. Then it was apparent that at intervals someone rattled the windows as if trying them in order to get in, or to frighten the people. But nothing could be seen of the person doing it as he dodged out of sight immediately after giving them a good shake. In half an hour Julius Eveleth, the son, came from the store and he heard the windows rattled and went out and tried to find the stranger, but could see nothing of anyone. Still later, after

Mr. Eveleth had come home, the performance was repeated. This window shaking happened at intervals until after 11 o'clock.

Albert Towns' house was another place visited. This is adjoining the Eveleth farm. The people there had retired early when they felt the house jarred by the shaking of windows. This happened several times and then the boy got up with his gun and went out to find the man but could see nothing of him around the place. Mrs. Towns says she saw the man's shadow just once, but could not see distinctly enough to tell how he looked.

A third house to have this mysterious visitation was that of Edwin Crockett, an eighth of a mile from the Bend. There also the windows were rattled violently. Mr. Crockett went to the door and shouted, "What is wanted?" But he received no reply nor was anyone visible.

Friday morning there was an early consultation of the neighbors and it was then that the investigators found the bare footprints in the sand near these houses and traced them back and forth. According to the story these footprints tell the man must be fairly mature, having feet that measure 11 inches each. These also indicate that he toes in considerably with his right foot.

The first sign of the tracks is where the man leaped over the gate from the Wesley Day pasture into the soft sand. This pasture is near the Congregational church. There he paid a visit and then the tracks show that he crossed the road to the grange hall opposite and tramped around that building.

Then the tracks lead up the road as far as the Eveleth house where they turn in and are seen in various spots near the dwelling. Thence they can be trailed to the town house and from there to the Crockett house.

From this point the footprints disappear, down the Mill road, [sic] so-called, which turns off about half way between the Eveleth and the Crockett houses toward the wooded tract where he is thought to be in hiding and where search was made by the officials and citizens Tuesday. About half way down this Mill road, the morning searchers discovered skins of cucumbers and apples where he had eaten the vegetables and fruit stolen from some of the farms visited. There also was found an old tin can in which milk had been put within a few hours previous. From this it is thought that the "wild man" must have found access to some cows during the night.

Inquiry among the farmers showed that no cows were at pasture and all the barns were securely locked except Mr. Crockett's, where it is concluded the stranger found entrance easily and probably secured his milk.

George Haskell, whose pasture is near the woodland where the "wild man" probably keeps in retreat by day, says that he is sure some of his cows have been milked from time to time for several days past.

Owing to the fact that residents of Durham have thought this man might be someone missing from the school for Feeble Minded at Pownal, inquiry was made at that institution, Friday. One of the teachers answered the telephone and positively denied the absence of any inmate there which might be this stranger now in hiding in Durham.

None of the citizens of Durham are missing and the only conjecture left is that he is someone escaped from an insane hospital or a traveler suddenly bereft of his senses, who has lost his clothing and from lack of sufficient food has been driven into the open to seek garden stuff and to milk the cows.

It is probable that so close a watch will be kept about the wooded tract where is thought to stay that his capture must follow soon. Some concerted effort may be made to effect this capture tonight if he shows himself again in this vicinity.[56]

[6] Durham Wild Man May Be Artist John Knowles—Durham people, who have lately heard much about a wild man in that town, are wondering if the headline maker is not John Knowles, the Boston artist. Their theory is that Knowles made a compass out of birch bark and spruce gum, or some other magnetic substances, and that these have led him to Durham.

Deputy Sheriff U. G. Harding of Lisbon, is as much at sea as ever. He does not believe that there is any wild man nearer Durham than a Portland baseball fan, and believes the story without foundation.

It has been learned that a lumber camp is situated in Durham and that some of the lumbermen, as lumbermen should be by tradition and Holman Day folklore, are not over particular as to their brands of shaving soap, and hence, perchance, often present an appearance more or less uncouth.[57]

[7] Hunt for Wild Man—Deputy Sheriff Stackpole and Posse of 12 men Searched Until 11 o'clock Last Night—Several Houses Were Visited by Wild Man Thursday Night. Tracked to the Old Mill Pond—The hunt for the wild man of Durham was begun in earnest last night by Deputy sheriff M.G. Stackpole and a posse of a dozen men this being the first systematic search that has been made for the strange acting nude individual. He is believed to be living in the woods near the old Seward Merrill place, the

buildings of which have been vacant for the past half dozen years or so.

Not until the story published in The Sun a few days ago of the hold up of William Welch was the Durham wild man taken seriously. Previous to that time reports of a strange acting, nude individual having been seen dodging through the woods or leisurely picking berries by the roadside, had been discussed by the townspeople, but few if any of the Durham folk took the matter seriously. Some thought a local joker was busy, while others were of the opinion that those who had been circulating the reports had been dreaming of Joe Dignard and the Auburn panther.

But now Durham citizens are convinced that the wild man is no dream, but a solemn reality. During the past week he had been seen by several different persons and many others who have not seen the man, have seen the imprints of his bare feet in the dusty roads, in marshy places or along the river bank. He has a very sizeable foot which unsocked and unshod measures 11 inches strong. It has also been discovered by Durham sleuths that the wild man toes in.

On Thursday night of this week he appears to have been very busy and judging from his freshly made tracks covered many miles, visited several farm buildings and farm gardens and milked at least one cow.

In the early evening he visited the homes of Marcus Eveleth, Albert Towne and Edward Crockett, tried the doors, rattled the windows and did other things to make his presence known. A woman in one of the houses saw his form indistinctly as it disappeared around the corner of one of the buildings. The wild man lingered in the vicinity of these buildings until nearly 11 o'clock.

Yesterday morning the tracks were found about the buildings, in the road, about the Congregational Church and about the Grange hall. The tracks led from the Grange hall to the home of Edward Crockett and then turning away continued to the old mill pond where a tin can containing a small quantity of milk was found. Beside the can were two partly peeled cucumbers. From the appearance of the cukes it was the opinion of the searchers that a ragged edge piece of tin had been used with which to do the paring. There were also some apple parings and other evidences that the wild man had stopped for lunch in this rather secluded spot. The milk in the can is supposed to have come from one of Mr. Crockett's cows. Mr. Crockett feels certain that during Thursday night someone entered his barn and milked one of his cows.

Deputy Sheriff Stackpole, Mr. Crockett and Wesley Day followed the tracks some distance yesterday morning, but did not catch sight of the mysterious being whose presence in Durham has caused consternation among the townspeople.

The wild man has been seen more frequently near the old Seward Merrill place. The buildings which have been vacant for the past six years or so, are widely separated from other buildings, and it is thought by some that the wild man has taken up his abode in the old farm house. However no one has inspected the house, so far as known.

There is a large track of wood and swamp land in this part of the town and it is conceded by the searchers that the wild man has every advantage over them. He can "see them coming" and the dense woods furnishes him ample hiding places.

The party of searchers under the leadership of Deputy Stackpole remained out until about 11 o'clock last night but heard no strange noises and saw no signs of the wild man. A few may continue the search today, but if the mystery is not solved before Sunday morning it is then planned to organize a big party and scour the woods for miles around.[58]

[8] Posse of 60 Men Hunt Wild Man of Durham—Traversed Miles of Wooded Country Sunday But Found No Tangible Clue—May be Escaped Convict. Talk of Reward for his Capture—A posse of more than 60 men headed by Deputy Sheriff M.G. Stackpole, Selectman H.J. Wagg and Constable Leon Bowie, scoured the Durham woods for miles around yesterday in search of some tangible trace of the wild man who has been terrorizing the community. They found his barefoot tracks, the parings of apples and cucumbers and a small quantity of milk in an old tin can, but got nary a glimpse of the much sought individual.

The townspeople are thoroughly aroused to the importance of clearing up the mystery that has been discussed pro and con for the past two weeks. Women and children dare not venture far from their homes, day or night, unaccompanied and some of the men folk are none too brave about facing the wild man.

The posse which was made up of Durham people and residents along the South River Road, Auburn, formed at South West Bend at 9 o'clock yesterday morning. Some of the searchers carried guns while others carried heavy sticks with which to defend themselves against the attacks of the wild man, for it is generally believed that when he is finally cornered he will make a desperate struggle and somebody is pretty likely to get hurt. Several dogs accompanied the searchers, it being hoped that they might hit the trail and rout the wild man from his haunts.

Divided up into groups of a few able bodied men each, a systematic search of the dense and almost impenetrable swamps was begun. Miles of territory were covered and while plenty of old tracks of the wild man were found not a fresh clue was discovered.

Almost every foot of the ground from the South West Bend to the Plummer Road, the Trask neighborhood and the entire section toward the Plummer Mill, about the old mill pond, the grange hall and other sections where the strange man had previously been seen.

The Seward Merrill place was thoroughly investigated as were also several logging camps and old shacks in secluded places, but no sign that they had been recently occupied, was found.

There is no doubt in the minds of the Durham people, that the wild man's diet is milk and Edward Crockett is of the opinion that his cows are furnishing the main supply. He believes one of his cows was milked as late as Saturday night but during the early part of that evening he heard no strange noises about the place.

Many of the searchers carried lunches which they ate with ravenous appetites at noon after three hours tramping through the woods. Those who had not taken their lunches were supplied with refreshments by the selectmen of the town.

Later in the day W.H. Ward and Also Merrill, members of the board of selectmen joined the party, Mr. Merrill having abandoned forest fire fighting on the Sylvester lot to join the wild man hunters.

The description of the wild man as given by the different persons who have seen him, varies but little. They say the hair on his head looks as though it had been clipped all over only a few weeks ago,

which leads many to believe that he may be an escaped convict. His face is covered with several weeks
growth of beard and his eyes look wild and his features drawn. He is nude except for a small brown
shirt which is twisted about his body.

William Welch who was held up by the wild man
and forced to turn back, got a close view of the
strange being. Mr. Welch says he was a vicious looking individual and that he said "You can't pass this
way" just as though he meant it.

Mr. Welch who has been in rather poor health,
was on his way to Lisbon Falls to see the doctor. The
mad man, armed with a heavy cudgel, jumped from
the bushes beside the road and faced Welch.

"Where you going" he demanded of Welch. "I am
going to Lisbon Falls to see the doctor" was the reply.
"Well you can't pass this way." growled the wild man
and drew his club back as though to emphasize his
words.

Mr. Welch was but a few feet away and got a good
look at him.

Mrs. Ralph Sawyer has seen the wild man twice
at least and has no desire to see him again. The first
time she saw him she was picking berries near the
Merrill farm. She heard a twig snap and looked up
just in time to see the wild man creeping toward her
on his hands and knees. He was partly hidden among
the berry bushes. She left the place in a hurry and
didn't stop running until she was a safe distance
away.

A few days later she was returning in a team from
Lisbon Falls and saw the same strange looking man
in the pasture with Edward Crockett's cows. Both
times that Mrs. Sawyer saw him he was not far from
the grange hall.

On Friday night William Thomas heard a loud knock on his door, but when he answered the call he found no one in waiting. He believes the wild man was prowling about his premises.

The same night Emery Jaba heard someone about his place and he took his gun and started to investigate. It was about 11.30 p.m. He had not gone far when he saw a man standing in the shadows about 50 yards away. He fired his gun into the air and the man ran for dear life down toward the Plummer Mill.

Everyone in town is wrought up to a high pitch of excitement over the presence of the wild man. It is understood that the town officials may offer a reward for his capture.[59]

[9] Fifty Searchers Out—Sunday to Find Durham's "Wild Man"—No Trace Found—More Cows Milked—Durham, Me., Aug. 25 (Special)—Sunday forenoon was devoted to a manhunt, in Durham. About 50 men from all over town, some of whom walked seven miles to participate, joined forces at The Bend at 8 a.m. In this party were the selectmen, Herbert J. Wagg, Also S. Merrill, and William H. Ward; Town Constable Leon R. Bowie and Deputy Sheriff Merton H. Stackpole.

Under their direction a line was formed, with the men but a few yards apart, and the woods on the Merrill place, where the "wild man" has been staying, were gone thru as by a fine-tooth comb. No trace of the stranger was found.

It is believed that the man either saw the men gathering or heard their approach to the woods and took to his heels and disappeared to some other retreat. That he seemed to have been frightened out of this particular vicinity is warranted by fact that no signs of him were seen or heard Sunday night.

On Saturday night when Ralph Sawyer drove home his cows he found one had been milked. This particular cow is the only one in his herd which will stand to be milked out of doors. Laforest Gale also reported, Saturday night, that one of his cows had been practically milked dry.

It will be remembered that Friday night, following the search made by 15 men under the leadership of Deputy Sheriff Stackpole nothing was seen or heard from the "Wild man" near Crockett's barn, where he had milked a cow the night before, although this was watched and the woods were watched until midnight.

These facts tend to show that the stranger is wily and knows enough to secrete himself when searching parties are out. When things are quiet he reappears.

Over at the Josiah Vining farm, where a German family lives, the man reports that on Friday night he heard someone trying to get into his barn. He fired his gun into the air and the man vanished. He thought the visitor was partly clothed, at least.

Durham citizens hardly know what to make of this peculiar visitation and now are simply awaiting developments.[60]

[10] A Wild Man—A strange wild man who has been living in the woods of Durham near the Hilltop, is believed to be a crazed Shilohite. Recently he held up William Welch, a farmer, and threatened to kill him. Welch escaped and gave the alarm, but when a party of armed men had been organized and visited the place where the wild man had been seen, the only evidence of his presence were the imprints of his bare feet in the dirt.

Welch said the man was nude except for a piece of burlap twisted about his body, that he had a beard, long hair, was lank and bony and looked half starved. This creature has been prowling about the Durham woods for days, and on several occasions has chased women and children who have been berry picking. A reward for his capture has been offered and a posse of fifty men has been organized to search for him.

The town authorities of Durham feel that this is only an evidence of conditions at Shiloh and that prompt action on the part of the state officials is necessary, not only to save the lives of the Shilohites, but to protect the property and citizens of the town of Durham.[61]

[11] Fired at Wild Man—Reben Hunnewell of West Durham Uncertain Whether He Hit Prowler or Not—Believed to Have Left his Old Haunts Last Sunday to Evade Army of Searchers—When Reuben Hunnewell of West Durham heard someone prowling about his place late Monday night, he seized his gun, slipped in a cartridge and went to the door. He saw the shadow of a person whom he believes to have been the wild man, dart behind a woodpile. Without waiting to make any inquiries, Hunnewell fired at the man, but thinks he failed to hit him.

While the home of Reuben Hunnewell which is located beyond the mill pond in a somewhat secluded part of West Durham is three or four miles from the previous scenes of activities of the Wildman, Mr. Hunnewell had read of the strange being and of his night prowling, pillaging gardens, milking cows, rapping at doors and rattling windows, and when he heard foot steps [sic] about the house and later raps on one of the doors, his first thought was of the wild man. That is why he took his gun when he went to the door.

It was late in the evening and quite dark in the Hunnewell yard, but when he opened the door gun in hand, a person who had been standing near the door started to run. Without losing any unnecessary time, Mr. Hunnewell raised his gun and fired. By this time, however, the wild man was fifty yards or more away and just dodging behind a woodpile which sets in the Hunnewell yard.

Mr. Hunnewell is uncertain whether he hit the man or not but he found no trail of blood and no dead wild man behind the woodpile yesterday morning. It is possible that the man carried a few scattering shot, but it is not thought that he was seriously hurt.

No more rappings on the door were heard that night and everything remained quiet about the Hunnewell place.

Mr. Hunnewell was unable to tell from the view he obtained of the wild man whether he wore any clothes or not. There is no question in the minds of the people in that section of the town but that the disturber at the Hunnewell home was the same strange acting individual that has been "doing things" near South West Bend. They believe that the wild man left his old haunts Sunday to evade the army of searchers.[62]

[12] May Search Sunday for Durham Wild Man—There is talk of organizing a posse for another systematic search for the Durham wild man next Sunday. While so far as reported, he has not been seen since the night he rapped at Reuben Hunnewell's door and was the target for Mr. Hunnewell's shot gun, fresh tracks have been seen and it is believed that he is still hiding somewhere in the Durham woods.

It is the opinion of some that possibly he was wounded when Mr. Hunnewell fired at him, but no traces of blood were found at the time and nothing has since occurred to lead the citizens to believe that the wild man is in distress.[63]

[13] Durham, Me., Sept 5 (Special)—The wild man has wandered to "pastures new," as he has left no signs of his presence in Durham for over a week. Many think that the crowd of men out hunting for him Sunday frightened him away.[64]

In the story below is found the oft-mentioned rationale for the presence of a Wild Man in the local vicinity: an attempt to scare off berry pickers from bountiful places of harvest.

[14] 25 Years Ago Today—'13. It is thought the mystery of Durham's wild man has been solved. He appeared at about the time people from miles around had learned that blueberry bushes were covered with berries. When stories of this mysterious, unclad person, appeared in the newspapers, berry parties were postponed.[65] The effect was as planned, and the schemers harvested their berries.[66]

[15] Durham Wild Man Now at Pownal—It is believed that the nude stranger acting individual who created so much excitement in Durham a few weeks ago, is now hiding somewhere in the woods in Pownal. His tracks have been seen and other signs of his presence in that locality have been discovered.

Charles Chadsey saw the bare foot tracks of a man near his house one morning recently and people in that vicinity are reported to have heard someone prowling about their homes at night.

The tracks seen near the Chadsey place are said to be of the exact shape and size of the bare foot tracks made by the Durham wild man. The wild man

has not been seen in Durham since the big searching party was organized and scoured the territory
where he had previously been seen. It is believed that
he became frightened and left that section.

Durham people are indignant that they should
be reported to have "framed up" a wild man story to
frighten the berry pickers. They also think they know
a wild man when they see one.[67]

[16] Durham Wild Man Left Clothes—Rubber Coat,
Hat, Belt, Rubbers and Satchel Found—Also Testament with Name 'W. H. Monroe' Marked Therein—
Graham Buns were Wrapped in Newspaper Bearing
Shiloh Postmark—Lard for Butter—Durham, Me.,
Sept. 22 (Special) While out gunning Friday, a
Durham man came across something startling, nothing alive, and not the wild man, but perhaps the wild
man's clothes. He found a rubber coat, sou'wester,
a pair of rubbers, leather belt, an old leather satchel
containing a testament with the name 'W. H. Monroe', some graham buns wrapped in a piece of newspaper, a can with a small piece of lard and a paring
knife in it with which to spread the buns no doubt, a
weather bureau report post-marked Shiloh, and tied
with a cord to use for a handle.

The hat was on a dead alder about six feet from
the ground and looked almost new. The lining inside was clean and white. The coat was as good as
new. The rubbers were right side up and full of water.
The satchel was hung on another tree. There was a
ten pound lard pail, also a five pound pail. One was
tipped over but the other had a few berries in it also
about a quart of water which the finder thought was
cider, but it had rained in it and the sun's heat had
made vinegar of it. He thought the berries were raspberries.

When the man got home he told his folks he had found the wild man's clothes and Sunday he and two others brought them home.[68]

[17] That the *Kennebec Journal* called the Durham Wild Man "that L.J. Wild Man", likely provides a hint of a patronizing position the Augusta daily had with regards to the *Journal's* long-running patronage of the story.

> That L. J. [*Lewiston Journal*] "Wild Man"?—While out gunning, Friday, a Durham man came across something startling, nothing alive, and not the wild man, but perhaps the wild man's clothes.[69]

> [18] The Durham people still believe in the wild man and are impatient with any criticism or expression of doubt. Geo. H. Hascall, the well known grange chorister, has even caught a glimpse of his flying form. Women and children are getting very uneasy and bolted doors may be found in many houses.[70]

> [19] (The Androscoggin Pomona Grange met with the Solid Rock Grange in South Durham.) And it was a jolly party that went from these cities. They went in the big truck auto owned by W. E. Leland and used as a delivery team for the grange store in Auburn. The 14 miles to the Solid Rock Grange hall was made without incident, save for the running conversation and discussion that was kept up all the way. A close watch was kept for the wild man of Durham, but beyond an occasional track no sight was caught of the bogie.[71]

> [20] Durham, Me., March 20 (Special)—Because his mother wouldn't take a boiler full of clothes off the stove and let him boil sap, Guy Towns of Durham,

aged 20, poured the sap on the kitchen floor. He remained ill-natured all day, and at night had some trouble with his brother, Ray, and threatened to shoot Ray, so it is charged.

Durham Wild Man? Some think that Guy Towns is the so-called "Wild Man" who terrorized this section last summer, and whose antics kept the country from being too quiet. It is recalled that farmers found their cows milked by some unknown person, and that vegetables and grains were garnered by hands other than those of the owner. Little children and women told tales of a naked man running at large thru the woods, but always he was too far away from any one to recognize him. Weird stories of this "wild man" were told from time to time all summer.

For several months suspicion has been growing that the wild man was Guy Towns, son of Mr. and Mrs. Albert Towns, who live near Southwest Bend, in the former home of Annie Louise Carey. The Towns family moved to Durham about a dozen years ago from Patten and are respected by their townspeople. The son, Guy, has a reputation of possessing an ungovernable temper, and when aroused is said to talk threateningly, often saying "I'll shoot, if you don't let me alone." and other like statements. Of course, no one has ever taken these threats seriously.

Several times last summer on hot days, farmers working in the fields, saw him running in the woods. He would strip off his clothing and then stark naked race back and forth in the woods behind the school house.

Later in the fall he was seen by a neighbor peering into her window at night. The moon was shining brightly and she says she had no difficulty in recognizing him. Thoroly [sic] frightened but putting up

a good bluff, she ordered him to go away and em-
phasized her command by picking up a rifle, which
by the way was not loaded. It had the desired effect,
however, for Guy went away immediately.

Guy is a tall, stout youth, and while many believe
him harmless, others are afraid of him.[72]

[1914]

[1] An eerie character, very scantily clad, is frightening the women and children in Westbrook. The
"wild man" is reported as being tall, dark, badly in
need of clothes and a haircut and shave, clad only in
an ordinary coat sweater. Several parties of berry
pickers in the woods around the tow path have been
badly frightened by the man, who persists in drawing attention to himself by loud shouts.[73]

[1919]

[1] Ocean-Going Wild Man in Maine Woods—Raspberry Profiteers are charged with story—Constables
chase nude terror from treetops to moaning surf—
Special dispatch to the Globe—Lubec, Me.—Aug. 6—
Hundreds of quarts of luscious raspberries have
gone to waste this year in the old cuttings in Morton's woods at North Lubec because the nimble-fin-
gered women and girls who usually gather the crop
have been scared of the "wild man" whose appearance there several weeks ago is still the mystery of
the town.

It is intimated that some people who are not
afraid of wild men have had great berry picking
there, and some of the women folks are beginning
to think that there was never any wild man at all,
but the berry season is about over. Reports came
from North Lubec that a man entirely nude had been
seen several times and gave chase to women along

the country roads, reports so circumstantial that an army of several hundred men and boys, commanded by constables, beat the woods and finally scared up the wild man. Then, they say, instead of running he shinned a tree and was lost from sight in the tree-tops, finally breaking away and making a dash for the shore where he dove off a high ledge and swam out to a dory moored some distance off shore.

He pulled in his anchor, got out the oars and made for the Lubec shore while some of the pursuers gave chase in motor boats and others went around the road in automobiles. There was no sign of any wild man when they got there, but some say they met an inno-cent appearing and harmless hobo who had evidently robbed a scarecrow for his wardrobe. He said he hadn't seen any wild man, but borrowed a cigarette and went on his way. Numerous expeditions have been made to Morton's woods since, but no signs of the wild man. But it is not a popular place for picnics.[74]

[1920]

Just in time for Halloween...

[1] Tramp Scared Residents—City Marshal Justus B. Cobb went down to the lower part of the Ferry road Tuesday afternoon, and with two other men searched the vicinity for a "wild man described as seven feet tall and wearing a beard two feet long." There was no sign of any such creature. The belief is that a tramp accidentally got into the neighborhood and lost no time in getting out.[75]

[1922]

[1] Maine "Wild Man" Sought in Woods—Mattawam-keag, Me.,—Dec. 2—Searching parties are scouring

the woods in the vicinity of Mattaceuk Lake looking for a so-called "Wild Man" while the county authorities are looking over their records of missing men which however, do not furnish much clue. For the past two weeks hunting parties have reported seeing an old man in the woods, who after some incoherent mutterings has disappeared. His clothes were ragged, beard and hair long and general appearance as of one demented. With the cold and rough weather of the past few days which would be dangerous to a woods wanderer, the sheriff's office has taken up the case and Sheriff O. B. Fernandez instructed his deputy in this section to locate the man if possible.[76]

[2] "Wild Man" Hunt On in the Woods—Mattawamkeag, Me., Dec. 7—Fear that the cold weather may prove fatal to the "wild man," seen on several occasions in the woods near Mattaceuk Lake, the authorities have taken up the hunt. Citizen searching parties have failed to locate his hut.[77]

[1929]

[1] No Trace of "Wild Man" Seen in Topsham—Sheriffs from two counties have been unsuccessful in two searches for a man who has been wandering naked in the woods of Lisbon Falls and Topsham. The fugitive has been spoken of as the "wild man" but whether or not this is true, those who have seen him say his lack of dress can be classed as ultra-primitive or perhaps ultra-modern.

He was seen near the home of George Roberts, Topsham, Memorial Day. Sheriffs from Androscoggin county [sic] went to Lisbon to search for him. On Sunday morning they received another complaint about him this time asking them to cooperate with officers under Sheriff Henderson of Sagadahoc. Sheriff [name illegible] and his men spent some time in an ineffectual search.

It was reported the man might be William Knight, who escaped from the Pownal State school,[78] but inmates there last night brought the information that Knight was missing only a short time and apparently he had never been off the school property also that no other inmate of the school is missing.[79]

WILD MEN AT THE FAIR

Wanted: Wild Men . . . who knew that 'Wild Man' was once a marketable skill set, putting fair managers in competition with farms and industry for engaging qualified candidates? The following excerpt is a tongue-in-cheek look at this competition exacerbated by across-the-board low unemployment. This article is from the *Biddeford Weekly Journal* from June 14, 1907:

New York showmen report a scarcity of good wild men. One who needed such an attraction in his business advertised his need and called out a single applicant when in ordinary times a dozen or more would have hustled for the job. And this sole applicant, it was discovered, had been next to civilization long enough to have one of his front teeth filled with gold, which rather disqualified him for the position. With the farmers everywhere calling for help, with a scarcity of laborers in almost every industry and with every wild man of reputation engaged, no one can doubt that we are at present in the very midst of an era of prosperity.[80]

In the present work, it is recognized that one of the earliest means for people to see a Wild Man up close and very personal

was the opportunity provided by fairs. Such exhibits did two things: it both played upon and reinforced the idea of a Wild Man, displayed for gawkers as an exhibition imported from some mysterious country or from the dark, steamy recesses of a far-off jungle. Fairgoers lured by the barkers' cries and tempted by brightly and garishly painted billboards and canvas drops understood what was lurking behind the tent flap. And Kiko, Moji, and their ilk spread this lurid and fanciful depiction of a savage, humanesque deportment embodying wildest nature, and perhaps realizing some secret desires of those living in (and wishing to escape) modern society.

Below is an excerpt from *A Busy Year at Old Squire's Farm* by C. A. Stephens (see Chapter XI for further offerings from C. A. Stephens) presenting the author's depiction of a behind-the-scenes encounter with a sideshow Wild Man.

> "It was not a full-fledged circus and menagerie, but merely a show on its way from one county fair to another. On the back seat was a gypsy fortune teller and a Wild Man, alleged to hail from the jungles of Borneo and to be so dangerous that two armed keepers had to guard him in order to prevent him from destroying the local population. As we first saw him, divested of his "get-up", he looked tame enough. He was conversing sociably with the gypsy fortune teller." "I think the proprietor of the show recognized us, for we saw him regarding us suspiciously; and we moved on to the cage in which the Wild Man sat, with a big brass chain attached to his leg—ostensibly to prevent him from running amuck among the spectators. Two of his keepers were guarding him, with axes in their hands. He was loosely arrayed in a tiger's skin, and his limbs appeared to be very hairy. His skin was dark brown and rough with warts. His hair, which was really a wig, hung in tangled snarls over his eyes. He gnashed his teeth, clenched his fists, and every few moments he uttered

a terrific yell at which timid patrons of the show promptly retired to the far side of the tent.

When Willis and I approached the cage, a smile suddenly broke across the Wild Man's face, and he nodded to us. "You were the fellows with the hogs, weren't you?" he said in very good English. I can hardly describe what a shock that gave us.

"Why, why—aren't you from the wilds of Borneo?" Willis asked him in low tones.

"Thunder, no!" the Wild Man replied confidentially. "I don't even know where it is. I'm from over in Vermont—Bellows Falls."

"But-but-you do look pretty savage!" stammered Willis in much astonishment.

"You bet!" said the Wild Man. "Ain't this a dandy rig? It gets 'em, too. But don't give me away; I get a good living out of this."

Just then a group of spectators came crowding forward, and the Wild Man let out a howl that brought them to an appalled halt. The keepers brandished their axes.[81]

In 1900, during a week where Maine was in the grip of many forest fires—including Ellsworth, Townships No. 8 and 14, Trenton, Lagrange, and Oldtown, the fair came to Lewiston. "There's everything that goes to make a good show of the rattling, ballyhoo order," proclaimed the *Lewiston Evening Journal* of the September, 1900 Lewiston Fair.[82] Attendees could see many things. "The Lewiston fair this year has a rattling good midway. It is full of life and ginger but it is clean."[83]

"The day of the hootchie-coochie has gone and in its place there is good, wholesome fun, but fun that doesn't make a man turn round and see if any of his neighbors saw him coming out of the tent."[84] Patrons could sample hotdogs and sip red lemonade. The big tent in the middle of the grounds housed the moving pictures show where fairgoers could see "The Magician's Budoir", "The

Battle of San Juan Hill," and still photographs from the play that had set New York "by the ears" the previous winter—"Sappho". There was a six-footed, four-horned heifer named Baby. Miss Blanche Coombs of Brunswick was receiving wide praise for her needlework display. There were three palmists on the grounds— Laquilla, Bright Eyes, and Juanita.

And there was not just one, but two wild men—Kiko and Moji in separate exhibits. Moji was billed as hailing from east Africa, "an unholy terror and refuses to eat anything which is not served up to him alive. At least, he never has up to the present time, but his owners are trying to get him down to a raw meat basis. It isn't always easy to get stray dogs enough to satisfy the pangs of hunger which Moji develops. Moji, like Kiko, is strongly guarded and may be seen without personal danger to the audience."[85]

> Moji is Safely Chained—But it Cost his owner $25 for the Damage he did when the Stranger Dropped Hot Ashes on Him—Corrected returns show that it was Mo-ji and not Kiko that did the scalp-opening act on the Midway yesterday.
>
> Moji belongs to Mr. Hallett, who went to large expense to bring him to this country from East Africa. And it comes rather hard to have to ante up the way Mr. Hallett did yesterday. It cost him $25 to settle for the damage which Mo-ji did, and even then there was the awful possibility that he might repeat the trick. Though he has been some time in captivity, Mo-ji has not settled down to civilization for a cent. Part of the time he lays quiet in his cage, and at other times he rises up and evinces a disposition to clean every strange white man off the face of the globe. There is nothing of the fake about him, and nothing goes to prove it better than the fact that he tried to brain a man Wednesday and came so near doing it that Mr. Hallett had to cough up $25 to square the bill for damages.

Bird's eye view of the grounds taken from the window
of the Exhibition Hall (Lewiston Fairgrounds).

It seems that it was a case of nicotine that ailed
him Wednesday. Some fellow came in smoking a ci-
gar and was so unfeeling as to try the experiment of
dropping hot ashes on the wild man to see what he
would do. The man found out. Mo-ji seized the side
of his cage and with superhuman strength ripped off
a piece and shied it as he would an assagil [assegai]
at the nearest foe. That person happened to be the
aged mail carrier of East Auburn, and the story of
his injuries has already been told to the Journal
readers.

Since then Mo-ji has languished in chains and
straps and at the present time can't get out to do any
damage, and the boards along the side of his cage
have been securely nailed down, so that even he cannot rip them off at his own sweet will.

It was really interesting to see the way in which the people shied when they went up on the stand to have a look at Mo-ji just after he had ripped open the East Auburn man's scalp. They all wanted to see him and yet they were all afraid that they might get served the same way that the first victim did. Every time that Mo-ji made a rush for one side or the other they would scuttle for the ground and then they would go back to look at him once more after the paroxysm had passed.[86]

The attractions at the station were the three fat women and Ki-Ko, the wild man, which will be at the fair this week. Ki-Ko complained of being sick and he certainly looked it. He stated to a Journal reporter that after the Central Maine fair he was going to some hospital and be treated. He has undergone this strain for such a length of time that he is a nervous wreck.[87]

The Death of Ki-Ko—Brings to Mind a Story of the Ter-Centennial Midway—Ki-Ko, the old negro, who for many years has been on exhibition at all of the fairs in this state as the wild man is dead. He was gazed upon with awe by hundreds of the boys and girls of the state who wondered what would really happen should he get loose. He really was a harmless and bright old negro. During the Ter-Centennial two years ago he was one of the exhibits on the Midway and stopping in front of the booth where Ki-Ko had been exhibited, stopped to look at the gaudy painting showing the fierce looking wild man apparently ready to break his chains and destroy all of the children in the community. One of the women remarked, "I wonder where the wild man is this morning," At just this time the old darky, who was standing near by, at work, looked up and with a grin, which displayed his ivories, remarked "I'se Ki-Ko, de wild man leddies." And then continued with his work until the time came for him to get into his cage and strike terror to the hearts of all who paid a dime to see him.[88]

Two Grange exhibits—Auburn and Norlands. Every inch of space under the big grand stand is taken by exhibitors of agricultural implements and besides what are shown there, there will be other exhibits nearly as large under private tents. There will be anything and everything a farmer wants in the line of modern improved machinery. George Wiseman of Lewiston submitted 100 poultry entries. "What is new in State Fair today?" inquired the Lewiston Journal of Makes J. Lowell of Auburn, secretary, Thursday morning.

"We're chock full, packed down and running over with new stuff," replied Secretary Lowell. With the prospects of good weather the secretary and fair organizers were confident their fair would be bigger and better than Bangor's. One area where fair organizers thought they could best Bangor was in the midway. "We're going to have all they've got and a lot more." [Attractions included a merry-go-round, a Ferris wheel, the Knabenshue airship and a Last Days of Pompeii exhibit. Three women weighing 400 pounds, snake eaters, snake charmers, fortune tellers were also available and on display.]

"One of our attractions will be a wild man. He's genuine, with hair all over him, ourang outing eyes, monkey tail and coon paw hands. I saw him at Bangor, and he's no fake."[89]

HERMITS

Bizarre behaviors, eccentric manners, murky intentions, and hidden purposes describe the colorful and enigmatic personalities that seem to be ever-attendant fixtures throughout Maine's history. The below anecdote of a Maine hermit comes from the *New York Times*:

> A Hermit in Maine—A paper published in Penobscot County, Maine, prints the following: "About ten years ago a traveling tinker named Kenniston, well known throughout this and Piscataquis Counties, mysteriously disappeared, and fears were entertained that he had been murdered. But in a short time he was discovered to be living the life of a hermit on a lonely island in Moosehead Lake. His dwelling-house befitted the fortunes of a retired tinker, being in Summer [sic] a large dry-goods box. In Winter [sic], Diogenes-like, he ensconced himself in a hogshead. He lived on the frugal fare which dame nature provides in that region and appeared happy. His clothing bill during the ten years of his voluntary exile did not amount to $5. His original garments were in a few years entirely gone, and replaced by patches innumerable and of all colors. His social visits were limited to occasional journeys across the lake to another island on which also lived a hermit of like habits of life. This Summer [sic] Kenniston's friends sought him out, and yielding to their persuasions he has abandoned his Summer[sic] and Winter [sic] 'residences' and returned to civilized life."[90]

The motivations of these characters past and present often remain secrets, enduring in time to become legends and folkloric fabric of hamlets, towns, and regions. Their very presence in Maine's tapestry suggests a possible link to Bigfoot, namely, could encounters with hermits be mistaken for encounters with Bigfoot? People on the fringes of society and living alone in the wilds could be interpreted as wild creatures, beings more of the hills and forests than town and square. The following story tells of a hermit, a veritable "Wild Man of Maine," who with his family eschewed the trappings of civilization:

> The other day, one of the oldest residents of Pittsfield gave your correspondent the following about a wild man who used to hunt and fish in the wilderness, near where now is the thriving busy town. Said he: "People used to say he was the wild man of Maine, Isaac M__. He was a Yankee hunter who dwelt in a log cabin at Plentoma Point, about three miles above the town, about the time the Maine lot was set off by Massachusetts. He lived all his days like an Indian chief, and had his war-paint and his tomahawk. The story is told of him that every Sunday at four o'clock in the morning he used to bedaub his face and arms with copper paint, and dance an Indian war-dance about his rough log hut. He never associated with the white man, and whenever anybody asked him a question he would only grunt, and say, "Me big Injun, me no talk to white man." He was a great hunter, he used to traverse the forests for miles about Pittsfield, bringing his peltry to the settlement once a month, and make an exchange for rum and molasses. He raised a large family of children, who were all wild, like himself. The story is told that he taught them the Indian tongue, and they never learned how to speak a word of English. They were hunters like himself, and used to roam over the forests trapping muskrats. Whenever they saw a settler, they ran from him like a deer, shouting a shrill Indian war-cry. Father and children lived in this wild, barbarian state up to the time of their death. The children were drowned while attempting to cross the river in an ice freshet, so the legend goes.[91]

Those unfamiliar with this particular hermit and his brood, living as they did in the environs of Pittsfield, could be forgiven if after a fleeting glimpse of a presumed child or adult flirting between the shadows in the wilderness believed they had seen something more animal than man.

William Dean of Chairback Mountain lived in a rustic camp whose stove was furnished by an inverted funnel of a wood-burning locomotive. As the *Hartford Courant* put it, "Dean is the jolly man on his township, and claims that he is constantly attended by a spirit guide, who directs him to game and warns him of danger. These warnings are usually given by raps on the boarding over his bunk, and when they come, Dean leaves his camp and hides himself in the forest."[92] One may wonder what Dean was hearing, and if indeed they were physical raps on the side of his camp, what could have caused the rapping?

From the same *Courant* story is the tale of Hermit Sprague who would have presented a conundrum to an unsuspecting observer's perceptions.

Hermit Sprague of Schoodic resembles a huge hedgehog when he rambles abroad through the forest. For fifty years he has tied to himself by strings and shreds of cloth all the old garments that he has been able to accumulate by begging them from chance passers in his lonely retreat.

He strips clothing up and ties it on his person—thatching himself as it were. He never takes anything off, but waits for it to drop off, and no one knows how many thicknesses of cloth cover him. His feet are wrapped in the same fashion. A more peculiar or forbidding object when he goes groping about the woods cannot be come up with outside a tropical jungle. It is a wonder that the strange old man has not been popped over by some astonished city hunter.

The interior of his camp is crowded by rag-tags and odds and ends that he has picked up. Sprague even observes this hodge-podge rule of life in his diet. Whatever is given him or whatever he secures for food he throws into a huge pot, and cooks the mess, meat, pit and all, in one batch.[93]

Christopher Knight is a modern example of a Maine hermit, achieving notoriety by his record of living alone for 27 years. His story garnered interest amongst some in the Bigfoot community, specifically for Knight's almost three decades spent as a hermit and his ability to remain so well hidden and undetected. If he could do it, a survivalist by life experience, why not a Bigfoot creature?

Knight was known as the North Pond Hermit, living concealed in the central Maine Belgrade Lakes region. Though many were convinced of his presence, he escaped definitive detection for decades. Also remarkable is Knight's shelter and living area, when found, was determined to be only 600 yards from the nearest camp.

6
BIGFOOT, OR SOMETHING ELSE—
ANALOGOUS, AMBIGUOUS, AND OTHER

In *Lizard Man: The True Story of the Bishopville Monster*, author Lyle Blackburn's research provides details of eight separate sightings of the eponymous Bishopville, South Carolina creature. Blackburn's comment on the sighting by a Colonel Mason Philips (name changed) reveals how the creature may have gained its famous reptilian qualities: "The only puzzling detail of this sighting is that he claimed to see a tail, whereas no one else had reported seeing one."

> One of the most promising scenarios is that the Bishopville Lizard Man is not what the name suggests. In other words, the creature is not actually reptilian but a more familiar primate-like cryptid. Yes, Bigfoot. Given the amount of Lizard Man witnesses that reported seeing hair or other ape-like characteristics, it does seem like a reasonable explanation.[1]

Blackburn raises the possibility that what may have been witnessed and reported in and around Lee County, SC, was a Bigfoot, rather than a tail-sporting reptilian. This suggestion is congruous with the majority of the Bishopville Monster reports and aligns with, as far as cryptids go, a more widely accepted cryptozoological candidate. The present chapter takes a brief look at several unusual events set against the question, *Could it be Bigfoot?*

A MERMAN IN MAINE

John Josselyn, an Englishman, sailed twice to New England, in 1638 and again in 1663. Based upon his travels in New England, Josselyn produced two books that were essentially travel guides, focusing on New England's flora and fauna and his observations on Puritan society. After writing of a *lyon* and relating an anecdote of a sea serpent, Josselyn writes of a Mr. Mittin and this particular gentleman's encounter with something in Casco Bay.

> One Mr. Mittin related of a Triton or Mereman which he saw in Cascobay, the Gentleman was a great Fouler [sic], and used to goe [sic] out with a small Boat or Canow [sic], and fetching a compass about a small Island, (there being many small Islands in the Bay) for the advantage of a shot, was encountered with a Triton, who laying his hands upon the side of the Canow, had one of them chpt [sic] off with a hatchet by Mr. Mittin, which was in all respects like the hand of a man, the Triton presently sunk, dying the water with his purple blood, and was no more seen.[2]

Whatever Mittin encountered, it had hands like a man which serves to discount known fauna. If Josselyn does recount an actual event, then the creature encountered by Mr. Mittin will likely ever go unidentified. Interestingly, one of the attributes of Bigfoot commonly agreed to by researchers and enthusiasts is that Bigfoot is often found near water sources and is in fact an avid swimmer.

MONKEY ON THE RUN

Those interested in Bigfoot may have been exposed to a possible explanation for the Florida Skunk Ape's identity—escaped monkeys or apes. This appears, on the surface, to provide a rational and believable explanation for the Skunk Ape, but could such an explanation apply to sightings in Maine? I wonder if anyone in Scarborough saw something surely unexpected over fifty years ago and left confounded at the experience.

> Runaway Monkey Likes Maine Air—Cordial to TV Men at Scarborough; Declines Ride—Scarborough (AP)—A meandering monkey named Joe, who escaped from an animal farm in September, smiled for the television cameras Friday but still eluded capture.
>
> Joe has been browsing around a neighborhood on Pine Point Road for the past several days, stealing food here and there.
>
> He boldly walked up to the back of Wendell Whitten's house Friday, broke down a bird feeding station on the ledge of the kitchen window and ate up all the bird seed. That's when Whitten called the Animal Refuge League and asked that Joe be picked up.
>
> The league workers had planned to try capturing Joe with a net, but they had too many other calls Friday and couldn't come out from Portland.
>
> A Portland TV station (WGAN TV) sent out camera crews and a newsman to cover the story. They found little Joe high in a tree in nearby woods. He was polite in his first encounter with the press but had nothing to say.
>
> Joe has been roaming the Scarborough countryside since he escaped from the animal farm in adjacent Old Orchard Beach. Officials there theorized at first that when winter winds started howling, he would go back home. But now, like the people on Pine Point Road, they're beginning to wonder.
>
> The Animal Refuge League will try to capture Joe Saturday. He weighs about 25 pounds and stands 2 ½ feet tall.[3]

THE LEACH'S POINT DEVIL

Emeric Spooner is from Bucksport, and the author of several books including *In Search of Sarah Ware* and *Ghost Hunters of New England*. On his website, *Maine Supernatural*, he includes a tale heard during his youth that he calls the Leach's Point Devil. An interesting facet of this story is the devil's aversion to light, a trait seemingly shared by the creature which confronted Eric Martin, mentioned below.

Two nuns on a sabbatical were walking along a road lit occasionally by a street light. It was a foggy night in the summer. One nun stopped at hearing a strange sound. It came again, a low guttural growling. It seemed as if the maker of the growling noise accompanied by snapping and breaking branches was following the pair. The pair quickened their pace knowing their retreat was only about three hundred yards away.

As they hurried to the next street light, one nun slipped on loose gravel and went down, her knee hurt, the flashlight she was carrying, the only light between the pair, went flying into the ditch. The other stopped to help her fellow and that's when they noticed the sounds following them had stopped. They decided to make a go for the house they were staying at.

Just as they were about to leave the circle of light from the streetlight, the nun with the injured knee pointed. There through the fog was discernible a large dark shape coming right at them. The thing had glowing red eyes and advanced on two legs like a man. The nuns retreated back into the center of the streetlight's light. The creature remained on the periphery of the light's glow seeming to circle the pair of frightened nuns. When it stopped and withdrew neither could later say, but the nuns remained where they were under the protection of the streetlight until daylight.

DOUROUCOULI—THE NIGHT MONKEY

Over the course of nearly three years, the *Biddeford Journal* ran stories of a strange creature disporting itself in the environs of Raccoon Gulley near Lyman. The mysterious animal was likened to a douroucouli and even a sapajou, both New World monkeys not known for attacking and killing animals their own size. There was never a final denouement to the identity of the creature haunting Raccoon Gulley. Whatever it may have been it was often described as not overly large and was understood to spend at least some time in trees. Perhaps the creature was more likely a pet monkey released by a sailor visiting one of Maine's port towns than a young Bigfoot but the very idea of the latter suggestion is appealing to the present chapter. The reader should also note the storytelling tone evinced by terms like *geezle* and *whingdingit*. Below are several stories printed by the *Biddeford Journal* concerning the douroucouli of Raccoon Gulley.

> Strange Animal Prowling About Raccoon Gully— People living in the neighborhood of the Raccoon Gully woods in the town of Lyman are puzzled over the appearance of a strange animal which prowls about the woods after dark and makes nights hideous with its hair-lifting screams, which are loud and sonorous, much like that of a jaguar. It also mews like a cat after it has had a good meal. Although a small creature, not over a foot and a half in length, it attacks and slays animals of about its own size. In daytime the critter hides between the branches of trees that are thick with foliage, such as firs and spruces, and keeps perfectly still.
>
> John Jay, who is something of a naturalist, expresses the opinion that the animal is none other than a douroucouli, but how in thunder it got into Raccoon gully [sic] is a puzzle. The nearest habitat of this animal has been supposed to be Brazil. It is a tropical animal and has never been known to wander from a region where it could obtain fruits and seeds the year round.
>
> A Douroucouli, or What?—One Reader Surmises it is a Sapajue or a Tarsius—Another Intimates That the Raccoon Gully Naturalist Needs Bone-Dry Treatment— and Another Harks Back to the Days of the Geezle.

Makes a Noise like a Sapajue—Editor of the Journal: I read with no little interest the story in Saturday's Journal about the strange animal prowling about the woods in Raccoon Gully and which the written [sic] thought might be a douroucouli. I never heard of an animal by that name. I have heard the cries of the animal which has alarmed the people around Raccoon Gully and the peculiar noises it makes, a cry as if the creature was in distress, and it sounds to me as if it were made by a sapajue, [sic] but I never ran across one of those animals in this latitude. It is possible that the strange creature is a tarsius, which is a nocturnal wanderer and noisier than a screech owl. J.C.W.

Discoverer Probably Been Drinking—Editor of the Journal: Wait until we have the real article in bone-dry prohibition and you won't hear any more complaints from Raccoon Gully, or any other quarter, about douroucoulis or other strange animals, terrifying people with their unearthly noises at night. The stuff some folks are drinking nowadays is enough to make them see a rubicundus in Kennebunk pond. Speed the day when even Epsom cocktails will be under the ban! Yours for teetotalism, Joseph Thumper.

Now Read This—Editor of the Journal—In the Saturday evening edition of the Biddeford Journal was an interesting item regarding an animal supposed by "John Jay," a naturalist, to be the douroucouli, an exotic, so to speak, from Brazil. It occurs to me that it must be closely related to the geezle, a strange quadruped which you will recall was often found about Raccoon Gully in the early 70's. If I remember rightly the then zoological authority of Biddeford tracked the geezle to its lair and from his pen came the only authentic (?) account of that fabulous beast. Natural history and York county certainly owe much to him, to his bent for science and his willingness to share the results of his study and investigations with the unenlightened public. One cannot but feel encouraged that it is not necessary to explore the depths of Darkest Africa or the mysteries of the Amazon jungle to make and record astounding discoveries.

The writer cannot forbear paying this slight tribute. Yours truly, Yura Jay.[4]

Night monkeys, also known as douroucoulis, inhabit South American forests. They have large eyes which improves their ability to be active at night.

Raccoon Gully Douroucouli in Limelight—Learned
Zoologist Skeptical—Thinks Strange Animal a
Whingdingit—Newt Newkirk, the "All Sorts" man of
the Boston Post, who claims to be something of a
naturalist, although many people believe him to be
a 33d degree nature fakir, writes in his usual know-
it-all vein about the Journal's story of the appearance of a strange animal in Raccoon Gully thought
by some people to be a douroucouli. The Journal
never ventured to classify this animal but merely
printed the opinion of Mr. John Jay, a naturalist of
more or less distinction, who pronounced him to be
a genuine douroucouli. Now comes Newkirk with his
fluent flow of misinformation, as follows:

"It is plain to be seen that the naturalists of Raccoon Gully went to night school and studied by a dim
light. To suppose that the animal above described is
a 'douroucoli' is preposterous. In the first place a
'douroucoli' is strictly terra firmaceous and non-
aboreal. In short, it couldn't climb a tree in a hundred years. Moreover, the 'douroucoli' is not nocturnal—it sleeps nights and works days.

"Judging from the screams this critter emits, together with the fact that it roosts in firs or spruces
during the day and that it mews contentedly after a
square meal, I have no hesitation in saying that it is
an adult specimen of the Whingdingit, which is quite
rare in New England, but by no menas extinct at the
present time, although it was extinct for a period of
about 50 years prior to 1873, when it made its reap-
pearance. Captured when young, the Whingdingit
makes an affectionate pet, but it is very difficult to
teach a matured Whingdingit to eat from the hand
without lacerating the fingers. The Whingdingit
wears a luxurious fur coat which is full of cooties,
and as a result it leads a pretty miserable existence."

Now in the first place Naturalist (?) Newkirk writes about a "douroucoli" and there is no such animal in Raccoon Gully or anywhere else. Even persons not well grounded in the most elementary principles of zoological science know that the creature which Newkirk is trying to tell folks about is a douroucoulo, or as it sometimes called, durukuli. Note the difference in the spelling. Then again Zoologist (?) Newkirk declares that the "douroucoli" is a terra firmaceous and non-arboreal animal. Assuming that he really refers to the douroucouli, he again betrays his dense ignorance. The douroucouli, scientifically known as the nyctipithecus trivirgatus, is arboreal in high degree. Neither the procyn lotor nor any of the rapotial birds of the Strigidae family are more so. Newkirk unblushingly tells his readers that the douroucouli "couldn't climb a tree in a hundred years." Newkirk would climb one a good deal quicker than that if he should ever meet one in the woods.

"Moreover," continues the distinguished (?) zoologist, "the douroucouli is not nocturnal—it sleeps nights and works days." As a matter of fact it sleeps days and gets in its work at night. Not satisfied with the foregoing display of what theologians would call his "invincible ignorance," Newkirk proceeds to say that in his opinion the so-called douroucouli is neither more nor less than an elderly whingdingit. He probably means a whinchat, but as they are harmless and never scratch or bite, he is on the wrong track again. If he had classified the strange visitor as an olyooly he would have come nearer the truth, but still he would have been wrong. Mr. Jay is too eminent a naturalist to make a mistake in classifying any denizen of the forest, and he is too wise a guy to be misled by the chatter of a fellow naturalist (?) of the Newkirkian school.

Probably in a few days the "All Sorts" column will be burdened with more misinformation, for it is only natural to expect that Newkirk will deny that a koo-doo, a dziggetal, or even a korrigum, has been seen recently near Dave Wilcox's ice houses at West-brook.[5]

Closing of Goldthwaite's Much Regretted. The One Place in Two Cities Where People Met on a Level and Settled Great Questions. It was Goldthwait's that the strange wild animal of noctural [sic] habits which invaded the Raccoon Gully woods a couple of years ago, was correctly classified as an East Indian dour-oucouli.[6]

Farmer Snow's Sweet Corn—John A. Snow, who carries on a farm at Pine Point, has lost a fine stand of sweet corn through the depredations of a coon. At all events, Mr. Snow believes it was a coon, but in all probability it was a duoroucouli, perhaps the same animal which created a scare at Raccoon Gully last year. This probability is strengthened by the fact that Guy Burnham, a noted huntsman of Pine Point, says he came across a female douroucouli and three cubs while picking raspberries near Mr. Snow's farm a few days ago. He was unarmed at the time or might have captured the whole family. According to Mr. Burnham the douroucouli he saw was a handsome specimen, with a fine spread of pink whiskers.[7]

Sanford Men See Bear—The Portland Press of Monday contained the following interesting yarn: Four Sanford men came to Portland Sunday in an automobile and on their way, while passing from Alfred to Biddeford, they think they saw a bear. At least they saw some large, ungainly object on the ground and

of a dark brown color and which was busily engaged
in getting out of sight in some thick underbrush.

As near as they could locate the situation it was
within the limits of the town of Lyman. One of the
four and one who was not driving the auto and so
had time to look closely at the object, said it was too
near the ground for a cow or ox and was too large to
be a sheep and bore no resemblance whatever to a
deer or moose. He describes it a big, fat, dirty,
brown-looking object which he should say would
weigh about 350 pounds, and that it bored its way
through the bushes head down and body working
ahead all of the time.[8]

It was not a Bear—Strange Animal Seen in Lyman
Woods by Sandford parties Believed to Have Been a
Duroculi. Lyman people who read the story in the
Journal reproduced from the Portland Press, telling
of the sighting near the Alfred road by several
Sandford men of a large bear, scout the story and
declare that no bear has been seen in the town of
Lyman during the past 50 years or more. The men
describe the animal they saw as "big, dirty, brown-
ish-colored creature, which weighed about 350
pounds, and which bored its way through the brush,
head down." The bear story receives little credence
among the residents in the vicinity of Raccoon
gulley. They express the opinion that the animal seen
was the same old duroculi which Abe Slupsky saw
last fall and of which he gave the Journal an accurate description.

He was of a reddish brown color, weighed 250 to
300 pounds, had a tail like a kitchen mop, one bright
red and dull green eye, a short pudgy neck, and ears
about as large as the ordinary palm leaf fan. When

he growled he made a noise something like that made by an auto when the driver is using his cutout.

Newton Newkirk of the Boston Post, who is accustomed to butting in on other folks' business and who labors under the delusion that he knows quite a lot about wild animals, as well as post-holes, denounced Mr. Slupsky as a nature faker because of his claim to having seen a duroculi at Raccoon gulley. He said there might be an abundance of hootch in Maine, but staked his reputation as a naturalist of the first order that no living man had ever seen a duroculi there. Newt expressed his belief that the strange animal seen by Slupsky was nothing more or less than a common dingbat. Be that as it may, it is safe to say that the Sandford bunch didn't see a bear. The strong probability is that if they saw anything they saw a duroculi, but it is bear-ly (don't overlook the joke) possible that it was a nutennewkirk [a play on the Boston Post gentleman's name] dingbat.[9]

Raccoon Gully Duroculi is Seen Again—Biddeford Man Gets a Glimpse of Him and Has Cold Shivers— Scouts Dingbat Theory. Frank A. Goodwin, manager of the local telephone exchange, is the latest person to see the duroculi in the Raccoon Gully woods. While motoring on the Alfred road near the gulley yesterday he noticed a disturbance in the thicket close to the highway and immediately thereafter a strange looking animal which seemed to him to be as large as a baby elephant went crashing through the brush at lightning speed and quickly disappeared into the deep woods. He did not get a clear view of the animal fore and aft, but after reaching home and reading in the Journal of the experience of the three Sandford men who saw what they thought was a huge bear, near the foot of the gully, but which is believed by experts to

have been a duroculi or a dingbat, he is convinced that
the strange creature which for a moment riveted his
attention and caused the cold shivers to run up and
down his back, was a duroculi. Mr. Goodwin takes
no stock in the dingbat theory, with whose habits
and general appearance he had long been familiar.
The dingbat, he says, is half bird and half beast. He
has five legs and a pair of wings, and always uses his
wings to make a quick getaway. The animal he saw
yesterday had no wings, but was a high jumper.[10]

Looking for the Duroculi. Chief of Police Robbins
and Probation Officer Spofford are among the many
Biddeford people who had a sharp eye out for the
duroculi as they motored up Raccoon Gully hill this
morning on their way to the county seat. No one got
a glimpse of the animal, the nearest approach to it
being a fine specimen of the genus mephitis, which
Chief Robbins observed disporting himself by the
roadside a little beyond half-way house.[11]

The mind can race in attempts to fill in missing pieces of information, no less than when encountering an unknown *something* outside the canvas of one's tent. G. W. Hinckley, the catalyst for Good
Will-Hinckley Homes for Boys and Girls in Fairfield, Maine, was a
believer in real camping, i.e. camping minus all the modern conveniences and luxuries of modern life. He wrote *Roughing it With
Boys* in 1913 to celebrate outdoors life and the lessons from living in
primitive conditions. The below story, like all the anecdotes from
Roughing it with Boys was portrayed as a true event and captures
the apprehension and wonderment when intimately close with the
unknown.

[At Wilson Pond, Greenville, Maine] July 16. The
very first night I ever spent on this point was three

years ago when Blake and I lived close to nature five weeks—he a college student earning one part of the wherewithal to carry himself through another year of studies. That night some wild beast was prowling about our tent; but after the first night we heard little or nothing from him. It was the same way last night but a different kind of beast.

Chick was asleep; he sleeps wonderfully whenever the mosquitoes and midgets [sic] let up on him. I heard twigs cracking and the steps of an animal coming toward the tent. There was a pause of a full minute; then the creature bounded away. I heard his retreating footfalls and was pleased that he had left us. But before I could get asleep he returned and came nearer. I didn't like it. I let my imagination—what I have of one—take the reins. I could see in my mind's eye, just outside my tent, any one of several terrible beasts, such as inhabit the Maine woods. Then the unwelcome visitor started on the run again, back into the woods, though all the while I had lain perfectly quiet and Chick was in the land of dreams, where no midgets [sic] are and where mosquitoes neither bite nor buzz.

The next time he returned, and it was inside of ten minutes, he made approaches of a few feet, stopping for a long silence between each advance, until he was not, to the best of my knowledge, more than a rod from the tent. The situation seemed uncanny. I had left the lantern out at Greenville. My gun and rifle were down home—for I take a peculiar satisfaction in leaving firearms at home as I go into the woods in the summer, when there is nothing to shoot except in the wanton destruction of life. But I wished I knew what it was, just outside the canvas and what his plans were. I recall Manly Hardy's recent and explicit contrasting of the Maine bobcat and the Canadian Lynx. I could see the broad wooly paws of the one and the smooth cat-like paws of the other.

I reached out my hand and laid it on Chick's breast it seemed full and strong and—too deep for a lynx to crush in with his jaws; just as a caribou's, according to Roosevelt, is too muscular for any such feat. I felt of my own chest; that, too, was lynx proof—bobcat proof—just like a caribou's. But as I happened to touch my own flesh, I was annoyed to find that it was not like human flesh at all—it was rough, cold, pimply; for the first time in my life fear had given me "goose skin".

I continued to think, but it didn't help me. I grew desperate. Wanting to say something, I raised the wall of the tent suddenly and shouted. I didn't expect the animal to understand my language, but I fancied he would take the hint that he wasn't wanted from the tone of my voice,—my general attitude.

"Get out of this," I yelled.

The intruder had no idea of what was wanted and didn't move. I rather hoped, too, that my shout would awaken Chick so that I could feel that I had company; but mosquitoes and midgets [sic] had suspended operations pending the break of day and Chick was oblivious. About three minutes after my yell broke the stillness, that wild beast—whatever it was—went rushing through the woods making a great racket, having started from the point where I had last heard him, about a rod from the tent.

A rabbit makes a lot of noise in the dark.[12]

A "serenading" bobcat was the antagonist of the following tale, identified by its voice rather than a definitive visual sighting.

Oct 28, 1901 The Loocivee's Yell—And How it Shivered the Nerves of a Lewiston Sportsman—Night Adventure in the Wilds of the Moxie Region—Would have been a Story by Fred Groves Wouldn't Talk for Publication—Chadbourne's Ranch, Moscow, Me., Oct. 28 (Special)—Some strange stories come down from Chase Pond camps these days, and parties here profess not to believe them all. A message from the camps this morning (by special clothesline wire) says that Groves has had an awful dream, a night-mare, or more properly speaking, a night-moose. It is probably the aftermath of a late fill of cook's dough-mullions, and quill-pig. Groves, (you know F. B. of Auburn, president of the common council who tells interesting yarns about Mayor Wilson's fishing exploits)—well, Groves is very fond of fricasseed quill-pig, quill-pig on the half shell, etc. Now, while quill-pig as a diet, is very strengthening, if persisted in will result in bad dreams. But whatever the cause the result is to be regretted because the cook says he doesn't like to be scared out of four Christmas dinners and an equal number of years' growth by such midnight alarms. He swears it was caused by eating quill-pig and Groves declares it was the effect of an overdose of cook's dough-mullions. As we are not yet in possession of the full details we will not attempt to make the dream public, in this letter, but it is coming over our special wire in a doubled and twisted code, as soon as it can be translated we will forward to the Journal. In the meantime we forward Grove's story about Milt Allen of Lewiston, which is no dream. Allen has been at the camps for some time and Groves and the Dotens (Lewiston jewelers) have been much alarmed about him because he was losing his appetite. They tempted him with the breast of partridge fried in butter, with deer steak, with Moxie beef garnished with Bald Mountain celery, and Groves even offered him some of that precious quill-pig. But nothing had the desired effect until an incident of one evening last week. Perhaps we had better relate it in Groves' own words. Just as they came in over the wire this morning and were recorded on the cylinder of the field roller at the ranch. He says:

"It is strange what simple things, when the combination is right, go to make up a startling incident. For instance a loocivee had been serenading us for several nights. We didn't mind and he had lots of fun trying his voice. Then there is the spring, a few rods back from the camp in the woods. Beside the spring is a knoll. Near the knoll was an old tin pail (taint there now). And last, but not least, there was Milt Allen.

"Well, you see, as these things have been, separate and attending to their own private functions they were all right, but when properly combined they created a sensation. And the way this combination was arranged is the whole story. It was this way:

"Allen had been out after partridges and came into camp late. It was dark. The bob-cat sat on the knoll by the spring. The spring was there the same as ever (and is now). The pail was sitting comfortably behind the knoll listening to the bob-cat, or, perhaps, watching for the water to spring. The bob was there just because it's a habit loucivee's have of getting somewhere, especially where they are not wanted.

"Now, all it needed was Allen to complete the combination and perfect the incident. And sure enough Allen came stumbling along! He was thinking how he would go in and look over cook's stock in trade for the space of an hour and a quarter and then get up without doing a thing to it. Then he would smoke about four pipes of tobacco and try to get up a little appetite for breakfast.

"The bobcat saw Allen coming but kept his sitting. The pail must have heard those hunting boots for there was its tin ears, and the spring kept on bubbling over with mirth at the prospect.

"Yes, Allen was blundering over old logs and saying things as the twigs nipped his nose, but he didn't know some things that the bob-cat and the spring and the knoll and the pail knew. He (Allen) could see a glimmer of light from the camp and was just mumbling something about being almost there when that bobcat let out a yell that made the spring jump and the old tin dish turn pail, and even the knoll look grave.

"Perhaps Mr. Milt Allen of Lewiston didn't jump, but the loocivee did and he landed in the old pail and both crashed up against a log. Then all thought bedlam has broken loose or Old Obadiah had arrived. The curl came out of Allen's moustache, his hair poked out through the top of his hat, and other things happened.

"Say, you ought to see some of Allen's tracks from the spring to the camp. Many of them would measure 40 feet. When he burst through the camp door his eyes stuck out so that Cook hung his hat on a hook and said:

"Dam-p night out aint it?"

That's all there is of it, except Allen's story of being chased by a sidehill lounger with feet as big as bundles of hay, etc. But you bet, he wouldn't go after water later than 3 o'clock in the afternoon after that. He says Lake auburn water is good enough for him now, so he has gone home, but he didn't dare go alone. The Dotens went with him and I hope there are no loucivees in Lewiston.

We tried to get Groves to say more about the case but he said he wouldn't talk for publication. The only corroboration of the story at hand is the fact that Messrs. E. C. and H. E. Doten and Mr. Allen have gone down river with a fine lot of game. Each of the Dotens had a fine buck and many partridges and ducks. Allen had a big bundle of remembrances with his other game and perhaps a Journal reporter could persuade him to tell more about Groves' dream.[13]

SUPPOSED BEARS

Bears must be accepted as a potential answer explaining some experiences people have shared with unknown creatures in Maine. Second only to the moose in total weight amongst Maine's fauna, the black bear also has the distinction of being able to stand up on its hind legs which could, through simple misidentification, explain some sightings of what people may perceive as Bigfoot. On my own coastal Maine property I've seen a bear stand up, but when it ran away, it dropped back to all fours. The next three stories all aver a bear as their object, though a measure of ambiguity in each account leaves questions. Bear in mind (no pun intended) that many Bigfoot sightings have the creature moving comfortably in a quadruped fashion, the normal mode of locomotion for a bear. The first account has the author enjoying a lengthy viewing of a bear catching fish, but a bear's paw deftly closing around its piscean quarry sounds like the dexterity of a primate's digits.

> "I came suddenly upon a very large bear in a thick swamp, lying upon a large hollow log across a brook fishing, and he was so much interested in his sport that he did not notice me, until I had approached very near to him, so that I could see exactly how he baited his hook and played his fish. After I stood a while seeing him catch fish and eat them, he discovered me and made off into the dense underbrush,

but I had no gun with me, perhaps if I had I should have missed him and maddened him and it would have been worse for me and I might not have been here now writing about it. But, as I was saying, he fished in this wise. There was a large hole through the log on which he lay, and he thrust his forearm through the hole and held his open paw in the water and waited for the fish to gather around and into it, and when full he clutched his fist and brought up a handful of fish and sat and ate them with great gusto; then down with his paw again, and so on.

The brook was fairly alive with little trout, and red-sided suckers and some black suckers, so the old fellow let himself out on the fishes. He did not eat their heads. There was quite a pile of them on the log. I suppose the oil in his paw attracted the fish and baited them even better than a fly-hook, and his toe-nails were his hooks, and sharp ones, too, and once grabbed, the fish are sure to stay.

They also catch frogs in these forest brooks, and drink of the pure water in hot summer days, and love to lie and wallow in the muddy swamps, as well as our pigs in the mire."[14]

The animal in the next story is never positively identified, but it is supposed by Mr. Owen Chase of North Brownville as a bear, mostly due to its upright posture.

In all my experience with bears, and as often as I have run across them in the woods, I never had a bear stand up to face me but once, and that was scarcely enough to be worth mentioning. I was watching an orchard where bears had been plundering, one night. I waited a good while without seeing anything, but at last, hearing a noise, I went through the orchard in its direction and suddenly came upon

a big, dark object as tall as a man. It was motionless, and for a minute or two I was in doubt what it
was, it sat so still. I listened to hear it breathe, to
make certain it was an animal, as I didn't want the
boys to laugh at me for firing at a stump. I was a boy
then and had nothing but an old smoothbore to shoot
with, but I made up my mind I'd shoot, stump or no
stump, and raised my gun to my shoulder. When I
did that, as quick as a flash the old fellow dropped
on to his feet and made off in such a way that I lost
him. He was one of the biggest bears I ever saw.
That's the only time a bear ever sat up before me.[15]

Donn Fendler was separated from the rest of his family and
became lost on Mount Katahdin (Maine's tallest mountain) in July
1939. Fendler ultimately survived for nine days in the elements
before he followed a stream and telephone wires back to civilization. The following anecdote is from Fendler's *Lost on a Mountain
in Maine*, which has become a children's classic in the state.
Fendler recounts hallucinating from lack of sleep, food, and water
during his ordeal. A hallucinatory memory could account for
Fendler's depiction of a bear he encountered, which if taken literally, was actively moving and feeding on its hind legs before dropping to a quadruped stance.

The next thing I knew I woke up and it was getting
dark.

I was sitting on a rock looking at my feet. They
didn't seem to belong to me at first. They were the
feet of someone else. The toenails were all broken
and bleeding and there were thorns in the middle of
the soles. I cried a little as I tried to get out those
thorns. They were in deep and broken off. I wondered why they didn't hurt more, but when I felt my
toes, I knew—those toes were hard and stiff and had
scarcely any feeling in them. The part next to the big

toes was like leather. I tried to pinch it, but couldn't feel anything.

My head ached and I didn't want to move, but night was falling and I had to go on, at least as far as some big tree. I got to my feet. Was *that* hard! I could scarcely bend my knees, and my head was so dizzy, I staggered. I had to go across an open space to the stream, and as I went along, I saw a big bear, just ahead of me. Christmas, he *was* big—big as a house, I thought—but I wasn't a bit scared—not a single bit. I was glad to see him.

When he got nearer, I knew he didn't see me, so I crouched down a little behind some bushes—but kept my head up so that I could watch *him*. I didn't want him to run right in to me. He ate berries, as he went along. He'd swing his head and nip one way and then he'd swing his head and nip the other. He kept making grunting noises. I don't see how he ever came so close to me without seeing me, but I kept mighty still. I hardly breathed. I knew I was a goner, if he took after me. I couldn't run or climb a tree— not with those feet or legs. Pretty soon, he dropped down on his front feet and I couldn't see him any more, but I could hear him, breaking down the bushes as he went away.[16]

Samuel Spalding's encounter with a creature near Springfield is curious for the profound change it apparently wrought in his life, a once avid outdoorsman turned homebody. This intimates that Spaulding met up with something truly remarkable, certainly from his point of view.

A Hunter's Encounter with the Devil—Mr. Samuel Spaulding, of Springfield, while hunting in the vicinity of that town, recently, encountered the devil face to face. This he persistently and with trembling

declares. In corroboration of his testimony there is
a collateral fact: Whereas he was formerly a veritable
Rip Van Winkle, passing a great part of his time in
the woods with his gun, he now secludes himself at
home, hardly venturing from his premises. All ques-
tioners obtain from him is the same statement accompanied by a description of his Satanic majesty.
For many years the story has been periodically re-
vived of some strange creature in the woods, near
Springfield, and lumbermen have seen it as far distant as the head of Musquash stream; but Mr.
Spaulding is the first man bold enough to declare
that it is the devil. If his story is true, and his manner and character are in his favor, he had a very near
view and might have bagged the creature, had he not
been afraid to fire his gun. If it was really the devil
there are many who will never forgive him his reminisces.—Calais Times[17]

HOW DO BIGFOOT HANDLE THEIR DEAD?

The account of Benjamin Cole's discovery of an unusual rock pile
titled "Indian Grave or What?" is prefaced by an exchange taken
from the movie *The Mysterious Monsters* (1976) between Peter
Graves and a Dr. Lawrence Bradley.[18]

> Dr. Lawrence Bradley, Anthropologist: "Well in re-
> ality, very few bones are ever found anywhere in
> the forest. When an animal dies, it is immedi-
> ately eaten by another animal. This is a disposal
> service of nature at work. I go up in the woods,
> in Oregon and Washington all the time, and I've
> never seen the remains of bears, or even moun-
> tain lions, and these animals are in abundance
> up there.

Peter Graves: Dr. Bradley, it's been suggested by some people that maybe that Bigfoot creatures bury their dead.

Dr. Lawrence Bradley: That's remotely possible. If Bigfoot is, as some of us think, a humanoid, a species of animal more manlike than beast.

Peter Graves: But it is possible?

Dr. Lawrence Bradley: Yes, but it's a proposition completely rejected by my colleagues. But then, my colleagues completely reject the possibility of Bigfoot.

Indian Grave or What? One Fall when David had a furlough in hunting time, we were at the camp hunting for a week. It was fine weather so we decided that we would go to the top of Philbrook Mountain. When we were about two-thirds of the way to the top of the mountain, which was about three miles from Wanigan Brook where we had left the boat, we came upon a sight that has always puzzled me. I have always thought that it was a grave yet I do not know for fact that it was.

Here, in a level little glade, was a mound about six or seven feet long and probably three feet wide. It was built up about one and a half feet higher than the surrounding ground, packed over with small stones as big as a quart dipper—some larger. They were rip-rapped into the mound quite close together. It was so old that moss had pretty much over-grown it. I have traveled the woods a lifetime and I have never seen anything like it. It had to be the work of human hands.

We believed it to be a grave but who would have gone into that hard-to-get-to, out-of-the-way place, and when, to bury anyone! I have always intended

to go back with a pick and shovel and dig into it but
have never got about it.[19]

In form like a person is what Bigfoot and the creature mentioned in the story below, as experienced by Aaron Jackson in the
Allagash region, share in common. Otherwise, this quick story does
not suggest Bigfoot as the being witnessed by Jackson, but it is a
strange encounter which Jackson apparently believed in all his life.

> Jackson, a 93-year-old manufacturer of canoes,
> poles and snowshoes, sat among a collection of shav-
> ings and recalled the night 83 years ago when his
> aunt died.
>
> "I was down there that night," he said. "I started
> up home and it was just the same as wings over my
> head. I thought it was a bird. I looked up and it was
> a person, I'd say three or four feet long, naked and
> flying. It went right over my head. I seen it plain with
> my own eyes. It never scared me a bit. That was her
> death angel."[20]

MAINE'S LOUP-GAROU

French fur traders and woodsmen brought their traditions and lore
with them to North America. "The lutins are the only widespread
example of fairy-like creatures found in French folklore in North
America."[21] Lutins were said to ride farmers' horses at night as well
as busy themselves in plaiting the horses' tails. An Aroostook
County Oral History Project interviewer was told by a Valley resident that lutins lived in caves and came out only at night.[22]

Particularly in the southern sections of Maine, it was once common in late summer to enjoy sweet corn toasted over the glowing
coals from bonfires. The cooked corn would be removed from alder poles, slathered with butter, and collected in piles on tables
arranged under the spreading limbs of stately elms. Potatoes baked
in the ashes were also gathered under the moonlight. Feasts like

this were once community affairs, to be topped off by desert of hot donuts and new milk. At gatherings like these, ghost stories and legends were often told, much to the delight of the youngsters and many of which were derived from Franco-American culture. There was the belief that sinful life caused one to be transformed into a black bear or dog. Those so afflicted were condemned to wander through each night in this form until someone draws blood from them, causing the curse to be broken.[23] Juliana L'Heureux of the *Portland Press Herald* is a writer on Maine's Franco-American population. The account below concerning her husband shows an occurrence of the French werewolf, the loup-garou, as a belief powerful enough to genuinely scare young boys.

> On long October nights, while coming home from altar boy practice, walking down Sandford's Berwick Street heading towards the Ridgeway area, it was customary to take a dark short cut through "the pines" and up over the hill towards Freemont Street. While walking home, in a typical show of sibling rivalry, my young husband was frozen with fear by the loup-garou legend when his brother stopped to alert him. "Shhhh! Qu'est ce sais? (What's that?), he whispered. They paused wide eyed to listen to a distant howling, probably a dog, or the wind or both, while his brother sang a spooky refrain in a low cunning voice. "C'etait la voix de loup-garou—aaaaa-wooooo" (It was the voice of the werewolf—howling!).[24]

Bigfoot Field Researchers Organization (BFRO) Report #859 [see Chapter 8, sighting 34] contains the report by a witness who along with his friend saw "a very hairy animal standing upright on the side of the road" while en route to a club in Turner. The long, light brown hair had the appearance of "hanging off of it." The initial report ends with, "When we got to the club we told some people what we had seen and they said that others have also seen something in the area and even had a name for it. Loopgaroo."

Shelley Martin had listened to author Linda Godfrey on the *Coast to Coast* radio show and reached out to her with a remarkable story of her and her husband Eric's experience in a small Maine town. Not a classic Bigfoot encounter, the couple met up with something displaying qualities more canid than ape. Eric and Shelley's residence was in Palmyra on the Madawaska Road, and the experiences she related to Godfrey occurred on Labor Day weekend in 2007. Their 18-year old daughter was asleep upstairs while Shelley and Eric were on the porch when suddenly Eric told Shelley she should go inside the house immediately. Before going inside, the couple saw three large creatures whose pupil-less eyes reflected yellow-green and ran on hind legs shaped like a dog's. Eric and Shelley were shocked. The creatures appeared to be about 7 feet tall and had short tails less than a foot long.

> The creatures were covered in short fur; one beast was light brown and the other two were dark brown. They had pointed ears that started at the jaw and ran up the side of their heads so that the inner triangle of the ear extended past the top of their heads.[25]

Two more creatures were noticed which when first seen seemed to be attempting to covertly approach Eric and Shelley on their porch. When spotted, these two creatures also rose up and ran on their hind legs like the first three. The couple now raced inside the house.

> Shelley said they looked "sleek" when they ran on two legs, not at all clumsy, although their posture was slightly hunched at the base of the neck. Although their body fur was smooth, around their heads it "looked wilder and stuck out."[26]

Eric had guns stored in a barn and wanted to try to get them, an idea that Shelley flatly negated. While Shelley went through the house closing windows, Eric went outside in an effort to get the family car backed up to the house in order for the family to leave their besieged home. At a point in the yard, Eric set off the motion sensors and a light lit on. Eric saw that he was face-to-face with one of the creatures and feeling extremely vulnerable. The creature appeared to try to move toward Eric but the light appeared to stop it. Eric returned inside.

The couple eventually went to sleep and in the morning they began to investigate their yard. They were able to find tracks described as "over-large dog prints with a 'hook' on the end that made them longer." According to Godfrey, Palmyra "appears to be another of those enigmatic window areas where the anomalies come in many flavors."[27] Eric and Shelley had seen strange lights emanating from a swampy area near their home preceding the night encounter with the five creatures. Also prior to that Labor Day weekend, Eric had been seeing the apparition of a girl dressed in old-fashioned clothing in the house.

The SyFy show *Paranormal Witness* aired "Wolfpack" on Wednesday, August 7, 2013, dramatizing the experiences of Shelley and Eric. The show was produced by *Raw TV*'s Mark Lewis and Lucy van Beek. *Raw TV* is also the production company behind the shows *Locked Up Abroad* (National Geographic) and *Gold Rush: Alaska* (Discovery). Shelley stated that the program's dream sequence of one of the creatures breaking through a window into the house was contrived and her permission was solicited to put it in the show. The following statement by Shelley is taken from *Dogman-Monsters-are-Real blogspot* which credits her quote to the Facebook Group *Dogman Mystery*:

> I did, however, learn a few things, not those ridiculous things they made up, but other things Eric did I had no idea he had done until I heard his interview! They were here from London for a full week and we each had at least one day of interviewing with their crew and some of us more than that. But one thing Mark asked me was, "How did you feel when

Eric went outside the second time?" I matter of fac-
tually [sic] stated, "He didn't do that." Then they
showed me his interview and I literally cried. My
poor friend Linda Godfrey, she called me right up
when she saw the sneak peak because some of her
other friends and fans were mentioning her book
didn't discuss what they were showing! Little did we
know that when the show came out there would be
so many embellishments?[28]

The Martins' dramatized experiences in Palmyra reveal the
creative license taken in depicting specific events for a television
audience; some events were exaggerated and some were contrived
specifically for the show. Also of note is the nature of the creatures the Martins described, namely humanoid in basic form with
several canid features including tails and tracks resembling large
dog prints. The encounter in Palmyra begs the question, could the
Martins have had an experience with a group of Bigfoot? But since
the Bigfoot-in-a-box goal has not been achieved yet, and science
has no phenotype for study, one could just as easily ask the con-
verse: could many other sightings of hominid creatures within
Maine be wolf men?

FOOTPRINTS

Clearly, *something* is leaving enormous humanlike
tracks on the backcountry roads and riverbanks of
North America's mountain forests. With the potential for misidentification of bear or human footprints
accounted for, the otherwise inexplicable footprints
that remain must be either hoaxed or humanoid. In
the absence of bones or body, the tracks constitute
the most abundant and informative data that can be
dealt with by scientific evaluation.[29]

> In order to evaluate the long-standing, complex is-
> sue of the reality of the Sasquatch, the most concrete
> evidence we have are tracks.[30]

Bill Brock (see Chapter 9) is a Mainer who is looking for Bigfoot. He has been involved with outings which have discovered and re-covered casts from tracks that he believes are Bigfoot. Detailed later in Chapter 9, Brock has found tracks (all mostly around 14″ long) from Moosehead Lake, Chain of Ponds, and in an area near the Brunswick/Durham line. Brock's casts have been on display at the International Crytozoology Museum in Portland.

From the *Bath Independent*, July 13, 1901: "Many years ago be-fore the introduction of rubber shoes, ladies in wet weather wore what was then known as 'pattens.' They were made of wood and brought the feet about six inches above the ground. A number of years ago a lady at Hardings station found a pair in her attic and putting them on one dark and wet night, tramped to several resi-dences in her neighborhood. On discovering the tracks the next morning there was great excitement in the vicinity, many suppos-ing they had received a visit from the devil, as only the round heel and sharp toe left as an impression on the ground."[31]

Bangor—The Maine Central built their railroad bridge over the Kenduskeag Stream using in part, one particular stone as part of the bridge's foundation. The well-known rock in question evinced footprints on it which some attributed to the devil. Of using this stone, "The antiquarians protested against an act to him so icono-clastic, but all to no purpose. 'Where then,' said he, 'will be your bridge when the devil comes to claim his own?'"[32] Fannie Eckstrom noted these tracks.[33] Not far above the spot was the Devil's Arm Chair, a recession in the riverbank which gave the appearance of a seat replete with two arm rests.[34]

Byron—Tony Martin was digging in a clay bed beside Coos Canyon while rock hounding in Byron, when he discovered two odd-shaped stones. Each weighed about 5 lbs. and 12 inches in length. After immersing them in the Swift River to clean them, he saw what could pass as a curved instep and a heel on each of the stones. His first reaction was that he had found Bigfoot tracks. "Based on a photo of Martin's find, Tom Weddle at the Maine Geological Survey said they're likely metamorphic rock with mica crystals and a mineral called staurolite, around 400 million years old (pre-Bigfoot or other bipeds)."[35]

Cutler—At Cutler there is a line of tracks, made as if by some passing giant, called the "Devil's Track." A legend sprung up concerning them that one night, the devil came out of the ocean surrounded by a sulphurous light that pierced the darkness. Carrying a heavy pack on his back, the devil made his way across the ledge leaving his footprints separated by a mighty length of stride.[36]

Manchester—connecting the Prescott road with Rte 135 is the Scribner Hill Road in Manchester. The Old Meeting House cemetery has a stone wall where a particular rock in the eastern section contains impressions resembling footprints. One is a large print with five toes and the other is smaller and three-toed.[37]

The story of this footprint goes back to a road commissioner from Manchester whose experience with attempts to fix the road gave the foundation for the tale. The road was infamous for its deep ruts and ubiquitous potholes, making passage by buggies and wagons a chancy thing at best. Filling in potholes and dragging chains were routine business for the road commissioner and his crew. But dealing with a particular stone jutting out of the middle of the roadway proved to be far more challenging a task. Oxen teams with chains were enlisted but their efforts couldn't budge the stone. As day gave way to dusk, the road commissioner—who was said not to be a very patient man—exhorted his crew to give one last effort. "We'll give it one more try, God damn it and I swear to you if that rock would be moved, I would sell my soul to the devil." As they had experienced all day, the chains again slipped and the rock remained firmly in place. The men retired from the arduous day to their dinners and sleep.

During the night, the road commissioner was heard to cry out. His brother heard him and rushed to his room. The road commissioner appeared to be in pain and begged his brother to fetch the doctor. The brother agreed, but when he opened the door to make the trek to town, he found a large black dog on the doorstep. With the door open, the dog raced into the house and made for the road commissioner's room. When the brother returned to the feverish road commissioner's room, he found the dog under the bed. Coaxing wouldn't budge the creature and seeing the road commissioner was still in pain, the brother decided to start out again for the doctor. He returned an hour later, doctor in tow and upon opening

The mysterious footprint

One of the Manchester impressions
resembling a five-toed footprint.

the door both were startled by the black dog, bolting past them to disappear into the night. The doctor made it to the road commissioner's bed, but it was too late, he was dead.

The next day the road crew returned to the rock in the middle of the road. What greeted them at the same site as their previous ineffectual exertions was a large crater. The rock itself was now off the road and by the cemetery wall. On its surface was the impression of a large splayed-toed footprint, which some said was the devil's own, made when the stone was kicked off the road.[38]

Another local legend tells the footprints' story as owing to a farmer who offered his soul to Diablo if he would but have a good crop to settle his debts. His harvest turned out to be successful and when the devil came to collect, a chase ensued. The farmer climbed a structure on his property with the devil close on his heels. Both of them leapt off at the same time, landing on the large stone which bore the marks of their landing ever since.[39]

Wallace Nutting the author of "Maine Beautiful" received religious instruction in the North Manchester church which borders the wall in which the two prints are located. Nutting contended the origin of the prints were not owed to Old Nick, but to an angel's step, accompanied by a little child as they walked hand in hand.[40]

Milo—In several places around Milo are impressions of footprint-like tracks fascinating generations and popularly known as "the Devil's Tracks." Found on Route 16 a mile and a half east of Sebec Corner, between Dover Foxcroft and Milo, there are also so-called paw prints accompanying the tracks attributed to the devil's dog.[41] The tracks attributed to Old Scratch measure around three feet long and up to a foot and a half deep. With regular spacing and accentuated by time, the tracks give the idea of some giant monster passing this way long ago.

Before the construction of the highway it was a hard scrabble into the woods to reach the tracks. There is also a cave near them, made it is said, by a powerful breath from the devil, created by him for protection from the cold. A similar trackway discovered in

1906[42] during the construction of the Coffer Dam on the Sebec River—and now under water—seem to lead in the general direction of the cave.[43]

Early in the twentieth century, Professor William Cooper offered his opinion that the impressions were the trackway of a mammoth, left in a malleable substrate which has hardened over time. The account offering Professor Cooper's mammoth opinion, states that he was a taxidermist and lifelong student of natural history and geology. He had previously worked with Professor Woods of Rochester, N.Y., with whom Professor Cooper constructed the mammoth exhibit for the World's Fair. Also to his credit is a gold medal presented to him in London by the Lord Mayor of London for exceptional merit in taxidermy during a national contest. He moved to Milo and opened up a shop where examples of his taxidermy drew many admiring and appreciating eyes.[44]

In the late 1950s, some of the tracks were obliterated by road construction. So when once the trackway seemed to lead to the cave, the new road separated the two, the remaining tracks on one side, the cave on the other. Elmer Jenkins created a sign in the mid 1960s to point out the tracks to tourists, but it was vandalized. A similar fate met a second sign erected at the spot.[45]

Prospect—Work crews blasting Route 1's new path connecting to the span of the Penobscot Narrows Bridge unearthed what looks remarkably like a human footprint complete with its toes. The print hangs high above the road, difficult to see while driving but visible from the new 420-foot high observatory.

Some speculated it was a mark owing to the legend of Glooscap, who climbed the rock face, leaving behind this print, to contend with Wuchosen, the Great Wind Bird.

A more mundane account for the print was offered by the State Department of Transportation: the impression in the rock face is an inclusion, or the remains of volcanic activity. The seam where the footprint is was once a crack which was filled by molten material which hardened into granite. This concept is the same attribution given to the well-known Buck Memorial grave stone in Bucksport, the side of which facing Route 3 sports what appears to be a booted or slippered lower leg.[46]

A NOTE ON MISSING PEOPLE

David Paulides is the author of two books on Bigfoot: *The Hoopa Project* (2008) and *Tribal Bigfoot* (2009). In 2012, Paulides published two books: *Missing 411—Western United States and Canada* and *Missing 411—Eastern United States.* This series continued with two additional volumes: *Missing 411—North America and Beyond* (2013) and *Missing 411—The Devil's in the Details* (2014). The *Missing 411* books provide details on disappearances of people in North America forming over 25 distinct clusters of missing persons.

Paulides' interest in this project began while he was visiting a national park on business unconnected with missing persons. During this visit, he answered a knock on his hotel door and met a man who introduced himself as one of the park's rangers. Paulides and his visitor talked of rangers who had gone missing within national parks and of a shared opinion held among many rangers of inadequate publicity or vigorous follow-up resulting in many missing person cases becoming ignored and forgotten. This discussion included the many non-rangers—tourists, campers, and park visitors who had also gone missing. Intrigued, Paulides started to explore the subject, to include sending Freedom of Information requests to several national parks asking for the number of people missing in their areas. Paulides' initial research convinced him that the ranger had valid concerns—there indeed had been many people missing under unusual circumstances, and not just solely inside the national parks. Paulides developed a list of factors which lent an unusual aspect to the cases he studied and which recurred throughout his research:

People went missing in rural settings.

Dogs are involved in many of the cases Paulides presents, either in the disappearance itself or during the search. A common thread is search dogs will often not track or cannot track.

Storms play a significant role in many cases. Inclement weather often follows a disappearance with deleterious effects to search operations.

Many of those missing who are found are located within or on the periphery of swamps and bogs and briar patches.

Paulides' research also reveals that for those missing who are thankfully found, they are often in a semi-conscious state.

Berries also came up frequently in the cases: berrying was the activity engaged in when the disappearance occurred or people were later found in berry patches.

Paulides' cases share a recurring theme of missing people found in a location previously searched.

Clothing is removed from the missing in enough cases for Paulides to consider this a significant and relevant factor. Significant articles of clothing have disappeared or the missing are found naked, with many articles of clothing never found. Paulides states, "Some search manuals indicate that children remove their clothing when it is extremely detrimental to their survival; however, the facts surrounding these cases do not seem to support the assertion."[1]

Early discussion of *Missing 411* within some cryptozoological and paranormal blogs as well as online radio shows dealing in such subjects concluded that Bigfoot could be the causal agent for some of the disappearances.[2] This is an assertion Paulides never makes. Paulides' work is included in this book not to assign a direct link between Bigfoot to the cases mentioned in Paulides' books or to missing people in Maine generally. However, there is little doubt that some of the accounts Paulides writes about do contain elements that dovetail with some commonly inferred characteristics of Bigfoot. Following are examples (none from Maine) from Paulides' books that provide some intriguing associations for Bigfoot enthusiasts:

> "They [the missing children] had seen some of the searchers, he said, but were afraid to yell because they thought they were big gorillas." Pg. 192, *Missing 411—Western United States & Canada.*

> Story titled, "Carried off by bear, an 18-month old Child taken into the Mountains but found alive and well the next day" pg. 256 *Missing 411—Western United States & Canada.*

> "At approximately 1:30 p.m. on July 4, Ida May Curtis and her nine-year-old brother, Cecil, walked off the camp grounds. Cecil told the family that Ida May had seen something in the forest near the clearing where they were standing. The kids returned to the camp and were playing around the tent when they claim that a bear entered the tent and then left hopping on three legs carrying Ida May. They stated that the bear ran into the woods with the little girl. Curtis confirmed that he thought he saw a bear just about that time running through a creek, and it appeared to be carrying something. He stated that he tried to chase the creature but could not keep up." pg. 281, *Missing 411—Western United States & Canada.*

> Mrs. Herman Bilgrien was horrified on emerging from her home to see a huge black bear making for

the woods with her child. Helpless from fright her hysterical cries attracted attention of a neighbor's daughter who was passing the Bilgrien residence and she immediately gave chase to the bear. Following the bear for about three blocks, on encountering a wire fence the bruin dropped his burden and disappeared in the forest. pg. 49, *Missing 411—Eastern United States*

The Katie Flynn story was again printed in the July 6, 1961 *Ludington Daily News.* In this copy of the story it had the following as to Katie's rendition of events.

"In answer to questions, the child said that after her father had left, she had played a little while in the sand when a big black thing came along and played with her. Then it held out its paw and she caught hold of it and it had walked away with her. Just before dark it had left for a while and when it came back its paw was full of winter green berries. The bear ate some of the berries and she ate some. Then it scraped a big pile of leaves close to her and lay down close to her and during the night had tried to cover her with its body." p. 57, *Missing 411—Eastern United States*

The following account from the *Lewiston Evening Journal* of a New Hampshire toddler gone missing in 1911 is not mentioned in Paulides' four *Missing 411* books. Hill, New Hampshire is 112 miles from Howe Hill, Maine where Elsie Davis (discussed below) disappeared in the same year, at almost exactly the same time. This account is included here due to its similarities with cases meeting Paulides' criteria in his *Missing 411* volumes, notably, a 3-year-old is found two miles from his home and on a mountain ledge—a distance and location seemingly at odds with a child's capabilities.

Heard Boy Crying—Missing Youngster Found on Ledge on Top of Mountain—Hill N.H., August 5—An all-night search in the woods and hills of this little town for three-year-old Donald H. Jones, who disappeared from his home yesterday noon ended today when the boy was found sitting on a ledge at the top of Huse Mountain, crying lustily. Little Donald started out to meet his grandfather, Dimond Colby, returning from his work at the New England Novelty Works yesterday noon but took the wrong road and got lost in the woods. When his grandfather returned home and found the boy missing, he gave the alarm and searching parties at once began. The Novelty Works and Needle Shop were closed and the employees joined in the search. The place where the boy was found is about two miles from his home. He was located by Alonzo Wadleigh, who heard him crying.[3]

Paulides' books mention the following cases from Maine. Names in italics are expanded upon in this chapter, fleshing out the respective accounts and providing additional details direct from primary sources.

NAME	DATE MISSING	LOCATION	OUTCOME
Unknown	09/07/1895	Bradford	deceased
Lillian Carney	08/08/1897	Masardis	found
Elsie M. Davis	07/30/1911	Howe Hill	found
David Brown	11/18/1922	Big Bog Dam	deceased
Mertley Johnson	11/18/1922	Big Bog Dam	deceased
Gerald "Terry" Cook	04/28/1948	Moosehead Lake	deceased
Teddy Bernard	06/17/1951	Gray	found
Gary Bailey	07/17/1954	Spears Mountain	found
Diane Abbott	07/6/1956	Hebron	found*

* Paulides has Diane Abbott's final outcome as unknown.

Douglas Chapman	06/2/1971	Alfred	missing
Kurt Ronald Chapman	08/31/1975	Chain of Ponds	missing
Geraldine Largay	07/23/2013	Sugarloaf Mountain	missing

The disappearance of C. H. Bordwell (p. 14, *Missing 411—Eastern United States*) is listed as Keewatin, Maine—the correct state is Minnesota.

LILLIAN CARNEY

Lillian Carney went missing while her family was berry-picking in Masardis in northern Maine. Two days following her disappearance, Burt Pollard found the girl two to three miles from the spot she had vanished from. Paulides includes her enigmatic mention of "the sun shined all the time in the woods,"[4] while she was missing. The August 12, 1897, *Lewiston Evening Journal* account included the following on Lillian's additional comments and her condition:

> She was perfectly cool and unconcerned and said she was looking for mama. She had in her hand some wild berries, but said she did not like to eat them for they might not be good for her. She said the sun shined all the time in the woods. (The moon shone brightly part of the nights.)
>
> The child said she heard people talking the day before, but she kept still for she was afraid they were tramps, and that she saw some little things about as big as her kitty (rabbits, probably) and she clapped her hands and they ran away.
>
> Guns were fired as signals, the steam whistle at the Stimson mill proclaimed the joyful news, and the tired but happy crowd went back to the usual business of everyday life, the better perhaps, for the exhibition of the kindness and goodness of their hearts.[5]

ELSIE DAVIS

The Elsie Davis story shares similarities with several *Missing 411* criteria, including disorientation, loss of some clothing, and eventual discovery in a previously searched area. The Davis story is laid out below from contemporaneous newspaper sources.

"She is a little more than of medium height—a little taller than ordinary, says one friend—weighing about 125 pounds, with dark, almost black hair, fair complexion and brown eyes. She is 24 years of age."[6]

Miss Davis left her home Sunday morning seen heading toward the Greenwood road from her home on Mason street. Her brother Moses Davis, upon his return from Church was the first to notice her absence.[7] The search was first taken up in that direction. "Miss Davis had not been feeling as well as usual for two weeks and did not attend church Sunday morning as was her custom."[8]

Miss Davis was the organist at the Methodist Episcopal Church. When last seen she was wearing a long light-colored coat. Area mills closed so about 250 men could aid in the search which was

Elsie Davis.

focused between Bethel and Lockes Mills. An initial report that she took the Sunday morning train to Berlin, N.H. proved incorrect.[9]

"Not in years have the people of this town been so wrought up over anything as over this case. From the moment that the news of Miss Davis' disappearance became public, until her recovery, men by the hundreds have been beating the woods of the vicinity, searching the river and stream banks, in an effort to find her."[10]

"'Find Elsie!' That has been the one watch word of the men of Bethel and vicinity since the bell of the fire house rang the alarm at 9 o'clock on Sunday night."[11]

"Whether Elsie M. Davis in a demented condition of mind is wandering over the hills and thru the woods of Oxford county or her dead body is lying in those woods or in Androscoggin river the people of Bethel believe that religion is responsible for her strange disappearance.

"That a constant brooding over a fear that she was not doing all which a true Christian person should do, which thought, so it is freely alleged by people here, aided and abetted by others, so preyed upon her sensitive nature that her mind became unbalanced to the extent that she determined either to take her own life or to go away, is the judgment of many well acquainted with the young woman and her mode of life. That she was deeply religious is unquestioned."[12]

"For nearly a month past Miss Davis has been in poor health and has been morose. Nearly a month ago she was very despondent. She dwelt at length on her duty to the family and church and appeared to be worried and possessed of the idea that she was falling short of what she ought to do along these lines."[13]

"Several years ago, while a student in the academy here, Elsie suffered a break down and for a time she worried over the same thing which troubled her at this time with the addition of a fear that she might be ill with consumption, the disease with which her mother died."[14]

"From the information available, it appears that the missing girl not only had premeditated suicide, but that she carefully laid her plans for carrying out this determination.

"Carefully choosing a time when the chances of discovery were the least, she left the house and went away, taking only a rain coat in addition to the clothes she was wearing. In declining to attend services at the church, suspicion was aroused. The fact that she did not come to attend to her duties as organist at the Methodist church might caused [sic] no comment for the suggestion had been made to her that in her feeble condition, it might be wise to put aside the work for a time. When she did not attend to these duties, it was accepted that not feeling well, she had decided to follow this advice."[15]

HER MEMORY
IS A BLANK

Elsie Davis Says She Went Walking In Woods and Lost Her Way.

Bloodhound Helped Find Missing Bethel Organist.

Miss Davis Discovered Perched High among Branches of Big Pine Tree.

A bloodhound was brought from Hartland by automobile on Thursday morning and by 11 o'clock was put to work finding Miss Davis' scent. By this time several days had passed, many volunteers and neighbors had tramped the countryside and heavy rains had fallen. At about 2 p.m. the dog picked up a workable scent. The dog eventually stopped at a pile of woman's clothing consisting of shoes, stockings, shirt waist and rain coat—all identified as belonging to Miss Davis.

The dog was put back on the lead and for a time headed back in the direction of the village of Bethel, but then turned and headed in the direction of Howe's Hill, the spot where Davis would be found.

Fred Edwards and Charles Stowell were searching apart from the party working the bloodhound. They had heard the news of the clothing's discovery and knowing that where they were found was an area previously searched, believed the girl would soon be found alive. At Howe's Hill, the pair discovered women's clothing hanging from a pine branch. Approaching the tree they looked up and saw Davis seated on a limb about twenty feet up and apparently trying to hide from them.

"They called her name, but she made no answer. Evidently dazed and frightened she kept quiet and more and more sought to shield herself from view"[16]

Davis was wet and exhausted. A ladder was procured and with great care she was eventually brought down from the tree, never speaking a word. The first sign of recognition she gave was the placing of her head on her father's shoulder during the automobile ride to her home.

"The physician stated that Miss Elsie Davis' mind was clear today and she could remember plainly that she started out for a walk and became hopelessly lost in the woods. She lost all sense of hearing and the more she attempted to find her way the deeper she penetrated the woodland, so that finally she had gone six miles from home. She slept some at night on pine boughs but had nothing to eat except berries. She did not see a single person during her wanderings, and as far as known, the searchers were not near her at any time until yesterday. The physician stated that she will be mentally well and out of doors in a short time."[17]

"A few persons are inclined to credit the finding of the missing girl to the information given Wednesday by a Portland clairvoyant because the medium said that she would be on land belonging to a man whose name began with 'Fred' and had an 'E' in it? [sic] Miss Davis was found on land owned by Fred Edwards. Those who discredit the clairvoyant idea, say that no one knows how much had been told the spiritualist before he gave his information and further that it would have been almost impossible to have found Miss Davis on land owned by another man than Mr. Edwards."[18]

TWO WARDENS

Chief Warden David Brown of Greenville and Warden Mertley Johnson of Patten headed up to Big Bog in November, 1922 to investigate illegal beaver trapping in the St. John Valley. Beaver at the time was worth a dollar an inch making it a very lucrative commodity. The two men went missing in November and their bodies were discovered five months later in the water near Big Bog Dam.

> David E. Brown and Mertley Johnson were two of the most skillful woodsmen in the warden service of the State and from the time that they disappeared on Nov. 15, last, it has been the firm conviction of all who knew them that they had been murdered. A reason for this, beside the knowledge that they were experts in woodcraft, is that it is known that, after getting into the woods on that trip, they changed their plan. When the men left Greenville they had no intent to go to Abacotnet, but that they learned that certain Canadian poachers, whom the department had long sought, were in that region and so went there to arrest the men.[19]

MERTLEY JOHNSON
First Published Picture of Game
Warden Whose Body Has Been
Recovered.

Mertley Johnson

David Brown

"The autopsies . . . apparently, they didn't find anything at that time, but then again, you're talking a couple of bodies that had been underwater all winter and probably not in the best of condition."[20]

> Send Organs on to be Analyzed—But Money in Pockets Give Belief that Deaths of Wardens Brown and Johnson were Due to Accidental Drowning—Rockwood, May 5 (Special)—That Game Wardens David E. Brown of Greenville and Mertley Johnson of Patten, died by drowning, was the decision of County Attorney Thorne and Sheriff Morris of Somerset County upon hearing the report of the autopsy performed Friday afternoon by medical examiners Sawyer and Stinchfield.
>
> No bullet holes, knife wounds, or bruises which might have caused death were found.
>
> No water was found in the lungs, but both medical men stated that this was by [no] means unusual in cases of drowning.
>
> While convinced at the conclusion of the autopsy that it was a case of drowning, the county officials ordered the medical examiners to remove the organs of the stomach and send them to Bowdoin for a complete examination.
>
> In the pocket of Johnson was found $32, his keys, knife and watch. The watch had stopped at 3:42. His revolver was missing from the holster, but as the flap was unfastened it might have slipped into the water. In Brown's pocket, was found $105, his watch and keys. This indicates accident rather than foul play.
>
> Brown's body was found by Deputy Sheriff Lee M. Smart and the men who went from here to bring Johnson's body down about a mile and half below Big Dam, it has evidently swept down stream and over the Dam without being seen by the watchers.[21]

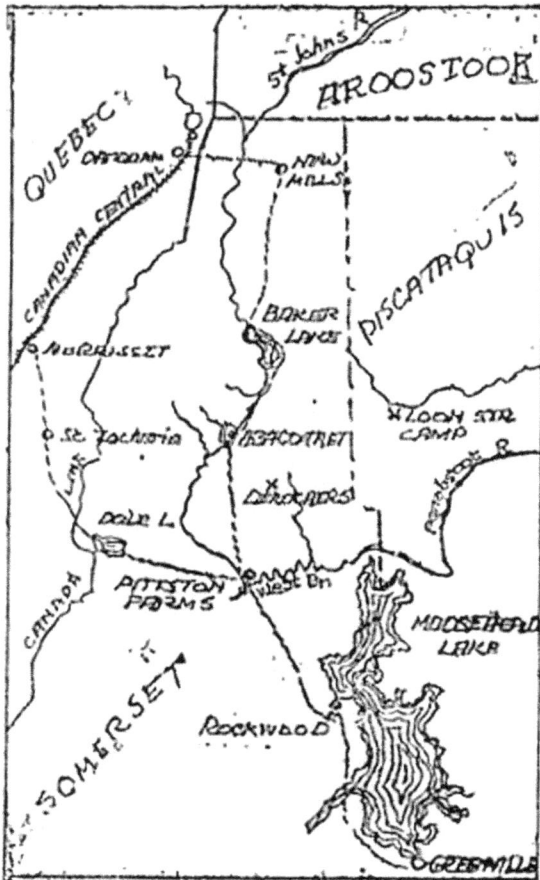

"Region where Game Wardens are
believed to have been murdered."

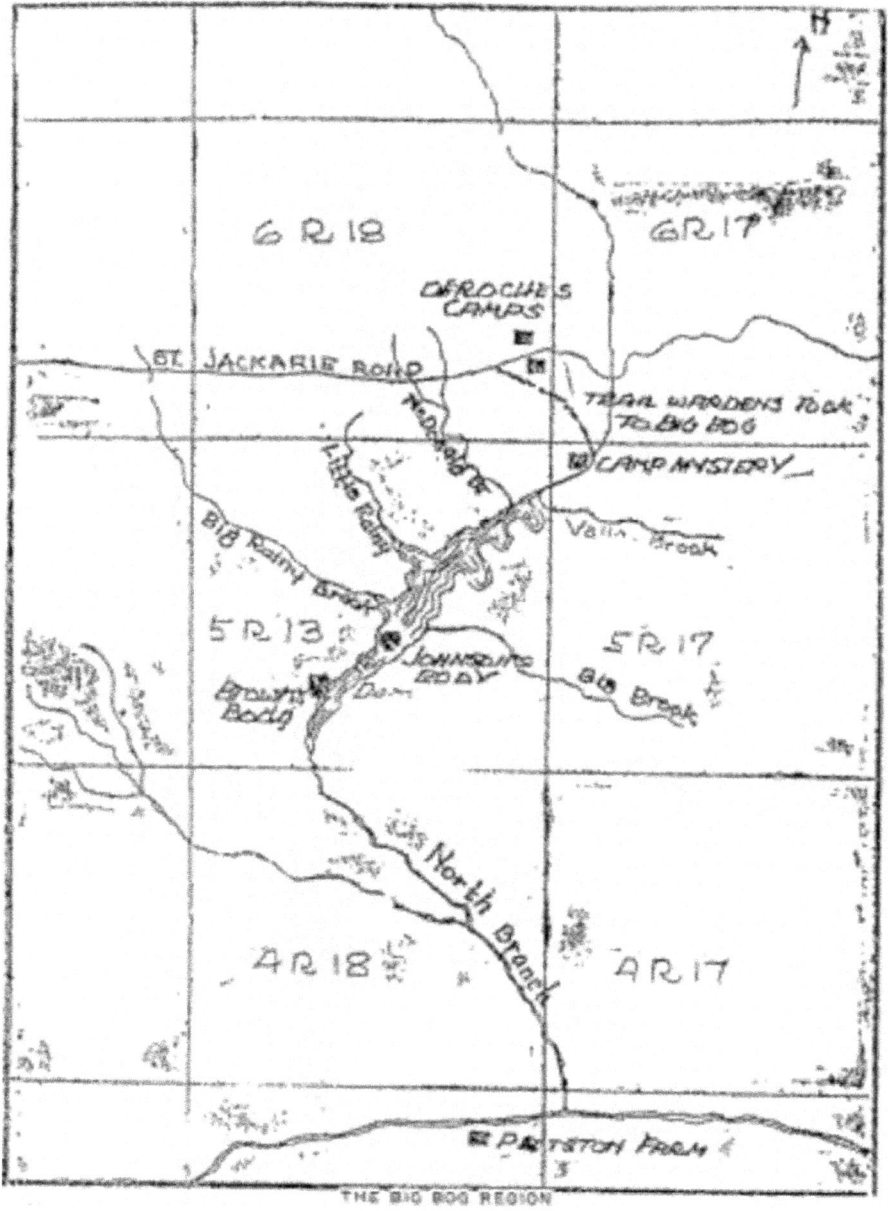

Big Bog Region

Critics of the drowning theory have raised several problems with that cause of death for Brown and Johnson. No water was found in either man's lungs. The water they were found in was fairly shallow—about waist deep—at the time the men went missing in November.

Retired Warden Bill Allen holds the opinion that "It would be pretty unusual for an experienced game warden supervisor like David Brown or an experienced warden like Mertley Johnson to just wander out onto a flowage in northern Maine on thin ice in the middle of November. Woodsmen of their caliber don't do that, period."[22]

Allen knows the story of a time when there was, informally, a suspect in the case. "There were two brothers from Canada that were poaching beaver in Maine around the same time Johnson and Brown went among the missing." One of the brothers returned to Canada when he learned of the Wardens' interest in looking into and stopping the illegal beaver trapping in the area. "The other one stayed. And he was later being given a ride back to the border in a horse drawn wagon and he mentioned to the driver that he'd got a couple deer. The driver asked him where they were and he said 'well, they weren't that kind of deer.' That kind of makes you wonder if he wasn't somehow involved."[23]

Warden Mertley Johnson was only 21-years old and is the youngest individual to die while on duty whose name is carved onto the Law Enforcement Officers Memorial in Augusta. Chief Warden David Brown was a widower who left three children and a woman he intended to marry.

TEDDY BERNARD

A posse of nearly 200 persons was searching a pasture and woodland area in East Gray last night, for a three year old Mexico boy, missing since early afternoon from a strawberry hunt.

Teddy Bernard, and his parents, Mr. and Mrs. Edmund Bernard had been visiting an uncle and aunt, Mr. and Mrs. Russell Ladd.

HUNT BOY, 3, IN WOODLAND AT EAST GRAY

Teddy Bernard, Mexico, Vanishes While Picking Berries—Is One of Seven Children of Mr. and Mrs. Edmund Bernard—Posse of Nearly 200 Aids in Search—May Use Bloodhounds

He and four young companions were about 500 feet from the Ladd home at 2:30 Sunday afternoon when the child told his friends he was going back to the house for a cup to hold his berries.

When he did not return, the children looked for him, and later his parents and Mr. and Mrs. Ladd joined the search. They estimated the child had been missing more than two hours before an organized party began the hunt.

Search crews, headed by Sheriff Phillip M. Dearborn of Cumberland County, and including police and firemen had combed the area near the home and down to the banks of the Royal River, two miles from where the children were picking berried.

The Civil Air Patrol sent Richard Carl of North Yarmouth over the general area in a plane, but he reported no trace of the youth, described by his mother as "small" for his age and wearing a blue shirt, brown sweater and shorts.

A call was sent out to Keene, N.H., for bloodhounds and crews, armed with powerful searchlights, were awaiting their arrival. A quick trip was made to the child's home in Mexico for clothing, which could give the dogs a scent, since none of the boy's clothes were found in the house where he was visiting.

The area where the child disappeared is on the north side of Route 115, between Gray and North Yarmouth, and about two miles from Gray village.

The boy's companions were his cousins, Kay Ladd, 12, Rene Ladd, 4, and David Ladd, 2, and Lyanne Spaulding, 6.[24]

MEXICO CHILD FOUND AT GRAY

Teddy Bernard, 3, Discovered Asleep After Night in Wilderness

Boy Unharmed Except for Scores of Mosquito Bites and Bruise Suffered in Fall

BY HAROLD R. SMITH
State Editor

East Gray—Lost nearly 24 hours in rainy wilderness, Teddy Bernard, 3, of Mexico, was found asleep, early Monday afternoon, beneath a clump of bushes. Hungry, suffering from scores of mosquito bites and a huge forehead bruise received in a fall, he was carried three-fourths of a mile to the small camp his parents, Mr. and Mrs. Edmund Bernard, were visiting at the time he became separated from other children while on a berry-picking expedition.

Wanted "Mommy"—The child was discovered by three searchers about 100 feet from a logging road

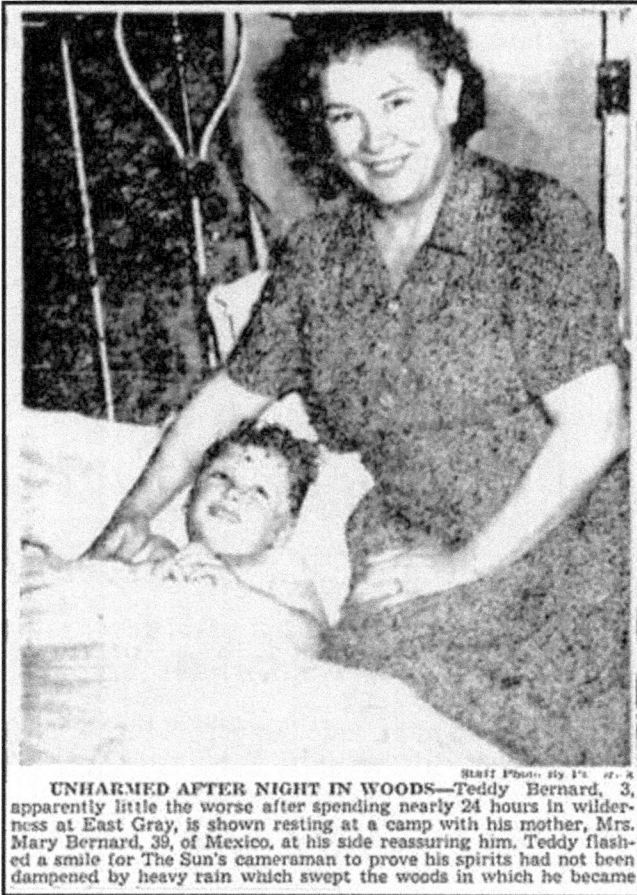

UNHARMED AFTER NIGHT IN WOODS—Teddy Bernard, 3, apparently little the worse after spending nearly 24 hours in wilderness at East Gray, is shown resting at a camp with his mother, Mrs. Mary Bernard, 39, of Mexico, at his side reassuring him. Teddy flashed a smile for The Sun's cameraman to prove his spirits had not been dampened by heavy rain which swept the woods in which he became

which leads from the Mosquito Road, so-called. He was soaked from heavy overnight rains and his shoes were missing.

Awakened by James Fruzia, Robert Sawyer, and Joseph Bernard, who found him, the boy cried, "I want my mommy, I want to go home."

Dr. Kenneth Russel of Gray was called. He examined the boy and said his condition was good. The doctor advised giving the boy only a small amount of warm milk because he had gone without food more than 24 hours. Two hours later, after having consumed two glasses of the fluid, the boy was crying for more food but the family was carrying out Dr. Russell's instructions.

Mrs. Bernard said that Teddy perhaps had not received his forehead bruise in a fall. She reported the child had a habit of pounding his forehead into his pillow when he went to sleep and she thought he might have pushed his head into the ground. There was a considerable amount of dirt in the wound.

Went in Opposite Direction—The spot where the boy was discovered was in an opposite direction from that which Teddy was believed to have taken. Authorities said he must have crossed two roads to reach the area.

The mother, Mary, 39, fainted when she learned the child had been located and was unharmed. She recovered quickly and was on hand to greet the lad.

More than 300 members of a posse had combed the section beginning late Sunday afternoon. The parents were visiting the home of an uncle and aunt, Mr. and Mrs. Russell Ladd. Teddy and his cousins, Kaye Ladd, 12, Rene Ladd, 4 and David Ladd, 2, and Lyanne Spaulding, 6, had been hunting strawberries. The Mexico tot became tired and started back for the Ladd camp. When the other children returned, it was discovered Teddy did not find his way back to the camp situated on the North Yarmouth highway leading from New Gloucester via Pownal.

Prints Found in the Morning—It was at first believed the child might have tumbled into nearby Royal River. Footprints found near a woodland spring spurred on the search, Monday morning.

Teddy obviously was none the worse for his experience when he and his mother posed for the Sun photographer. He said he was all right, "only hungry."

Mrs. Bernard said Teddy told her he got tired, Sunday afternoon, and went to sleep.

Involved in the search which became a community project in nearby Gray village were Portland National Guardsmen, Civil Air Patrol planes, Pownal Boy Scouts, highway department workers, deputy sheriffs, and foresters. Clayton Weymouth of West Newfield and Philip Barton of Gray, both experienced foresters, and Deputy Sheriff Gerald E. Davis of Cumberland took leading parts in the search.

Teddy was dressed in a T shirt, a jacket, short pants, stockings, and shoes when he disappeared. He had all these articles of clothing on him when he was found except for the footwear.[25]

GARY BAILEY

Find Missing Thorndike Boy Safe—(AP) Knox, Maine—Jubilant siren peals and a thankful mother's rush up a hillside signaled success yesterday in a search of mountainous bear country for a 3-year-old boy lost for 24 hours while picking blueberries.

Donald Bradstreet of Hampden and Stephen Fowler of Albion, members of a search group aided by a helicopter, located Gary Bailey unharmed except for scratches. He was found a mile and a half from where he disappeared Saturday.

The child's mother, Mrs. Kenneth Bailey of Thorndike, searched throughout Saturday night with the group and was trying to rest when woodsmen trotted out of the brush with Gary. Catching sight of the lad she sprinted up a rugged slope and hugged the youngster to her.

Bradstreet and Fowler said they heard the lad crying and found him on the opposite side of Spear's Mountain from which he disappeared. They said he

Gary Bailey after his night in the woods.

apparently roamed over the 1,400 foot peak's crest, resting overnight beneath a tree.

The searchers were spurred by reports that bear were roving the area for food which they can't easily find until the summer's berry crop ripens.

About 500 persons, including game wardens, state police, firemen, Boy Scouts and volunteers scoured the dense woods. The lad's father, almost blind and unable to search actively, kept a vigil close by a phone in a forest fire watch tower.[26]

DIANE ABBOTT

Paulides finishes the Diane Abbott account with, "In the lengthy archive search I performed, I could never find any information that Diane was found." Here is the original United Press story:

2-Year-Old Girl Missing in Woods—Oxford, Me, July 14, (UP)—Four hundred persons searched through rain-drenched woodlands of this central Maine town today for a two-year-old girl missing more than 17 hours.

The child, Diane Abbott, disappeared yesterday, while playing at her home. Her mother said she had heard a "faint cry", went out to look, and could not find the child.

Clad only in a blue play suit and red shoes, she had been playing not far from two brother, Richard and Charles Williams. They said they did not see her wander off.

State police and Oxford County officials organized the giant search which was carried on through the night, using floodlights.

The child's mother, Mrs. Helen Abbott, said she was feeding Diane's sister, Holly, 3, and a nine-month old brother when Diane disappeared. Their father is a patient in a Lewiston Hospital.[27]

A check of online newspaper databases revealed the following happy denouement to the Abbott case.

> Missing 2-Year-Old Found in Maine—Oxford, Maine (UP)—A two year old girl missing 24 hours in rain-drenched woodlands was found Saturday three miles from home while 600 persons searched for her.
>
> The child, Diane Abbott, walked out of woods near the farm home of Mrs. Margaret Tobey who was sitting on a porch.
>
> "She was crying 'Mommy come, Mommy come'," Mrs. Tobey said. "She only had on one of her red buckled shoes."
>
> "I couldn't understand anything else she said," Mrs. Tobey said. "I didn't know it was a lost child. I thought she was a neighbor's girl from down the road."
>
> Mrs. Tobey wrapped the child in a blanket and took her to the home of Mrs. Lloyd Davis. Mrs. Davis called the sheriff.
>
> The child appeared in good health though covered with scratches and insect bites.
>
> She had disappeared yesterday about 2 p.m. while at play. Her mother at the time was in her home, feeding a 3-year-old daughter, Holly.[28]

One of the interesting aspects of this disappearance is that when she was found, Diane Abbott was wearing only one shoe. This is one of Paulides' chief criteria for the unusual disappearances he collects in the *Missing 411* books—the absence of clothing from those who are found.

8

BIGFOOT ENCOUNTERS

The New Year was only two score hours or so old when Loren Coleman posted his January 2nd *Cryptomundo* blog entry, the title of which asked, "First Blobsquatch of 2010?" The posting was in regards to a YouTube clip contributed by video poster webtech88. The video was titled, "What is it? Sasquatch, Big Foot, monkey, ape!" Webtech88 provided further explication on this post: "NOT a FAKE!!! REAL VIDEO!!! This is a video taken by some friends hiking through the woods in upstate Maine. What could it be? Sasquatch? Monkey?"

The video opens with the image below, the camera just capturing a person in the bottom of the frame before zooming in on a dark object in the trees.

Viewers can hear people exclaim the object resembles a monkey. The object does not appear to move during the filming.

Webtech88 responded to Coleman's blog post. "Hi Loren. I am the original poster of this video but I assure you I did not create this as a hoax. I simply added it to my youtube (sic) account after it was sent to me by the person who took the video. She sent it to me in hopes of trying to figure out what it was. It is a strange video but nothing in it has been modified in any way. To my surprise when I awoke this morning and checked my email I was bombarded with comments from my youtube account."[1]

The BFRO concluded the video depicted a porcupine. "The footage is relevant to Bigfoot research, though, because a porcupine in

Arrow is pointing to the object (prior to the camera zooming in) which would become known as the Maine Tree Creature.

Close-up of the from the Maine Tree Creature video.

THE Sun

NEWS

Got a story? Text: - Email: talkback@the-sun.co.uk

Bigfoot: Is this the proof?

IT'S definitely big. It's certainly hairy. Could this mysterious beast be the elusive Bigfoot?

Shot in snowy woods in the US state of Maine this video shows a creepy creature lurking in the treetops — and the footage has taken the internet by storm.

Speculation is rife about the identity of the fearsome animal.

It has been likened to a monkey and even a bear. But many of the 20,000 people who have already watched the clip are convinced it could be the legendary sasquatch.

The video was posted on YouTube by American user webtech88 five days ago.

MULTIMEDIA

TOP STORIES

NEWS | SPORT | BIZARRE

Labour blasted in Sun survey

MOST READ STORIES

NEW YEAR DEBT CLEAR

A typical over-hyped sighting in the media.

a tree can look like a dark-furred primate in a tree, especially at a distance."[2] The so-called *Maine Tree Creature* is an object lesson showing how certain events may be both proposed, as well as received, as possible Bigfoot encounters.

Misidentification, distortion, fabrication, and misrepresentation explain many sightings. But there are reports that do not subscribe themselves to easy dismissal. This chapter collects in chronological order, hairy hominid events occurring within Maine. Information, when known, is included for month, year, town, and county. These sightings, by their chronological number, are aggregated on the county maps at the end of the chapter.

#1   1855 * Waldoboro, Lincoln County * winter

Chad Arment's *The Historical Bigfoot* includes a report of a creature resembling a Bigfoot, though at the opposite side of the size-spectrum which one might expect. The *Thomaston Journal* is given as the source for an account of the capture of a small monkey-like creature. C. H. Paine started the *Thomaston Journal* in Thomaston on March 9th, 1854, a paper which professed to be "neutral in nothing," and "independent on all subjects." It lasted four years under the *Thomaston Journal* masthead before a title change to the *Lincoln Advertiser*.[3]

In the nineteenth century, Maine had very important ports which brought a lot of people to the state's coastal communities. One of the fashions was to keep exotic birds as pets. The Waldoboro story may refer to a monkey brought to the state by some seafaring person which then escaped into the coastal environs. This report can be compared with the "douracouli" from Chapter 6.

> The following communication relative to the discovery of a wild man in Waldoboro', (Maine,) was handed to the editor of the Thomaston Journal by Mr. J. W. McHenri. The Journal says there are several persons who can vouch for the truth of the statements which it contains:

Mr. Editor: On the morning of the 2d of January, while engaged in chopping wood a short distance from my house in Waldoboro', I was startled by the most terrific scream that ever greeted my ears; it seemed to proceed from the woods near by. I immediately commenced searching round for the cause of this unearthly noise, but after a half hour's fruitless search, I resumed my labors, but had scarcely struck a blow with my axe when the sharp shriek burst out upon the air.—Looking up quickly, I discovered an object about ten rods from me, standing between two trees, which had the appearance of a miniature human being. I advanced towards it, but the little creature fled as I neared it. I gave chase, and after a short run succeeded in catching it. The little fellow turned a most imploring look upon me, and uttered a sharp shrill shriek, resembling the whistle of an engine. I took him to my house and tried to induce him to eat

Postcard image of the Waldoboro waterfront, circa 1900.

some meat, but failed in the attempt. I then offered him some water, of which he drank a small quantity. I next gave him some dried beach nuts, which he cracked and ate readily. He is of the male species; about eighteen inches in height, and his limbs are in perfect proportion. With the exception of his face, hands and feet, he is covered with hair of jet black hue. Whoever may wish to see this strange specimen of human nature, can gratify their curiosity by calling at my house in the eastern part of Waldoboro, near the Trowbridge tavern. I give these facts to the public to see if there is any one who can account for this wonderful specimen. J. W. McHenri Marysville, Ohio, Tribune, January 31, 1855.[4]

This story also appeared in abbreviated form in the *Georgia Telegraph*, February 20, 1855. J. W. McHenri became J. W. McHenro. Otherwise the Macon, Georgia, paper's story is essentially the same.

#2   1856 * Greenville, Piscataquis County * season unknown
"Cascade," the handle of a contributor to the *Finding Bigfoot in Maine* website, mentions a newspaper article telling of an "ape-man" found dead on the railroad tracks near Greenville. Further research by this author did not produce the original source.[5]

#3   1886 * North of Moosehead Lake, Aroostook County * autumn
This encounter received excellent research by No-Noi Ricker, staff writer for the *Bangor Daily News*. The account below was taken from the October 8, 1886, edition of *The Industrial Journal* of Bangor, which cites the story's source as *The Waterville Sentinel*.

A New Kind of Game (*Waterville Sentinel*). An affrighted Frenchman from over the line had his fellows in town all by the ears with a story of a gigantic wild man killed a week or so ago in the dense woods above Moosehead. The Frenchman's story, which is implicitly believed, is that three men were camping out in the woods about a hundred miles north of Moosehead Lake. Two of the campers were away from the camp for a week and came back to find the dead body of their companion. They went for help and reinforced by a dozen others searched the woods for the unknown murderer. It proved to be a terrible wild man, ten feet tall, with arms seven feet in length, covered with long, brown hair. The party fired several shots into him and finally succeeded in reaching a fatal spot, laying the monster low. The story has caused great excitement among the more ignorant portion of our foreign population.[6]

According to Ricker's research, the *Sentinel* story also ran in *The Wilton Record*, making a total of at least three Maine newspapers that ran the story. Beyond Maine, the story is known to have been printed in *The St. Albans Daily Messenger* of Vermont under the title, "Odd Gleanings" on October 12, 1886. The *Messenger's* version began with, "The wild man is coming to the front this fall." This is the same opening line as used in a paper on the opposite side of the country in early 1887, demonstrating the ability of some stories to be propagated:

The wild man is coming to the front this fall. A Maine newspaper of repute says that one ten feet high was recently killed 100 miles north of Moosehead Lake. He had previously killed one of three hunters, and the other two got reenforcements [sic] and slew the giant, who was covered with long, brown hair. (*Los Angeles Times*, May 5, 1887)[7]

#4   1895 * South Gardiner, Kennebec County * summer
This 1895 account has several classic, time-transcending hallmarks: "unearthly" shrieks, witnesses engaged in berry-picking,

the "monster" described as an African monkey, and mention made of the creature's footprint.

> Strange Monster Seen in Maine—A South Gardiner, Me., correspondent of the Globe writes that people living on the outskirts of that town have been startled by unearthly shrieks lately. Two women and three boys, who went into the woods for blueberries, came upon a hairy monster which walked upright on its hind legs toward them. They were badly scared, but the animal, which looked like an immense African monkey, walked past them, leaving a footprint round like a saucer. A posse will start on the hunt Sunday.[8]

#5   1895 * "North of the State", presumably north of Bangor * Penobscot County * autumn

This is the second Californian paper to carry a story of a Bigfoot-like creature in Maine in the latter nineteenth century. This story posits its own rationalization of the antagonist—a sportsman who has reverted to an animal state. Notable are the creature's speed, hairiness, and murderous nature.

> A Terror in the Woods—Great Consternation Caused by the Deeds of a Wild Man. Unlucky Lumbermen Slain and Their Bodies Mangled in a Shocking manner. Bangor, Me., Nov. 26—A lumberman who returned today from the forest in the north of the State brings the most harrowing intelligence of the doings of a wild man in the lumber region of the west branch. He states that great consternation has been caused and a large number of lumbermen have left the camps and returned to their cities rather than face the monster.
>
> For over two months quite a number of men have disappeared from the camps and when found bore

the semblance of having had an encounter with some wild animal, their bodies in every instance having been terribly mangled and torn. A lumberman who returned to a camp a little north of this city a week ago startled all by stating that while at work he had been attacked by this wild man, and it was only by the help of his ax that he had been able to defend himself from the murderous attacks. Since that time he had been seen by the crews several times, but on their approach he fled into the deep woods with the speed of a deer.

He is described as being so nearly like an animal that it is almost impossible to detect him from one. He has a long, shaggy beard, and is covered with a huge, skin coat. The general belief is that he is a sports-man who has become lost in the deep forests, and after wandering around for weeks has gone hope-lessly crazy, and already there have been over half a dozen instances of a similar character in the State.

The crews of the lumbering camps are out hunt-ing for the man and hope by shooting him in the leg to effect his capture.[9]

It is interesting at this point to note the gap in Bigfoot accounts (recognizing the author's discrimination between Bigfoot and Wild Man). During this 47-year gap between 1895 and 1942 there are 52 separate Wild Man entries noted in Chapter 5.

#6   1942 * Meddybemps, Washington County * season unknown
Jackson and Tulane Porter shared two Bigfoot stories with Bigfoot Encounters.com concerning a creature called the Meddybemps Howler.[10] During a Thanksgiving dinner in 2006, two ladies re-lated stories of their childhood and the times their family would go to Meddybemps Lake to fish for smallmouth bass. While at the lake, the girls slept under the night sky. The girls enjoyed a favor-ite rocky island where the fishing was good and would at times

remain on the island for the night as to begin fishing at their favorite spot with first light. On one occasion, a howling was heard to come across the lake, like a "monotonous singing from someone with a husky voice." When this was heard a family member canoed out to the island to police up the girls.

Another time the girls were left to fish on the island while the rest of the family pushed out onto the lake to fish another spot. They hadn't known what the howler was. The girls would be left with biscuits from breakfast, some worms and crickets for bait and would fish into the late afternoon. The girls' parents did not return before sunset and the howler started again. Anxious for their parents to arrive, the girls tied up the days' catch, three smallmouth bass, and placed them in a bucket and waited for their parents.

The howling grew louder and then it suddenly stopped. The lake was silent. The girls looked around and standing behind them was a pair of "giant Indians" dripping wet. The giants stared down at the girls and reached over and took their catch from the bucket. The bigger one handed the fish to the smaller one. Then the giants, black in color, went back into the water and swam away with the fish. A gentleman at the Thanksgiving table asked the Native American ladies how big they were and they responded, "two fathers high."

The "two-fathers-high" giants were described as covered in black shiny hair and both were dripping wet as if they had just stepped out of the lake water. They had huge hands with fingers as large as the fish they took. They had big feet and wore no clothing. When the parents finally came back in the canoe, they were not surprised at the girls' story and said it was the Meddybemps Howler, who were known to steal fish. After that the girls were left at the campground area and never were permitted to fish alone again. The Meddybemps Howlers were also purported to make off with little girls as well as fish. The ladies said they never returned to their favorite island.

The second anecdote from the Porters concerned a gentleman at the same Thanksgiving meal. During desert he shared he had heard of an entire family of the hairy giants living in the vicinity of

Stud Mill Road, north of Lower Sabao Lake. The story this gentle-man had heard stuck with him not only for the ostensible subject matter, but also of the detail that one of the giants was holding a jack hare by its ears, which was still alive, kicking and squealing.

Another individual, Tyler Smith, shared a family story with *Bigfoot Encounters*. His great aunt told him about the Meddy-bemps creature including an account where a pair of them threw rocks at a fisherman on the lake and sometimes stormed the canoes, swamping them by churning the water and howling.[11]

#7   1946 * Bigelow Mountain, Franklin County * season un-
         known
This account is believed to have occurred on Bigelow Mountain. Four dowsers were surprised to see an enormous man in the dis-tance and coming their way. This "man" was thought to be about ten feet tall. He had a complete covering of short black hair over his body, the hair on the head noticeably longer. He was carrying what was thought to be a large rock. The group, who were pre-pared to stop for lunch, instead left the area.[12]

#8   1949 * Bigelow Mountain, Franklin County * season un-
         known
The road mentioned may be Routes 16/27 which head southeast from Stratton toward Carrabassett Valley and skirt the southern edge of Bigelow Mountain.

While driving by Bigelow Mountain at 1 a.m. on a rainy night, the witness passed a 7-foot tall creature standing in the roadway. The witness later met a local Frenchman named "Old Pete" who had seen the creature before, and said, "him don't smell good."[13]

#9   1951 * Biddeford, York County * spring
The respondent recounted an incident which occurred when he was a boy with his parents on holiday in Biddeford. The occasion was during the month of April, and he had taken a walk down an estu-ary leading toward the ocean. He saw a figure 15 feet away floating

on its back with the tide. The figure was in the shape of a man, about 6 feet tall, covered in 6-8 inch grayish hair which the boy could see swirling about its body. Its face floated out of the water and there was about half an inch of water covering his body. The facial skin looked wrinkled and was pinkish-red. The boy saw no movement of the hands or arms, the figure was just drifting easily. He ran to tell his parents what he had seen who in turn told him he had probably seen a seal.[14]

#10   Early 1970s * Livermore, Androscoggin County * season
        unknown
In Livermore, Androscoggin County, a man was driving on the Strickland Ferry (Road) AKA "Tolly Wolly". He stopped his jeep to observe a large hairy bipedal creature.

> "As I kept staring, I knew it ain't no bear. This thing
> was standing and walking on its hind legs, and it was
> at least over 7 feet tall. It was huge."

The man started his jeep up again with the intention of encouraging the creature off the road. Instead, the creature advanced toward the jeep stopping in the middle of the road in front of the vehicle. The man revved his engine. The thing bent forward and ran off the road on two legs disappearing into the woods.[15]

#11   1970's * County Unknown * season unknown
Bigfoot enthusiast and researcher Rip Lyttle collected a Bigfoot anecdote from the Maine coast involving a family witnessing something walk out of the forest near their farm. At first taking the figure for a hunter, the creature kept coming towards them causing some family members to get guns before the figure headed back into the woods. The family was sure what they had seen was consistent with descriptions of Bigfoot.[16]

#12   ~ 1970 * Baxter State Park, Piscataquis County * autumn
During November three friends were camped south of Long Pond

in the Trout Brook Mt. Quad of Baxter State Park. While hunting,
one of the party encountered a musky smell and noticed the woods
were extremely quiet. He soon found foot-long black scat and
smaller scat. While continuing on, the witness was "confronted with
the sound of thumping like a partridge on steroids thumping. The
ground actually shook." The smell became rotten and got worse. A
hundred yards in front of him a dead tree came crashing down. He
returned to camp where his tales was guffawed by two companions. The next day one of his friends joined him and the pair saw
the previous day's scat. The drumming sound came again followed
by a large rock, its flight audible, hurtling over their heads. They
read this as a warning and proceeded no further and returned to
camp. The hunt for all three men was a failure because there were
no signs of game. "It was the scariest thing I have ever encountered in my hunting career."[17]

Another version of this tale was told by ParaBreakdown on a
YouTube video titled, "Death by Sasquatch." Billfish Mountain
(1,880 feet) is in the northern part of Baxter State Park (about fifteen miles north of Mt. Katahdin) and was listed on old maps as
both Bill Fish and Fill Fish. The latter name was probably a mis-
spelling, as Bill Fish Ponds is, ironically, quite barren of fish. All
other accounts and maps from the late 1800s list the mountain as
Bill Fish, a nickname for William Fish, an early lumberman in that
area.[18]

Casey Walton and three of his friends came for a camping vacation. The men made their camp northwest of Billfish Mountain
along the southern shore of Long Pond. Casey's friends included
two veterinarians: Dr. Mel Hallings and Dr. Donald Fox. Pete
Ridgeway from Montana rounded out the group.

Casey was heading along the south edge of Long Pond when he
picked up a musky scent in the air. Further south of camp, he noticed huge, black scat. Casey became aware of the woods around
him falling silent. The stillness was broken by repetitive and rapid
thumping sounds lasting several seconds—this in turn was followed
by a tree falling. Drs. Hallings and Fox were not impressed with
the story Casey relayed back in camp. Dr. Fox accompanied Casey

back to the area the following day. They saw the thirty-foot tall tree that had been pushed down and Dr. Fox agreed that it would have taken great strength to push it over. The thumping Casey had heard the previous day returned and a rock was thrown at them.[19]

#13   1971 * Bridgton, Cumberland County * presumably summer

This entry is based upon Ken Buckley's 1971 *Bangor Daily News* story titled, "A Sasquatch Search in Bridgton, Maine—Does Bridgton really have a Sasquatch. Or, is it something dreamed up by a couple of guys over a glass of beer somewhere?"

Buckley's piece tells of Ted Bradstreet, a graduate student at the University of Maine, and his intention of investigating the possibility of a Bigfoot in the Bridgton region. Bradstreet, a glacial geologist, along with his wife will spend part of their vacation "hopefully catching one and making a real contribution to anthropology."

> Says Bradstreet, "Everything I learned in anthropology and geology points to the possibility that Sasquatches exist. But I can't say they do because I've never held one by the hand. And I probably won't, because I'd run like hell if I came face to face with one."

Bradstreet cited a report by Portland's John Shelley Jr. as evidence of a Bigfoot's presence in the Bridgton area and the catalyst for his planned excursion. Shelley reported on the discovery of large "primate" prints in Bridgton as well as smaller prints indicating the presence of a smaller Bigfoot.

Bradstreet referred to reading Ivan T. Sanderson's *Abominable Snowmen* in 1961 and was familiar with Patterson and Gimlin's "Patty" film.

> Looks like Man—What intrigues him, though, is that *Argosy* and *True* magazines reproduced sections of

the film, which has been decried as nothing more than a fake—a man in an ape suit.

Strangely enough, Bradstreet says, the so-called man in an ape suit walks erect like a human, but not like a man in an ape suit. He admits that the walk is something that would take considerable rehearsing to accomplish. At the same time, he says that if someone had wanted to go to such pains to fake the shots, why didn't they go all the way and reproduce the Sasquatch's reddish-black hair, rather than leave the hair totally black.

Bradstreet maintained that creatures like Bigfoot could offer information on the glacial period.

He feels that such primates hid in the dense woods after leaving Asia and fled after feeling that civilization was catching up with them.

Besides a bag for plaster casts, and his wife, Ruth, Bradstreet doesn't intend to carry anything else on his expedition into the Bridgton forests, much less a camera. He just thinks that if he is fortunate enough to come face to face with one, he'll not have time to say "watch the birdie."[20]

#14   1973 * Piscataquis County * spring or summer (spring annotated on county map)

Philip Pruitt and his brother in June, 1973, were driving through a heavily wooded area north of Kokadjo on a private road, when they saw what they took to be a black bear, "stooped over or on all fours" running toward the road. When it crossed the road it stood upright and ran swiftly like a human, moving its arms like a human runner. The creature was 6-feet tall and covered with dark hair. They watched it cross the road, enter the forest on the right, and then look over its shoulder at them.[21]

#15   1973 * North of Bangor, Penobscot County * summer
This encounter occurred during a summer evening on I-95's south-bound lane just north of Bangor. Jean Williams observed a large, hairy creature, either dark brown or black in color, moving swiftly from the median to the woods on the right hand side. Its arms were too long for a bear. Williams' first thought was that it was a gorilla. Williams was the only one to see the creature, her other family members in the car were sleeping.[22]

Durham, Maine has an interesting history of unusual occurrences.[23] The Durham Wild Man as reviewed in an earlier chapter is one of the best examples of a Wild Man story in Maine's history. In 1966, a Durham man reported an unidentified flying object at 8 p.m. on Sunday, April 16. The object was reported to be at treetop level and emanated blue and green lights. While investigating, two deputy sheriffs from Androscoggin County Sheriff Office reported hearing "beeping sounds which moved as they approached." The object was originally spotted on the Shiloh Road after the witness heard a beeping noise outside his house. The object, traveling southward, displayed flashing green and blue lights. According to one of the deputies, a beeping noise came from a wooded area in the respondent's backyard. The sound seemingly shifted away from the Deputy when he approached to investigate.[24]

In the early 1970s, an entity that became known as the Durham Gorilla, would come to the fore and bring notoriety again to Durham.

In the *Lewiston Daily Sun* of April 20, 1973, an article noted a gorilla costume had not been returned to Drapeau's Costume Shop of Lewiston. It had been rented on March 15, apparently under false pretenses.[25]

> Residents are asked to keep an eye out for a suspicious-looking person who might be seen making an extremely large purchase of bananas, or swinging from tree to tree.[26]

The missing gorilla costume preceded the rash of Durham Gorilla reports. The costume was reported as realistic and designed to fit a large person.[27]

#16   1973 * Brunswick, Cumberland County * summer
This event occurred in the Shiloh-Lisbon Falls Road area in Brunswick, a half-mile from the Huntington home, in Durham. The three Huntington children (Lois, (13); George Jr., (10); and Scott, (8); and their friend Tammy S., (12) were bicycling in July, about a half-mile from the Huntington home. They reported an encounter with a creature resembling a chimp. Mrs. Huntington told reporters: "My 13-year-old daughter fell off her bike about three feet from him and all he did was cock his head and look at her." It was described as upright and chimpanzee-like. Lois reportedly told the *Maine Sunday Telegram*: "I fell right down in front of him and all he did was look at me. I would have known if it were a hippie or something. But it had a regular monkey face. You have seen a monkey before, haven't you?"[28]

#17   1973 * Durham, Androscoggin County * summer
This sighting occurred near the Jones Cemetery, close to the Durham Road in Durham. A gorilla-like animal standing on its hind legs was observed two or three times by a James Washburn. A subsequent search of the area later by officers found moose and deer tracks.[29]

#18   1973 * Durham-Brunswick line, Androscoggin County *
         summer
The following encounter occurred on the Shiloh Road,[30] near the Durham-Brunswick line. Mrs. George (Meota) Huntington, 33, of Lisbon Falls Road, was driving home from a baseball game when she saw the "ape" peeking out from the bushes on the Durham Road. It was twenty feet away, and made a "mad dash" on two legs into the heavily wooded area. The exact description of the "ape" was that it was a little over five feet tall, with a shaggy, black coat, and weighing about 350 pounds.[31] Mrs. Huntington reported the

creature had a "monkey face." When the creature saw her vehicle it ran into the woods. Mrs. Huntington had the opportunity to spy the creature twice, the second occasion during a return to the area with neighbors. During this latter incident she remained in her car as others searched the woods. While sitting in her car she saw it again, peering at her from behind a tree.

Reportedly over thirty officers searched for the animal for two hours, involving the Androscoggin Sheriff's Department, the Cumberland County Sheriff's Department, the Maine State Police, and State of Maine Game Wardens. Tracks found near a cemetery had the appearance of human-like prints but had claws—likely those of a bear.[32]

#19   1973 * Durham-Brunswick line, Cumberland County * summer

This encounter occurred "behind the Jones Cemetery," at the Durham-Brunswick, line. On Friday, 27 July 1973, several witnesses came forward saying they had seen "it." Peter and Jean Merrill found a footprint behind the Jones Cemetery, the scene of several of the Durham Gorilla sightings. The Merrills said the track looked like a chimpanzee print. Androscoggin County Deputy Sheriff Blaine Footman examined a series of prints, casting one of them which was described as "rather deep." Footman said it was about five inches wide with "the thumb part broken off. Whatever made it weighs 300 or 350 pounds and I can't tell you much more. It's definitely not a bear track. I don't know what's going on here and I'd rather not express an opinion."[33]

Durham Gorilla witnesses would describe a creature like a gorilla, a bear, and a big chimpanzee. Auburn Dog Officer Louis Pinette joked with a reporter that it could be "a hippie out looking for a free meal."[34]

#20   1973 * Durham, Androscoggin County * summer

A beast resembling an ape or gorilla was said to be prowling through Durham on Thursday night, July 26, 1973. The Androscoggin County Sheriff's Department dispatched several units to look for the beast which was reported by several residents. A spokesman for the department said similar reports originated from Brunswick Wednesday, 25 July, but local police hadn't found any trace of the beast.[35]

#21   1973 * Durham, Androscoggin County * summer
Robert Huntington, of the Shilo Road, Durham reported to the Androscoggin County Sheriff's Department that someone wearing a gorilla costume scared his wife and daughter while brandishing a knife. Huntington said he chased the creature into the woods and thought that its face mask fell off during the pursuit. He tried to locate it but could not in the fast-approaching darkness. Deputies Dufresne and Footman and Dog Marshal Louis Pinette were dispatched to the scene but they were unable to find trace of the creature.[36]

During the rash of sightings, the critter earned the moniker, Osgood the Ape[37] from locals. As the gorilla's fame spread, so did the interest in finding the creature. The County Sheriff's Department said that persons seeking the creature were causing a significant amount of property damage, due to the number of complaints from property owners which included breaking down shrubs and ornamental plants and tramping over lawns. The Department said that the creature, believed to be a man in a gorilla costume, had scared residents in Durham, Lisbon and Brunswick.[38]

> Gorilla Spectators Cause Durham Property Damage—Spectators were in the news in Durham Tuesday, being the cause of complaints to the Androscoggin County Sheriff's Department.
>
> A spokesman for the department said the complaints stated persons seeking to pick up firsthand any clues in the so-called "gorilla" case are trespassing on private property and causing damage.
>
> According to one complaint, these persons are tramping over lawns and breaking down shrubs and other ornamental plants as they roam about the

Shadows in the Woods   277

Shilo Road area where the "gorilla" reportedly has
been seen.[39]

About the middle of August, 1973 the Durham Gorilla sightings
abruptly stopped.[40] Kinship with the possible donning of a gorilla
suit in Durham may have inspired a political protest that occurred
later in the decade at the seemingly coincidently named Durham,
New Hampshire. A man in a gorilla costume mocked President Ford
and his main challenger, Ronald Reagan, during their University
of New Hampshire addresses. When Ford was there the gorilla was
called "Bozo", a nickname some liberal editorial writers had given
to Ford. At Reagan's appearance, the gorilla was dubbed "Bonzo"
for obvious cinematic associations with the former actor.[41]

#22   1974 * York or Kennebunk, York County * spring
In the spring of 1974, as reported to the Bigfoot Field Researchers
Organization (BFRO), a man and his girlfriend were returning from
a motorcycle rally in Laconia, NH. Somewhere around York or
Kennebunk, they decided to camp out for the night and pulled off
the highway about a mile down a gravel road. They set up a two-
man tent in an area thick with scrub pine growth.

An hour after setting up camp, the man could hear movement
outside the tent but it did not concern him as he knew there was
lots of wildlife in the forest. The sounds of movement got progres-
sively closer and he noticed a pungent order. His thoughts turned
toward a bear. He felt that the black bears in the area were usually
just curious more than dangerous.

Whatever it was that was outside and making the noise suddenly let out a scream "that made every hair on my body stand
straight out and nearly gave me a heart attack." He stopped, at
that point, in thinking the thing outside was a bear. The scream
awakened his girlfriend who became upset. The thing moved so
close to the tent that the couple "could actually feel its movement
through the ground and hear it breathing." The thing could be
heard moving about the campsite for about twenty minutes and
then it went away.

The man didn't sleep that night and at first light he crawled out of the tent to look for signs and tracks. He found impressions in the grass but no distinct tracks. The impressions created a trackway that he was able to follow until they ended at a cliff about 30 feet in height. He wanted to investigate but his girlfriend was adamant on leaving the area and getting far away from their campsite.

The man's account included that later the same year, he and his girlfriend were watching a Bigfoot documentary which played some screams reputedly from a Bigfoot. The man again experienced his hair standing straight up, and they both recognized the scream as what they heard that night when they camped out on their way back from Laconia, New Hampshire.[42]

#23   1975 * Manchester, Kennebec County * autumn
Debbie and Donna Adams were walking along the Camp Road near Cobbosseecontee Stream at 5:30 p.m. on September 22, 1975.[43] It was as they neared Benson Road they saw a hair-covered, man-like creature, swaying back and forth and watching a man chopping wood. The girls screamed and the creature, which had an ape-like face,[44] turned and fled across the road and into the woods. It had a white patch on its chest area and a bushy appearance around the jaw line and the top of its head.[45]

#24   1975 * Manchester, Kennebec County * autumn
Manchester, Maine; November 21, before 10 p.m.[46] Mrs. Paul Adams was sitting in the car waiting for her husband, who was tuning a guitar in J. Dupont's trailer. She noticed what looked like the figure of a man 6 to 7 feet tall standing by a nearby tree. She watched about 10 minutes as it crouched down with one arm around the tree, apparently listening to the country-western music coming from the trailer. She touched the car horn and it got up and started towards her. She leaned on the horn producing a longer blast and the creature stopped its advance.[47]

#25   1975 * Somerset County * winter
During brutally cold conditions in mid-February, the respondent

recalled a feeling of being watched while noticing two ponies acting agitated and distressed. The witness had the firm impression something unseen was lurking in an old barn. Feeling a sense of danger, the respondent did not enter the barn but did notice a pungent smell in the air like a dead animal. Upon hearing a scream or screech, the witness entered a trailer. Later that night there was a bang on the trailer's side as if someone had punched it and this was followed by another scream similar to the one heard earlier that day. Looking out a window, the respondent did not see anything, but did notice the ponies were running again. In the morning, the witness noticed several large foot prints around the trailer.[48]

#26   1977 * Lake Moxie, Somerset County * summer

> Was an unusually large animal akin to Bigfoot driven
> out of its lair by the Lake Moxie forest fire? That
> question is being circulated from Canaan to Jackman
> these days with reports that an awesome animal has
> been sighted since the fire.
>
> None of the sightings has been officially con-
> firmed but some have come from reliable sources,
> like a farmer in Canaan who described the animal as
> "bear-like" but taller than any he's ever seen and
> covered with a strange texture of hair rather than fur.

1977's Lake Moxie fire burned about 2,000 acres and forced many families along Lake Moxie, located about 25 miles from Greenville, to flee their homes. The Lake Moxie fire was the largest of about 100 scattered fires that broke out across Maine in May, 1977. Over 80 firefighters, rangers, and volunteers and some 20 pieces of heavy equipment built fire lines to contain Lake Moxie blaze.

> Crews fighting the Lake Moxie fire saw a number of
> moose being driven by the flames but none saw bears
> fleeing. They didn't sight anything that might be
> mistaken for 'Bigfoot' either.[49]

#27   1977 * Kezar Falls, Oxford County * winter

A Kezar Falls man heard a sound like a baby crying outside, opened the door and was confronted by a tall black creature. It promptly screeched and ran off. Footprints were followed across the road and back behind a camp for about a mile. The immediate terrain was hilly with brush and a layer of snow on the ground.[50]

Another version of this sighting was proffered to the G.C.B.R.P. The respondent recalled a visit with a Native American couple in Kezar Falls. The couple told of hearing a baby crying earlier that morning followed by a sighting of a *giant of the woods*. The couple pointed out a set of tracks still visible in the snow. One of those whom the couple showed the tracks to stood six-foot, two inches and he could not replicate the stride length evinced by the track-way. This telling of the story included mention of a wet-dog odor while following the tracks.[51]

#28   1977 * Deblois, Washington County * autumn

The following incident was relayed by a woman who heard the account from her husband and two of his friends as they camped and hunted in Deblois. The events started just before Thanksgiving, 1977.

> Incident Location: Upstate Maine—Deblois, ME (Washington County) behind the Wyman & Son Blueberry Barrens (7k acres of blueberry barrens alone) stretching from Deblois to Cherryfield Maine, on a tributary of the Narraguagus River off RT 193. GPS coordinates are unknown at this time. This is behind the now defunct Deblois Flight Strip area.

The party of friends, who were 19 years old at the time, planned to spend a week at an old logger's bunkhouse that one of the boy's parents retained a lease on. After their journey culminated with a stretch of canoeing, they arrived at the bunkhouse where they built a fire to warm up. A game bird was shot and anticipated for the morrow as their Thanksgiving meal.

The bunkhouse was a one-level rectangle of approximately 600 square feet serviced by an outhouse 40 yards away. The woman recounted her husband's need for the outhouse in the late afternoon as dusk was settling. Snow had begun to fall and at the time he planned on his outhouse excursion between ½″ and 1″ had already fallen. Snowfall total was expected to reach 1 foot. About halfway on his trek he stopped noticing "something didn't look right on the ground."

> He saw large tracks and thought for a second quickly replaying who had been out before him or after it had begun to snow (no one) he was looking down at huge, very huge tracks and became immediately alarmed. He thought it must be one big ass bear to make tracks that size, so he looked closer and saw separations between what looked like toes and no claw marks, these footprints were huge and in one line, not left/right like a bear or a person would make, and they were toes forward to the cabin then stopped. He estimates they were 18″.

The woman wrote that her husband went back to the bunkhouse and alerted one of his friends he had seen something unusual—big tracks right outside the bunkhouse—which must have been fresh based upon the falling snow. The friend went out for his own examination of the tracks.

> Joe was outside which was only for about 20 seconds when he flew back into the cabin and locked the door, my husband remembers thinking in his head 'oh man Joe, please tell me those are bear tracks'.

Instead, the friend said "We got a big problem," convinced the friends now had the company of something very big in the area.

My husband actually saw the toes spaces clearly; they were bipedal like human feet, one in front of the other too (not like humans or bears left/right) they came forward to the cabin and stopped 20 yards out and seemed to stop. They were 100% fresh tracks.

The remaining friend decided against checking out the tracks for himself, seeing that his compatriots were visibly shaken and utterly serious regarding the possible implications. The young men began pushing bunk beds up against the front door for a barricade.

Just as they were starting to lose that immediate intense feeling of high octane fear, it happened. At about 10:00 p.m.—came a blood curdling scream that lasted for about 20 seconds. It was *very* loud, clear and close. They could not identify it since none of them had ever heard anything like it before.

The account averred the woman's husband was then convinced he and his friends were experiencing a Bigfoot. Though they wanted to leave, they wished to have no part in trying to retrace their way through woods and stream back to their van parked off of Route 193. They armed themselves and settled into an all-night vigil focused on the door and a large, conspicuous window.

At about 1:00 a.m. Joe thought he heard something and blasted the door with his 12 gauge which almost caused my husband and Barry to have heart attacks.

After remaining awake all night, the group stayed within the bunkhouse the following day. Outhouse needs were done indoors and the game bird shot at the start of their adventure was cooked for a Thanksgiving Day meal. The next day, November 25th, the group ventured outdoors again but remained in sight of each other.

>They never heard another peep, no screams, no
>tracks, no sightings and no smells, no rocks thrown,
>nothing. They were there for 7 days total and went
>back 6 months later for another 7 days and nothing
>happened.

Years later while questioning her husband about the incident, the woman included he was certainly "freaked out" by the tracks and offered he hadn't examined them thoroughly but did recall there was one set, toes pointed toward the bunkhouse. He admitted there could have been additional tracks leading away from the bunkhouse but he hadn't noticed them.[52]

#29   Before 1978 * vicinity Rockwood, Somerset County * presumably autumn

Don Wilson and Joe Smith partnered on a hunting trip 20 miles from Rockwood in Somerset County. Their camp was on the edge of a big lake. The friends woke at 2 a.m. to see a big hairy head, light gray in color, looking in through a cabin window. It stayed for about five minutes. The pair retrieved their guns and followed its trail until the tracks were lost on hard snow. A Maine Guide later told them the monster had been known for 50 years in the area and had never hurt anyone.[53]

#30   1980 * Sebec Lake, Piscataquis County * winter

Footprints estimated to be about 16" long were found at 1:30 a.m. during snowy conditions, by a group of friends riding in a truck. Their find occurred near Sebec Shores Road near Sebec Lake. The party saw that the footprints crossed the road and headed up over a bank. Unnerved at what they were seeing, the group headed home without further investigation. The closest city is Dover Foxcroft.[54]

#31   1982 * Mount Katahdin, Piscataquis County * season unknown

A bird-watching group observed two upright, hairy creatures walking across a clearing on the north side of Mt. Katahdin. The two unclothed figures had their backs to the bird watchers and were reported to have backs and buttocks covered in long hair.[55]

#32   1984 * Sullivan, Hancock County * spring or summer
         (spring annotated on county map)
Herb Hatch says that while trout fishing in June 1984, he and his uncle saw a 7- to 8-foot, dark brown creature, with an estimated weight between 400 to 500 lbs. The creature was in plain view less than 50 feet away. No other specifics of the sighting reported though the respondent added he knows of other sightings in the same general area.[56]

#33   1985 * near Acton, York County * summer
While gathering cut firewood in the summertime,[57] the witness heard a noise causing him to turn and look. He expected a moose but saw nothing. Resuming work, he presently heard the noise again and turned to look and saw "it," presumably a Bigfoot creature, watching him. The respondent hurriedly left. This account is also found in *The Track Record* #96: During the summertime, near Acton, a single witness was gathering firewood when he heard a noise behind him. He turned around expecting to see a moose, but instead saw a "creature" only 30 feet away. Deciding not to stick around, the witness left the area.[58]

#34   1986 * Turner, Androscoggin County * autumn
This sighting occurred October 15 during a mild fall night in the 50°s. Two people were heading to a club in Turner. As they drove off Route 4 onto Route 117, they saw an upright hairy creature estimated to be 6' and 200 lbs. Telling their tale at the Circle Electric club, other people confirmed they had seen the thing too. The creature's hair was described as light brown, long and "hanging off it."[59]

#35   1987 * east of Bangor, Penobscot County * autumn
Two hikers east of Bangor, encountered a fecal, sweaty smell in the month of October. As the pair neared the top of a hill the smell became stronger. They decided to backtrack the way they had come while feeling a strong sense of being watched.[60]

The account below is similar, if not directly related to *IBS Report* #3822,[61] which relays the story of fishermen who had stopped at a cabin in the northern Maine woods where they absorbedly read a journal-like account of the Penobscot Ridge Monster. The Penobscot Ridge Monster and the Maine Ridge Monster which immediately follows appear to be the same encounter/story.[62]

#36   1988 * "northern Maine woods" * summer

The Maine Ridge Monster—The respondent relates a story from about ten years ago (1988) when he, then 13 years old, went on a fishing trip with his father for brook trout in the northern Maine woods. It was summer and the pair found the fish to be biting. While wading down the brook they came upon the half-eaten remains of a hare lying on the bank. The slow current betrayed that something had disturbed the water and spun up and agitated debris in the water.

The pair caught their limit of trout and headed back to their truck, parked about five miles away at the end of an old woods road, last used for logging about 30 years ago according to the reporter's estimate. The road was grown over but still easier than walking and slugging through the thick woods. They stopped at a cabin to have a rest. On a table they noticed a log book which was full of entries concerning a "Ridge Monster," so called, replete with detailed experiences with the creature.

After their rest and perusal of the journal the pair headed out again on the old road toward their truck. The father suddenly froze. The boy believed his father had stopped for a rest until he saw his father's face and realized his father was motivated by something completely different.

The father was focused on a small opening in the forest. The boy wondered what could be holding his father's attention so, and his thoughts turned toward a protective cow moose with her young.

His father suddenly grabbed him, urging his son to hurry. The pair made the rest of their walk out of the woods in record time.

On the ride home the father told the boy what he had seen. At the time the pair had rounded a small corner in the trail the father caught a flash of movement. Looking over, he saw what could have only been a Bigfoot. The father described a creature eight feet tall with an estimated weight of 500 pounds. It had one long arm grasped round a tree, one foot up on a rise in the forest floor, and it stared straight back at the man. With an apparent sense of being bored the creature melted away into the woods.[63]

#37   1988 * Mount Katahdin, Piscataquis County * autumn
A group of Boy Scouts were camping on Mt. Katahdin in September. They witnessed a big, triangle-faced creature covered with reddish-brown hair. The creature had a large build, broad shoulders and it seemed to be the source of a stench that permeated the area for hours after the sighting. The Boy Scouts observed the creature rooting in the groundcover as if searching for food. They heard it vocalize several times, making what the report calls "frightening" sounds.[64]

#38   early 1990s * Aurora, Hancock County * summer
The following encounter was remembered to have occurred in the early 1990s, taking place during July at Rocky Pond in Aurora. The witness at the time was right out of high school and enjoying camping for three or four days a friend and the friend's dad. The group was canoeing, heading to the north end of Rocky Pond. While not quite out in the middle of the pond, near the southwest section, the witness caught sight of something strange on the rocky shore. A big, black or dark brown figure, slender in build, stepped up onto a rock and looked towards the canoe before disappearing into the woods. The witness and his friend both saw the figure. When the figure was described to the friend's father—darker in color than Chewbacca but shorter hair—he said it had probably been a game warden. Investigating the sight, a little trail was noted encircling the rock the creature had stepped up on. The rock in question was about three feet tall. The witness estimated the creature's height at about 7 or 8 feet. The witness recalled seeing the creature's shoulders and the rear-view of its legs, but didn't recall seeing arm swings or specific muscle movements. The hair looked longer to-wards the upper portion of the body and a uniform length on the

legs. The creature was estimated to weigh, given its height, about 600 pounds. The witness said it was similar to the Patterson Gimlin film but slimmer, lacking the girth of "Patty." The creature's head had an inhuman shape, described by the witness as "longer." No facial details were recalled in the interview. When the party walked up the game trail they did not notice any footprints, or smell.[65]

#39   1990s * Aroostook County * spring
The Allagash is shielded from urban expansion by its location in northern Maine. The term Allagash refers to a region as well as a river. "This vast watershed springs from a half-hundred lakes, countless streams and rills, forming the heart of the wild land that blankets northwestern Maine."[66] The Maine North Woods is the northern part of the state encompassing an area of 3.5 million acres of forest. Along with tracts of cultivated timber for lumber production, the Allagash is also a favorite destination for campers, hikers, hunters, and fishermen.

Michael Merchant conducted and posted an audio of a Skype interview with a Bigfoot witness who had an encounter in the Allagash region of Maine in May in the 1990s (the exact year and the witness' name withheld by request). The interviewee reported that the occasion was dipping for smelt with friends.

When the group arrived at their smelting point, they got out their nets and milk jugs. As the witness was walking down to the water, he realized he had forgotten his milk jug (used as a scoop) at the car. The witness recalled a moon hung in the sky. Upon retrieving his milk jug he started back down the embankment and down a path to join his counterparts.

During his trek he heard a whistle and a crack like a tree break from the forest off to his left. Looking in that direction he didn't see anything, but when he turned back, about 20 feet away he saw a shape standing next to a tree only a few feet off the path. The thing turned its head and shoulders and looked right at the witness. The witness reported that it rolled its upper lip in a snarl revealing yellowish eye teeth.

Worried that the creature would think his advance down the path would be perceived as challenging, the man put his head down in a naturally submissive posture, though still keeping the thing in view through the top of his eyes. The thing, which the witness took to be male, started rocking back and forth. The description of the thing was that it was maybe four feet wide and when viewed from the side the barrel-chest was two and a half to three feet thick. It stood about seven and a half to nine feet tall. When it turned, its coat of hair flashed light for a moment before turning black, like it was sprayed with oil. The hair was knotted and when it was rocking he could see the four inch hair on its arms swaying back and forth. The hair length was consistent all over the thing's body but the hair on its head gave the impression of having had a hairdo. The hair on its head went down its back, across the face, and across the chin like a beard.

After a few moments it took a couple of steps into the forest and disappeared. The witness believes that he was close to two creatures, the first whistle he heard being some kind of communication to the one he clearly saw. He felt the creature he saw was actively observing his friends smelting at the river. He noticed no eye shine or odor. It had a rounded head, wide mouth, and a flat, wide nose. The head hair which came down over the shoulders, gave the impression of being messed and knotted with a wet or oily look. The witness said it was definitely a creature in its prime. Overall, the creature sported a massive build with pronounced shoulders and an overall thickness. The witness said it was like encountering a monster. It had a prominent belly similar to a Bigfoot purported to be captured on video by researcher Paul Freeman.[67]

#40   1990 * Stud Mill Road, Hancock County * spring
The witness related an encounter on the Stud Mill Road, about 20 miles further west near Pickerel Pond and an old airstrip. Occurring about 1990, the respondent was showing his girlfriend the spot of a recent ROTC exercise. Having to go to the bathroom, he went to an old gravel pit on the north side of the road. When he walked into the pit, he saw movement near the rim, about 50 yards away. Something

large, about 6 feet in height, and reddish-orange in color, moved away from the pit's edge and into the tree line. Thinking about what he had fleetingly seen, the witness decided it was the wrong color for a moose, and there had been no white flag from a bounding deer. The thing moved swiftly and noiselessly. The witness had the urge to get out of there, experiencing a "leave here now" kind of feeling.[68]

#41   1990 * Durham, Androscoggin County * autumn or winter
Something was noticed crouching over the site of a buried deer, apparently engaged in digging and making a faint grunting noise. The lone witness, a man on foot on a Durham road, grunted back a response. The creature turned toward him and "upped the ante" by growling. The witness reciprocated the growl and received an unexpected reaction from the unknown creature. This time the thing stood on its hind legs and advanced, growling as it made its way to the man. The man ran to a house, bug-eyed and claiming a seven-foot Bigfoot was chasing him. Later examination of the site where the thing had been observed showed a deer had been extricated but only the liver was missing. Broken branches, about ten feet off the ground, were also noticed in the area.[69]

#42   1992 * near Andover, Oxford County * spring
This encounter took place during May, in the early afternoon. The location was Lower Richardson Lake near the South Arm Campground. One member (the witness) of a three-person group caught a glimpse of the creature, happening to be the only one looking at an area of high ground where the creature was seen. The witness, though admitting the creature may have been a bear, was taken by the size of the thing. Thinking the group had encountered a black bear looking for food, one of the hikers fired a 22-caliber pistol twice to scare the animal away. The creature was described as brown, almost black, large, but it was hard to determine exact size. It moved rapidly producing a large shadow.[70]

#43   1993 * Solon, Somerset County * summer
Wilfred Judd, driving on South Solon Road beside the Kennebec

River in July, saw a hairy, dark brown or black creature, estimated as 6′5″ tall, standing in profile on the road. Hearing the car it turned and dashed to the side. Jamie Robinson, in a following car, saw the creature standing amidst some alders. When he came back for another look he heard loud growling.[71]

#44   1993 * Fryeburg, Oxford County * summer
One of the obvious problems with hoaxes is that incidents not directly related to the hoax become conflated with the deception, thereby remitting any potential value or significance to oblivion. The inclusion of this undeniable hoax's papier-mâché details is intended to alert and inform anyone who may believe they saw something Bigfoot-like in Fryeburg, Lovell, or Stow in 1993.

> Big Foot Sightings turn out to be Hoax—by Jennifer Sullivan, *Sun Journal* Regional Staff Writer— Fryeburg—Attention Sasquatch fans and Yeti watchers: That hairy 7-foot creature who's been roaming the banks of the Saco River is a fraud.
>
> Yes, it's true, no matter what the *Weekly World News* might say. If you spotted Big Foot, you played a part in an elaborate hoax concocted—with a bit of papier mache and fake fur—by a 30-year-old Gorham man who says he did it all for rock 'n roll.

Martin Kade used his Bigfoot creation to startle motorists because he wanted a recording contract.

> "I don't have any money to get a manager to get a record company to listen to me, and I can't get any manager to take me for less than $10,000," he said. "Unless you're backed by a record company, then you can get a manager."
>
> So when Kade heard that Maury Povich was coming to Maine to play in a charity golf tournament,

the idea of the tabloid TV guy and the hairy guy of legend fused in his mind: This was how he would get the national attention he needed to catch some record executive's eye. He would stage fake Big Foot sightings, get on The Maury Povich Show and, figuratively speaking, sing his song.

Kade would often shake his creation in pseudo-marionette style to give it the appearance of life and movement.

"I've been on just about every road there is in Fryeburg, Lovell, Stow. I was doing it down around the Saco River so the campers would see it," Kade said, adding that he'd lie in the bushes with his creature, shake it a bit, then hide it. An hour or so later, he'd casually approach the campers.

"I'd say, 'Hey, I heard you guys saw Big Foot. Did you report it?' They'd say, 'Yeah, I saw it, but people will think I'm crazy,'" he recalled.

The mannequin's ultimate appearance occurred when Kade brought the papier mâché figure to the Falmouth Country Club with a sign asking Maury Povich to call him. Overall, Kade was underwhelmed by the response he was generating.

In fact, Big Foot generated so little attention that a friend of Kade decided to help him out earlier this week by pretending to have spotted the creature on Route 113 in Fryeburg. The friend told his tale in Portland television station WGME which sent a crew out to get the story."[72]

#45   1994 * Coopers Mills, Lincoln County * winter
The following encounter took place in the early morning in February on a dirt road near Coopers Mills. The observer had gone off

the road in his pickup after encountering icy driving conditions. Being about three miles from work the witness started to walk the rest of the way. It was still dark out (about 4 a.m.) but there was some moonlight. The observer approached his place of work when he saw a figure appearing to be a person heading toward him. The man was not afraid but rather curious who it could be at that early hour. Eventually the two walked right past each other, each occupying an opposite side of the dirt road. The observer could make out the figure's silhouette. After their passing, the figure crossed to the observer's side of the road and kept going right into the woods. This behavior frightened the observer who ran the last half mile to his place of work. "Whatever it was I never saw it again."[73]

#46   1995 * Prong Pond, Piscataquis County * season unknown
"Prong Pond is a moderately large body of water located a few miles north of Greenville. The shoreline of Prong Pond is quite diverse. The southeastern portion of the pond consists of two long shallow arms. The shoreline in these 'prongs' is floating-mat, covered with grasses and shrubs. The shoreline of the main basin is boulder-strewn with occasional ledge outcroppings. There are many small boulder and ledge islands in the main pond."[74]

The witness was an avowed skeptic before the related incident. The party was made up of three including the witness' father and a close friend. The group had decided to try their luck at Prong Pond for smallmouth bass. After setting up camp at a boat launch, the group paddled to the other side of the pond to start their fishing.

> We were fishing for maybe a half hour when we all experienced a cry coming from the shoreline which appeared fairly close but could not pinpoint an exact location because of heavy fir boughs. At first I had thought it was a bird of prey of some sort then the screams took on a more human form sounding like a laughing hyena and they were very powerful.

The group continued to fish wondering what it was they had heard. As darkness approached, they paddled back to camp for dinner and talk around the campfire.

> My dad had mentioned that a good fishing guide friend some years back had told him that he was fishing the same pond and was close to the location where we had been that day and saw a huge hairy creature stooping on the shoreline watching him inquisitively, then the creature knew that he had noticed it and stood up and turned to exit in the firs behind him banging its head on a limb.

After waiting for a period of time, his father's friend pulled his canoe over to the spot where he had seen the creature. He measured the height of the branch the creature banged its head on and determined it was approximately eight feet tall.

In the early morning, the respondent was awakened by a "horrendous scream," that was similar to what the group had earlier the previous day. The father and the friend had already been awake listening to earlier iterations of the screams. The screams went on for some time and gradually sounded from further distances. The group packed up the next morning and left the scene.[75]

#47   1996 or 1997 * Bucksport, Hancock County * season unknown

Two sisters were driving to school on Route 46 near Bucksport one morning, when they saw a creature standing upright on two legs. It was covered in dark shaggy hair, and hunched over with long arms hanging down. It was a couple of feet onto the roadway but quickly moved off the road and vanished into the trees. "When we tell people they never believe us because most people don't believe in 'bigfoot.'"[76]

#48   1997 * Hancock County * season unknown

This G.C.B.F.R.O. report is essentially the same as BFRO Report

#1188 above. The witness provides information on another sight-
ing she had as she was riding with her father when she saw something walking on two feet. "Dad, turn around, quick. I think I just
saw bigfoot." Her father turned the truck down the road in question but there was no evidence of any creature. The witness added
that many neighborhood pets had gone missing at this time.[77]

#49   1998 * Parmachenee Lake, Oxford County * summer
The witnesses were a woman, her boyfriend, and two sons (5 and
10) driving in truck leaving Parmachenee Lake on July 3rd between
3:00 and 4:00 p.m. The encounter took place on the road heading
back to Route 16 in Wilsons Mills. The creature came out of a
wooded area at the top of a hill. It was described as approximately
7 feet tall, skinny, covered with long brown hair or fur, and standing on two legs. It never seemed to acknowledge the approach of
the witnesses' truck. Its strides were huge, requiring three steps
to cross the road, which was wide enough for two vehicles. The
group did not observe tracks and they did not stop. The creature
was last seen heading towards Aziscohos Lake.[78]

#50   1999 * Poland, Androscoggin County * summer
This incident occurred on August 15, at about 03:00 a.m. Route 11
and North Raymond Road in Poland, Me. Off the road about 100
feet, big and hairy, really big. 2 people in car. "My brother-in-law
was turning off Route 11 onto the Raymond Road when he said 'Do
you see that?'" When the respondent said "What?" his brother-in-
law said, "That big hairy thing in the bushes!" The early morning
was dark, and no moon. "This is more common that you may think,
people up hear [sic] see this stuff often but the locals keep it
quiet."[79]

#51   2000 * Kennebec County * winter
Multiple sightings of tracks discovered in January—some found in
snow—with specific mention of prints in Sidney, Topsham, and
Manchester.[80]

#52   2000 * Richmond, Sagadahoc County * winter
February 15, 2000, a Bigfoot investigator found two sets of tracks
in Richmond. These tracks were 1 mile apart, one set heading east—
a 16 inch track just off the White Road in Richmond. The other
was 16″ also, heading west to White Road. Both sets of tracks were
over a mile long.[81]

#53   2001 * Frenchville, Aroostook County * summer
The reporter and two 17-year-old identical twin nephews decided
to go fishing in a part of Frenchville in June, 2001, where they
found three fresh prints. The prints looked to be partial and clearly
showed the toes. No visible claw marks like a bear print. Prints
were impressed an inch into the mud. While walking to a fishing
spot, the man thought he had heard a "low-volume 'grunt'." They
covered the print with a shirt and that night it rained. The next
day they took pictures. Rough measurements suggested the print
was 6 inches wide and 13 inches long.[82]

#54   2001 * Somerset County * summer
In late July, while heeding the call of nature, a camper heard a
noise which he attributed to a moose. Later, and before daylight,
the witness woke to a loud moaning and screaming. Upon exiting
the tent with a flashlight the witness caught a glimpse of an 8- to
9-foot-tall creature melt away into the darkness.[83] This account
does include mention of aircraft performing unusually.

#55   2001 * Andover, Oxford County * summer
This account occurred in July, vicinity of Elephant Mountain near
South Arm campgrounds in Oxford County. "I heard the most un-
natural sounds of my life. My eyes are welling up just writing this."
The incident occurred during July while the witness was camping.
One night the witness heard a "deep groaning/barking noise." It
reminded the witness of recordings heard on old shows about
Bigfoot. The next night the same "scary, deep groaning noises" was
heard again. The witness packed camp and left the next day, never

feeling the urge to investigate or look for tracks. "Those noises were enough to scare me out of my wits."[84]

#56   2002 * Topsfield, Washington County * summer
Date and time of this encounter was July 28, 2002, in the afternoon, between 1 and 3 p.m. The witness was driving from Fredericton, New Brunswick, to Topsfield, Maine. The witness saw a big, dark, upright man-type thing walk on two legs left to right across the road about 400 feet away. The creature never looked at the witness and it disappeared into a wooded, swampy area. The creature reportedly took about 6 steps from the middle of the two-lane road into the woods. Its steps were huge but the thing didn't seem to be in a hurry. The creature was dark brown, covered all over with hair, and had a thick muscular build.[85]

#57   2003 * near Baxter State Park, Piscataquis County * summer
July 12, 2003, near Baxter State Park. Saturday evening. The reporting witness, his wife, and son were hiking and were invited to the camp of a couple they met who were returning from Roaring Brook via the Russell Pond Trail. The botanist perked up during their conversations at their camp. "Hear that?" the witness' wife asked. "Sounds like the whistle of a hawk circling nearby." They all listened and heard the shrill cry another 2-3 times from opposing directions giving rise to the thought there were at least two red-tailed hawks nearby. After awhile another whistle came, this was much closer and louder. Later, a small rock landed in the midst of the campers. Another small rock hit a tent. A flashlight beam illuminated a man-like thing with reddish-brown hair and no clothing about 40 feet away from the campers. The thing stared into the light and raised its forearm to shield its gaze; the long hair at the forearm was recalled as "shaggy." The creature slipped behind the tree and out of line of sight of the party. The rest of the night passed uneventfully and the morning produced no sign of the hairy creature.[86]

#58   2004 * Bemis, Franklin County * summer
September, Labor Day Weekend—The respondent and a friend
hiked the Bemis Valley Trail. They took a side trail leading to
Mooselookmeguntic Lake. On this trail they saw a dark figure wade
into the lake and stand very still. Then with its arms by its sides, it
plunged into the lake creating a huge splash.

> The ripples faded along with the last rays of daylight
> and yet we kept staring into the darkening lake with
> wonder. How truly bizarre, I'm not used to seeing
> anything like that, I'm a city girl. Jim however was
> raised in the country, is nature savvy and seemed
> floored at what we just witnessed. I asked him if that
> could have been a bear, he answered with a resound-
> ing "no, that was no bear, it was no moose." It was
> the biggest man I've ever seen or it was a bigfoot,
> but too dark to see details.

The next day the two found a track near the lake, "one clear
half of a barefoot print in the mud," suggesting a total footprint
estimated between 14-15 inches long and 6 inches wide. The track
was oriented toward the water.[87]

#59   2005 * Gardiner, Kennebec County * winter
R. Brown of Sidney reported finding tracks on January 12, 2005,
one mile north of Route 201 in Gardiner. They measured 16" long
and 7 inches across. Stride was 7 feet. It was 2:30 p.m. He followed them for a mile before quitting.[88]

#60   2006 * Canton, Oxford County * winter
January 12, 2006. A trucker was driving his timber-loaded 18-
wheeler headed for Rumford, two miles out of Canton on Route 108.
At about 10:30 p.m. he saw three large creatures, each on two legs
cross the road in front of him. The trucker estimated that one was
eight feet tall and hairy, one was tall, thin with grayish colored hair
and limping and a third was short with a fat, protruding stomach.[89]

#61   2006 * Jackman, Somerset County * summer

This encounter occurred in the summer of 2006. The contributor concluded he did not know what he and his father met, though several of the story's details do align with inferred behaviors of Bigfoot.

It was while on vacation that Greg B. and his father traveled the many logging roads in the Jackman area to enjoy nature and hopefully catch sight of a moose. The encounter in question happened at night about 16 miles into the woods from Route 201, and was preceded by Greg B.'s use of a rabbit-in-distress call. He and his father had arrived at a spot where they had previously left pizza crust only to find it gone. At the same location a freshly killed rabbit, its stomach torn open, lay where they had earlier left the remnants of their pizza supper. The men remained at the spot and after about ten minutes Greg B. heard a whistling coming from the woods to his right.

> The chirping increased in volume for another 10 minutes. That was when I started hearing the rustling of whatever was walking in the woods. I should have stopped using the predator call at that point and left.

Greg B. wrote of his curiosity getting the best of him, and he increased the rabbit call to once a minute.

> After each sound there would be silence, than a very deliberate movement toward me.

Listening to the sounds of movement caused the contributor to realize it was not a moose or any animal he could recognize.

> Whatever was walking in the woods did not rustle the leaves on the ground, it was taking steps as if it was lifting its feet up and putting them down. Whatever it was it was very big, because I could hear

branches snapping off of the trees and leaves falling
once the branches snapped.

Scanning the woods provided no clue to the identity of the unknown animal. The brush and tree growth immediately off the road
was an impenetrable screen that offered no glimpse of the animal
walking close to them.

> I figured I'd try using the call once more. I sounded
> it. Within one second of the end of the call I heard a
> VERY loud guttural grunt/growl in the woods to my
> right. I told my father to put his hands on the keys
> and get ready to get us out of there. My eyes started
> watering and I started to shake because I knew what-
> ever it was had to be huge to make that noise.

After another use of the call, a "deafening" growl came from
the woods. Greg B. noticed a large birch tree begin leaning toward
the car as if pushed. That was enough for the men; the car was
started and the pair drove quickly down the road. At a point roughly
three miles from the leaning birch tree, the pair came across a tire-
sized rock in the road at a spot where they had stopped earlier in the
evening. They were absolutely sure it had not been there earlier.

> The next day my father and I went back to the spot.
> The rock was no longer in the road . . . in fact it was
> gone. We came around the corner where the rabbit
> was sitting. It was gone. There was no blood, either.
> The birch tree was standing upright as if nothing had
> happened. There were no footprints as it was too dry.
> The only thing that indicated that anything had hap-
> pened was the grass matted down on the side of the
> road. The area that was pushed down was about the
> size of a pickup truck so something big was sitting
> in this spot.

At the time of the incident and confirmed by later rumination, Greg B. was convinced the animal was not a bear. Though the men did not go back to their back woods haunt again that vacation, they did have the opportunity, the night after they had encountered something strange, to pull off the side of Route 201 to watch a moose. But the strangeness was apparently not over for the pair as from off in the distance, from the location of the previous night's encounter, they heard what Greg B. described as a "haunting" sound that echoed through the valleys.

> It had the tone of a fog horn but I could tell this was definitely some sort of animal or human making the sound. I'll never forget that noise.[90]

#62   2006 * Eagle Lake, Aroostook County * autumn
The following anecdote takes place in November. A hunter near Eagle Lake heard grunting noise and saw a large, eight-foot tall, hair-covered creature. After staring at the hunter for about twenty seconds, the creature ran off into the woods.[91]

#63   2006 * Knox County * autumn
Date: September 26th. A tractor operator's grinding of gears may have caused a strange animal to reveal itself. Its height was 7 or 8 feet tall, and the hair over its body was long and gray and brown in color. "I just had the weird feeling the entire time. I am just baffled that it would even come around the place with all the dogs."[92]

#64   Late 2006 * Ellsworth, Hancock County * autumn
Ellsworth is the county seat of Hancock County. Historically, Ellsworth was a shipping center and spot of boat building. A flood in 1923 and the Ellsworth Fire of 1933 helped change the commercial direction of the city. Today, Ellsworth is one of Maine's fastest growing cities; it is home to the Grand Theater, a widely-used public library on the high bank of the Union River, and is known by many visitors and tourists as the gateway to Mount Desert Island and Acadia National Park.

A young couple were driving home to Ellsworth on Route 1A in October, before Halloween, after having caught a movie in Bangor. They made Ellsworth Falls sometime after 9 p.m. where the couple glimpsed on the left hand side of the road a figure standing on two legs. The figure moved to the center of the road where it was back-lit by a sherbet-orange streetlight. It was tall and very thin and had very long arms.

Initial reactions for the couple (who were actively scanning for deer during their nighttime drive) was that it could be a human and possibly some form of practical joke. The thing looked back behind it and then fluidly hunched forward putting its hands to the ground (maybe its knuckles) and it bounded across the road on all fours. The female witness has degrees in marine biology and secondary science education, a point she mentioned during the interview to establish herself as a creditable witness and capable of accurately describing what she saw that night. The fluid movements were contrasted with humans, for whom the act of getting down on all fours and trying to ambulate across the road would not look natural. After about four quadrupedal bounds across the road it was joined by an identical, smaller creature following it in similar bounding fashion. They likened it to a deer being followed by a fawn across a road.

The female witness said the larger creature had a look of annoyance when it turned and bounded across the road as if saying, "Come on, there's a car coming." Both witnesses stated the creature's arms were way too long for a human, and the ability to run on all fours with so much power and fluidity, so lean, lanky and gaunt, all made it hard to imagine a human dressed up playing a prank. The interviewees believed it behaved consistent with what an animal would do: displaying a moment of supposed hesitation before deciding upon a course of action—bound across the road.

If the morphology of the larger creature was reported to be overall thin and lean, its estimate of health, supposed to be fine, was based upon the power and capability to move the way it did. The incident was about seven seconds in duration, and there was no thought to getting out of the car to investigate further. One of the first things uttered was "We are never talking about this."

Looking at maps later the couple saw that the Union River waterway runs by Ellsworth before flowing into Leonard Lake and eventually dumping into the ocean. The couple surmised the river may have acted as a chokepoint creating a corridor of movement for the creatures if they were trying to stay in the woods. The hair was thought to be uniform in length all over the body and of brown or orangutan-like color, though they made the point that the streetlight backlighting the creatures had an orange hue to it. No details of the face, including eye shine, were able to be discerned. When asked about their prior thoughts of the Bigfoot phenomenon, they both said they were skeptics, and would have attributed the phenomenon to hoaxes and general bunk. The weight of the adult was estimated in the high 400s. The male witness stands 6 feet and weighs about 275 and he said the larger of the two was bigger than him. The smaller creature was about four feet tall.[93]

#65   2007 * Freedom, Waldo County * winter
This account occurred on February 28, 2007, at about 9 a.m. A school bus driver reported seeing a bipedal creature while on her way home in Freedom. The creature had a huge, hairy body. It was about to step over a snow bank but suddenly turned quickly and disappeared into the woods heading uphill. The witness said it was at least 8 feet tall with hair color ranging between light brown to gray. It reportedly stepped in the same tracks going back up the hill as it made when it had come out of the woods. She later returned with family members to examine the tracks. The footprints were sixteen inches long and spaced about 7 feet apart.[94]

#66   2007 * York County * summer or autumn
Very early on a September morning, the witness who was at home, heard a whooping call from outside. The witness returned this call out the bedroom window and noticed the call emanating from outside got closer. Feeling apprehensive, the witness next made a noise like a hissing cat and the whooping noise from outside receded.[95]

#67   2008 * Piscataquis * autumn
One hunter, among a party of three camping out in October, heard
two instances of sounds like wood on wood knocking in the early
morning. The wood knocking reminded the witness of a documentary claiming such sounds as Bigfoot behavior.[96]

#68   2009 * Freedom, Waldo County * autumn
Nearest town to this sighting is Freedom, on Route 137. On December 1, the reporter was slowing down to look at a couple of trucks
for sale. Pulling in front of the trucks, he noticed a tall, hairy creature cross the road and enter the woods. The witness proceeded to
drive to the spot where he saw the creature enter the woods and
got out to investigate, walking into the treeline. He could hear distant branches breaking and caught glimpses of movement but was
too far in to make it out clearly. He estimated that what he had
seen was about 7 feet tall and covered in dark hair.[97]

#69   2009 * York County * winter
In the waning days of 2009, a group out for a walk caught sight of
something high up in a tree that looked unusual. They videotaped
the apparent animal and loaded it onto YouTube in early 2010
under the title "What is it? Maine Tree Creature! Sasquatch, Bigfoot, monkey, ape!"

The hikers can be heard discussing what they've found, including wondering if it is a bear and one individual exclaiming that it
looks like a monkey. *The Sun* ran a story on January 12, 2011, with
the headline, "Bigfoot: Is this the Proof?"

The North American porcupine, (*Erethizon dorsatum*) has an
extensive range across much of the northeastern United States and
the vast majority of Canada, and was often cited as the likely subject in the Maine Tree Creature video.

#70   2009 * Aroostook County * summer
A bicyclist observed a "black hairy ape" walk across the trail about 50
yards away in early July. The biker turned around after watching the
thing, described as seven to ten feet tall, and quickly pedaled home.[98]

#71   2010 * Somerset County * spring
On May 17 the respondent was engaged in looking for a dog in the
woods, when chancing to look to the right, saw "a huge black musky
smelling thing about 10 feet away with big yellowish red kinda
orange, [sic] with long arms with what looked a lot like hands, and
long hair probably two inches long and was wide—probably two and a
half feet wide was on its knees first looking at the ground then
looking straight at me with a weird smile like it was trying to tell
me it was going to attack me . . ." The respondent ran back home.[99]

#72   2010 * Leeds, Androscoggin County * season unknown
A girl and her friend reported seeing a 6-7 foot tall creature completely covered in fur cross the Line Road moving north to south
in a bipedal fashion. The creature's arms hung low and remained
low while they swung. The creature did not appear to look at the
pair as it moved off the road heading in a southerly direction.[100]

#73   2010 * Leeds, Androscoggin County * winter
Leeds is a town in Androscoggin county Maine. It borders the
Androscoggin River, Androscoggin Lake, and is north of the
Lewiston Auburn area. Leeds boasts the only French-speaking summer camp in the United States, Camp Tekakwitha. Greene is south
of Leeds and the Line Road is between these two towns. Livermore
is to the north and Turner is to the west.

    The Line Road stretches from Quaker Ridge Road to the south-
east to connect at a three-road intersection with Wiley Road and
Old Lewiston Road and is approximately a mile and a half in length.
Traveling on I-95, Leeds is southwest of Bangor by 94 miles and is
about 55 miles north of Portland.

    On February 8, 2010, a couple reported something like Bigfoot
crossed the Line Road, near the Greene-Leeds town line, in front
of their vehicle. The man reported the creature as something hairy,
about seven-feet tall and bipedal. The woman in the car estimated
the creature's height at 6 ½ feet. The couple was interested in pro-
tecting their identity, afraid that they would receive ridicule for
reporting such a thing. The Red Roof gas and convenience store in

Leeds was a focal point for gossip about the sighting. Loren Coleman visited Line Road to investigate the couple's claim that something matching the popular description of Bigfoot walked in front of their car. Coleman has referred to the boggy area as the Turner Triangle.[101] It contains the towns of Turner, Turner Center, Greene, Leeds, and Durham, where several cryptids have been reported over the years, from mystery cats to Bigfoot. Of the recent Bigfoot sighting, Coleman dubbed the creature the Leeds Loki.

Coleman and investigator Jeff Meuse visited and investigated the area, interviewing five individuals who discussed what they had heard or seen along Line Road concerning unusual animal activity, including reports of Bigfoot-like hominoids.[102]

This section deals with plotting the various sightings and encounters from Chapter 8, all of which appear marked on a state map for a single holistic view of sighting locations. Maine's sixteen counties are then broken down denoting seasonal encounters reported within their respective borders for further specificity. Observation of the full Maine map shows that many Bigfoot sightings are reported at, in, or near water.

Table A shows Maine's counties by population. Table B shows the number of Bigfoot sightings by county. Piscataquis County, the least populous in the state, has one of the highest number of sighting reports. Table C shows, when specific information is available within a report, the number of Maine sightings by season. Summer is significantly in the lead as the top season for sighting reports. Summer coincides with several wild fruits ripening in the state, including blackberries and blueberries; several reports reveal berrying as the activity of the witnesses.

Lastly, Table D notes sightings by year in which they are reported to have occurred, showing the increase of reports as a trend moving upward over time. The spike in the mid-1970s is owing to the number of encounters with the entity that became known as the Durham Gorilla.

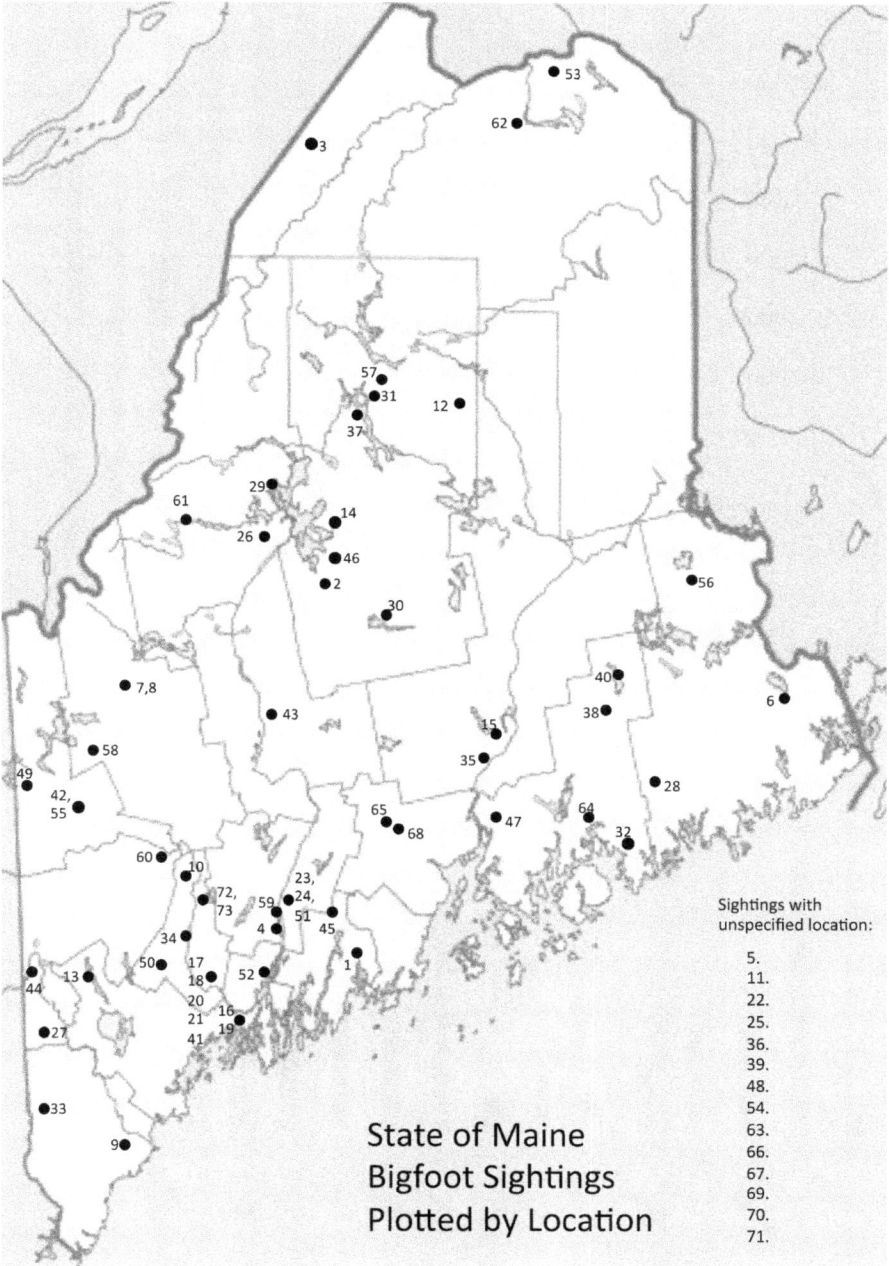

● 53

62 ●

● 3

57
● 31
12 ●
37

29 ●
14
61 ●
26 ●
46
2

30

7,8 ●
40
38 ●
6 ●
43 ●
56 ●

58 ●
15
35
28 ●

49 ●
42,
55 ●
65
68
47 ●
64
32

60 ●
10

72,
73
23,
24,
59
51
4
45

34 ●
50 ●
17
18
52
1

13 ●
44
20
21
16
19
41

27 ●

33 ●

9 ●

Sightings with
unspecified location:

5.
11.
22.
25.
36.
39.
48.
54.
63.
66.
67.
69.
70.
71.

State of Maine
Bigfoot Sightings
Plotted by Location

Maine Bigfoot Sightings Map

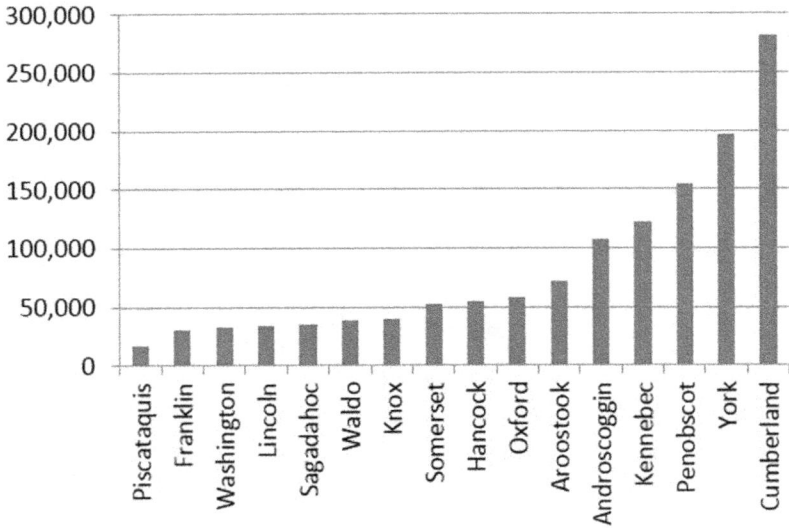

Table A

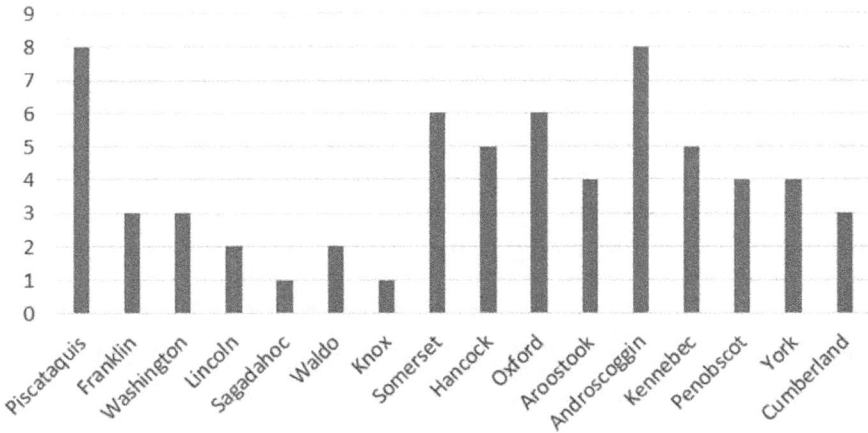

Table B

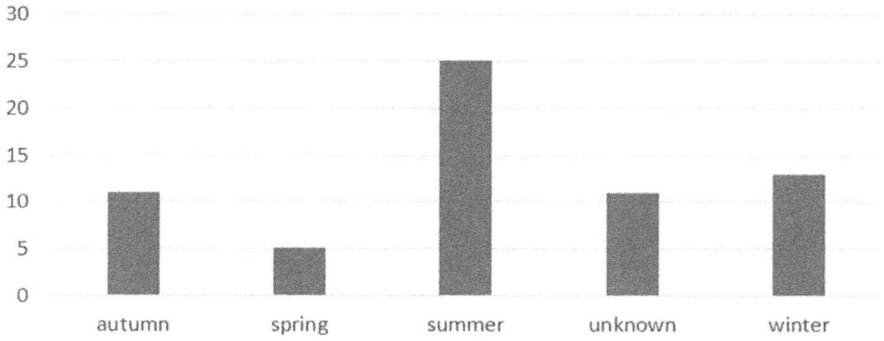

Table C

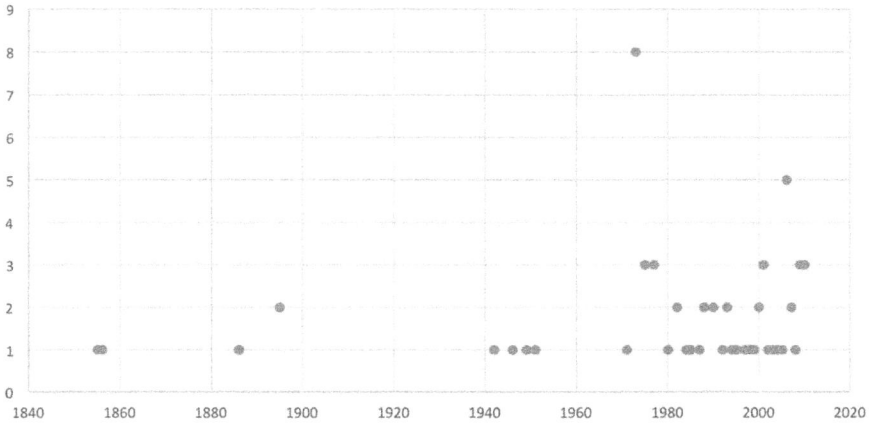

Table D

ANDROSCOGGIN COUNTY

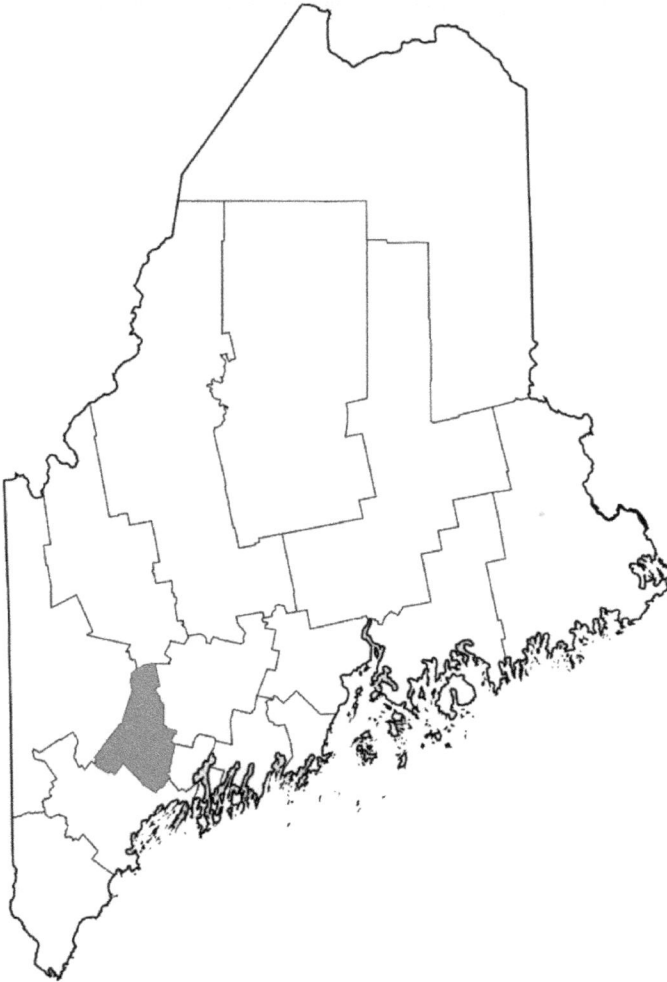

Androscoggin County is in southern Maine, and though one of the State's smaller counties, it has over 8% of the population. Referred to as the *Heart of Maine*, Androscoggin County is home to the Lewiston Auburn metropolitan area, Bates College, and the Androscoggin River on its way to the ocean. Androscoggin is an Abenaki term meaning "river of cliff rock shelters."

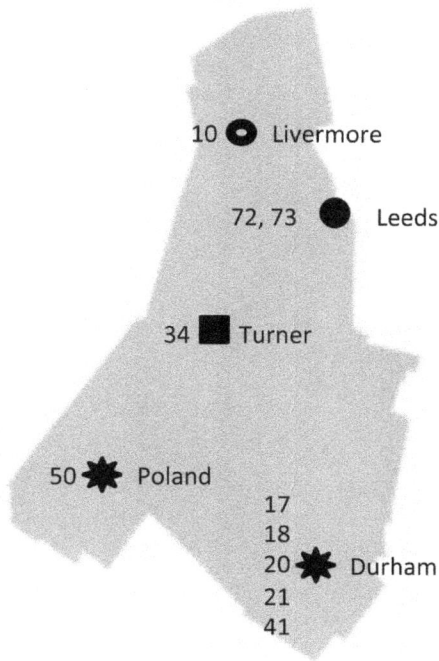

SEASON LEGEND

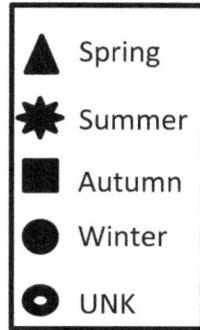

▲	Spring
✹	Summer
■	Autumn
●	Winter
◉	UNK

Sighting #	Year
10	Early 1970s
17	1973
18	1973
20	1973
21	1973
34	1986
41	1990
50	1999
72	2010
73	2010

Daniel S. Green

AROOSTOOK COUNTY

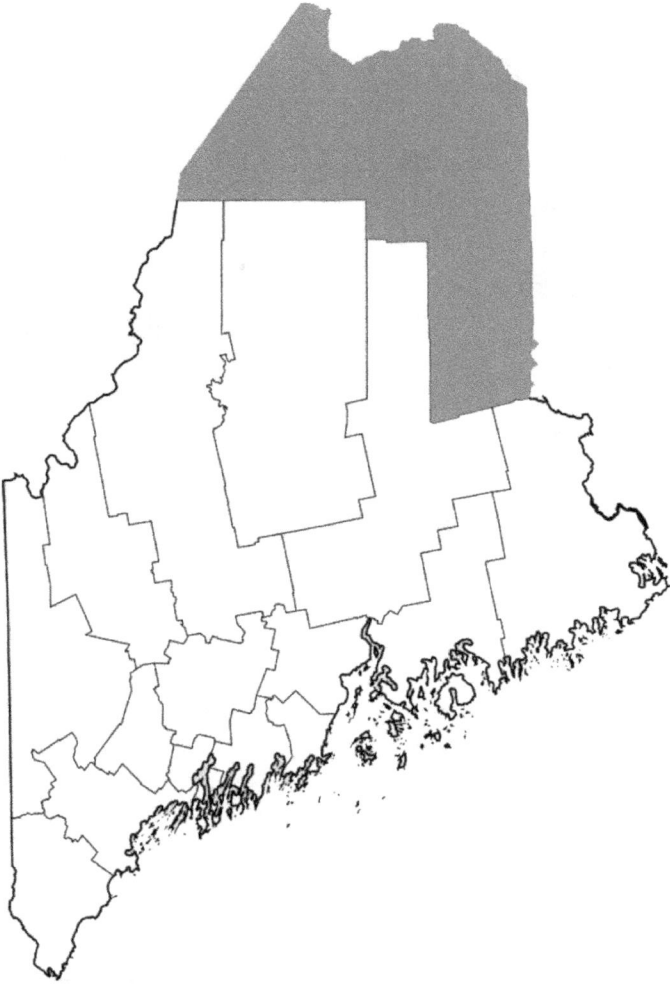

Aroostook County, known in Maine as "The County," is the largest American county by land area east of the Mississippi River. Aroostook was once known nationally for its potato crops. Aroostook has more than 2,000 lakes and streams and is home to the Allagash Wilderness Waterway.

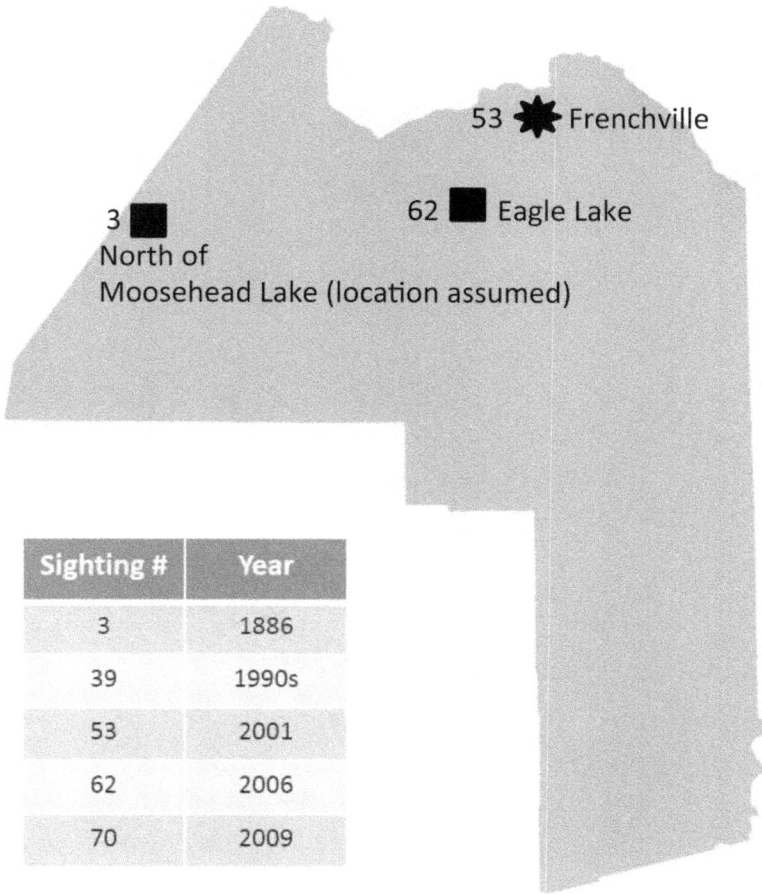

53 ✳ Frenchville

62 ◼ Eagle Lake

3 ◼
North of
Moosehead Lake (location assumed)

Sighting #	Year
3	1886
39	1990s
53	2001
62	2006
70	2009

39 ▲ Merchant's interview – precise location withheld

70 ✳ Precise location and time of year not given

CUMBERLAND COUNTY

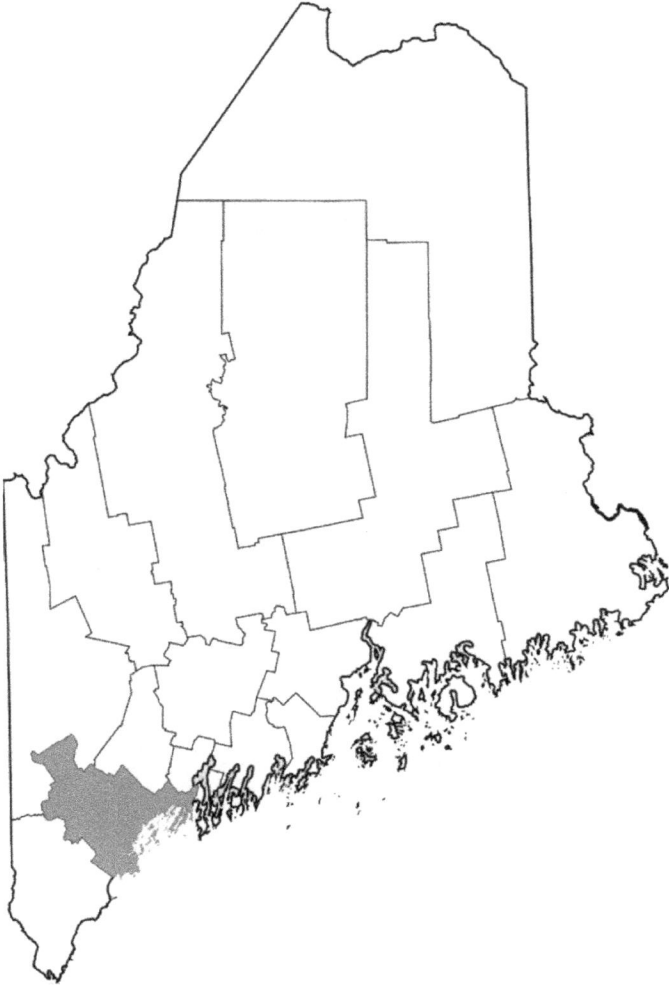

Cumberland County is considered one of the most affluent of Maine's sixteen counties. Over 1/5 of the state's population resides here. Cumberland is home to Portland and is the economic center of the state.

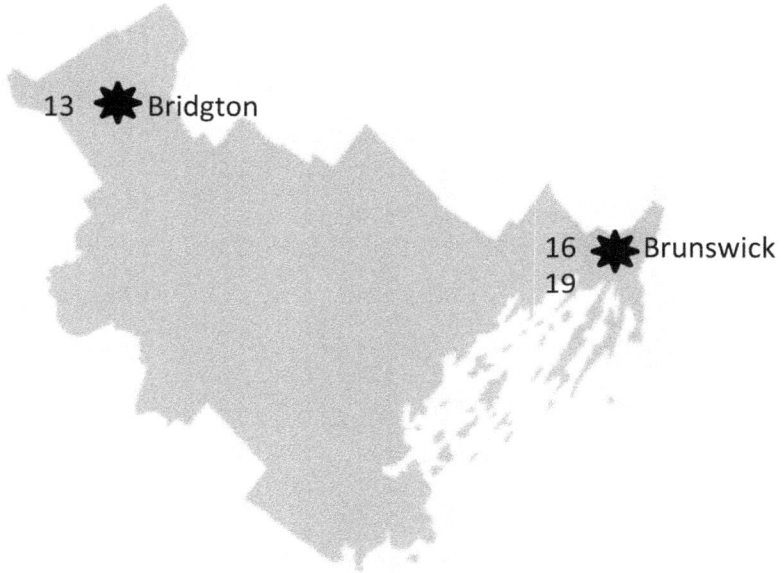

Sighting #	Year
13	1971
16	1973
19	1973

FRANKLIN COUNTY

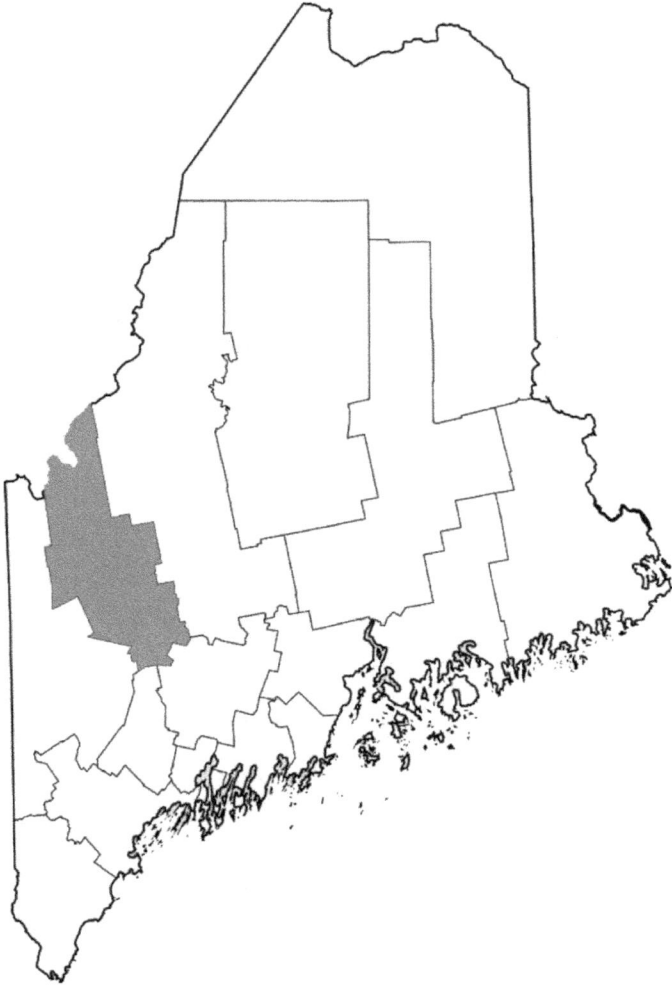

Sparsely populated Franklin County is known for its mountain ranges, glacial lakes, and streams. Farmington is the largest town and home to the University of Maine at Farmington. The Rangeley Lakes region is located here below Saddleback Mountain, whose 4,000 elevation provides unique alpine skiing for thousands every year.

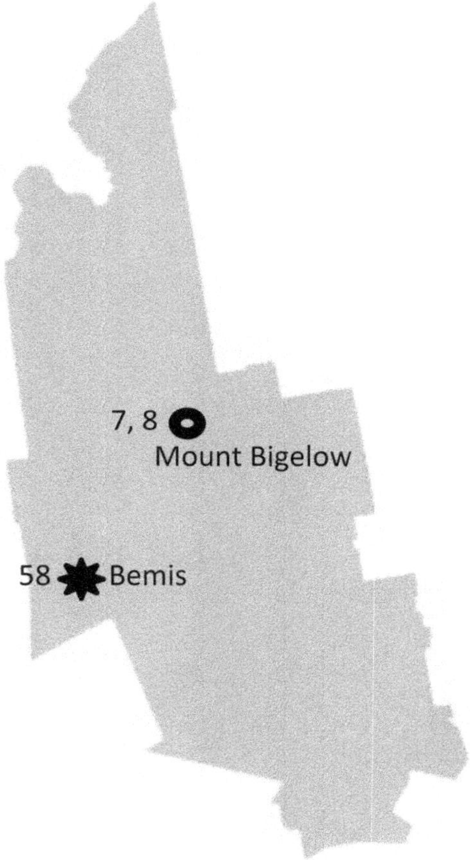
7, 8
Mount Bigelow
58
Bemis

Sighting #	Year
7	1946
8	1949
58	2004

HANCOCK COUNTY

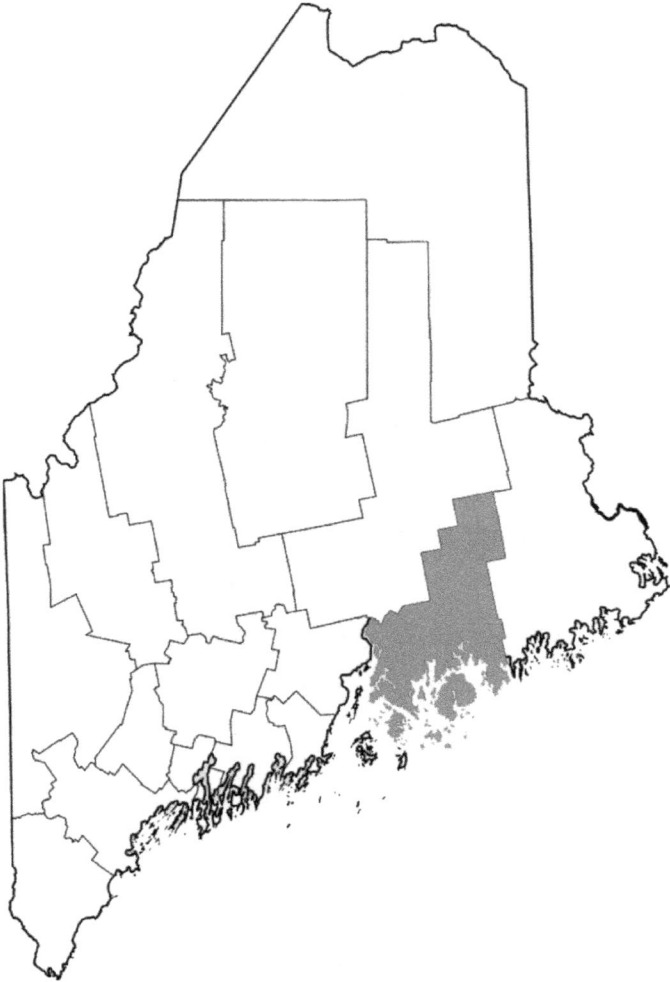

Hancock County's coastline is the longest of any Maine county. Acadia National Park, the only national park in New England, is here as well as Maine's Maritime Academy at Castine. The county is named for John Hancock, Declaration of Independence signer and first Governor of Massachusetts.

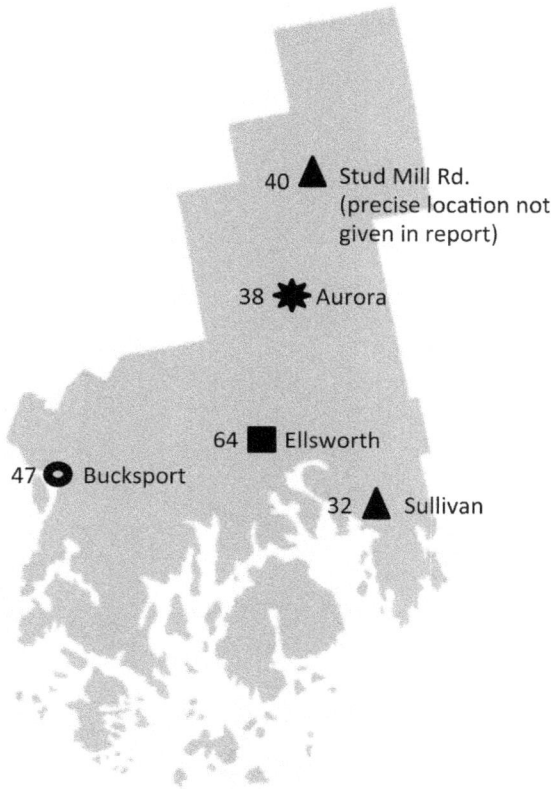

40 ▲ Stud Mill Rd.
(precise location not
given in report)

38 ✳ Aurora

64 ■ Ellsworth

47 ◉ Bucksport

32 ▲ Sullivan

48 ⬤ G.C.B.F.R.O. report; precise
location not given

Sighting #	Year
32	1984
38	1988
40	1990
47	1996 or 1997
48	1997
64	2006

KENNEBEC COUNTY

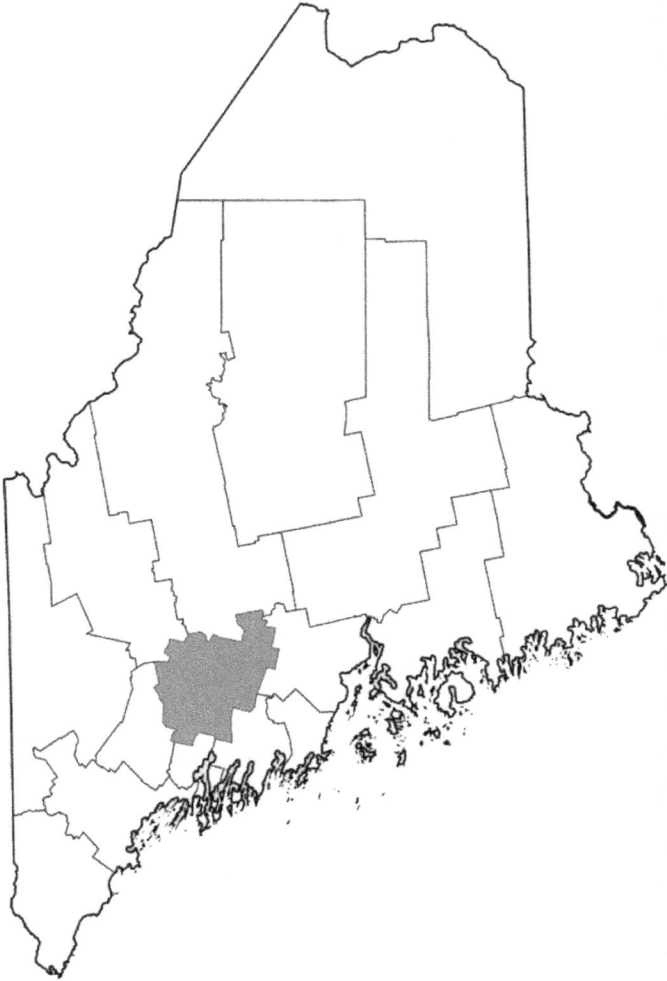

As early as 1607, the area of Kennebec County was explored by English settlers from the Popham Colony and later by English from the Plymouth Colony in 1625. Augusta became the county seat in 1799 and designated the State Capitol in 1827. Fort Halifax Historic Site is located in Winslow and Old Fort Western is located in Augusta.

23
24 ■ Manchester
51
59 ● Gardiner
4 ✹ South Gardiner

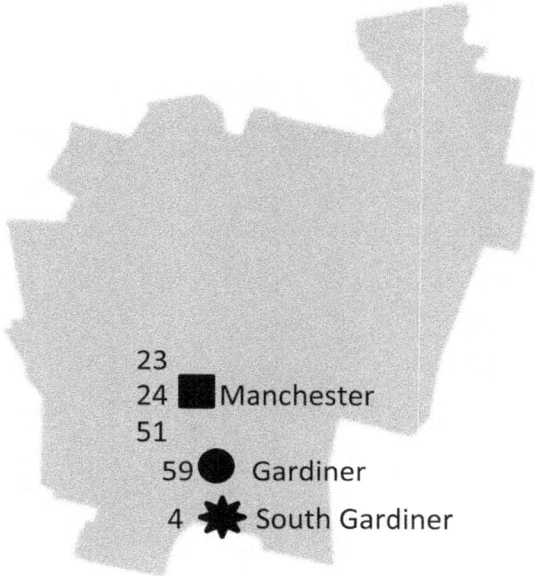

Sighting #	Year
4	1895
23	1975
24	1975
51	2000
59	2005

KNOX COUNTY

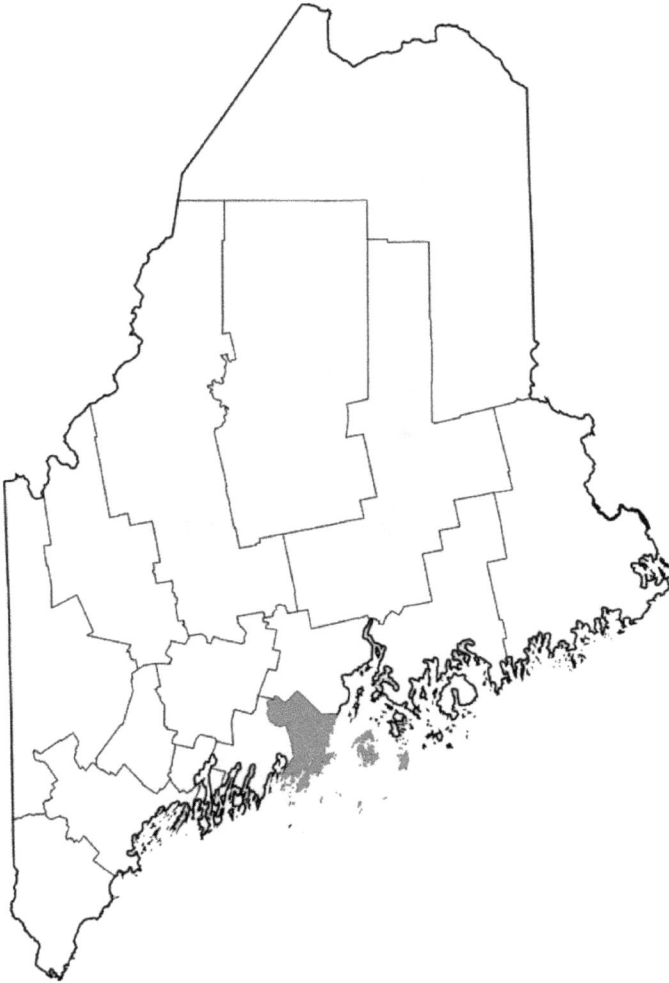

Named for Revolutionary War general and Secretary of War Henry Knox, Knox County has the distinction of being the newest county in the state, having been created in 1860. There are numerous islands in Penobscot Bay which are part of Knox County and home to many historic lighthouses.

63 ● Precise location and time of year not given

Sighting #	Year
63	2006

LINCOLN COUNTY

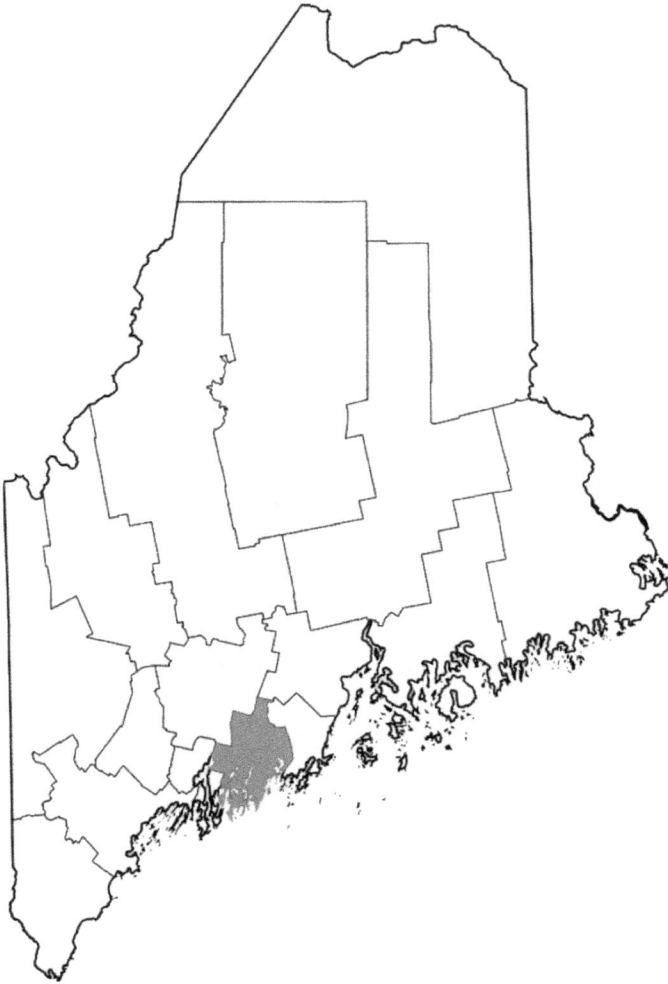

Lincoln County looks out over Boothbay Harbor and its many lobster boats, fishing trawlers, and windjammers. Wiscasset is the county seat, a historic center of shipbuilding which was the busiest seaport north of Boston until the embargo of 1807 halted significant trading activities with England.

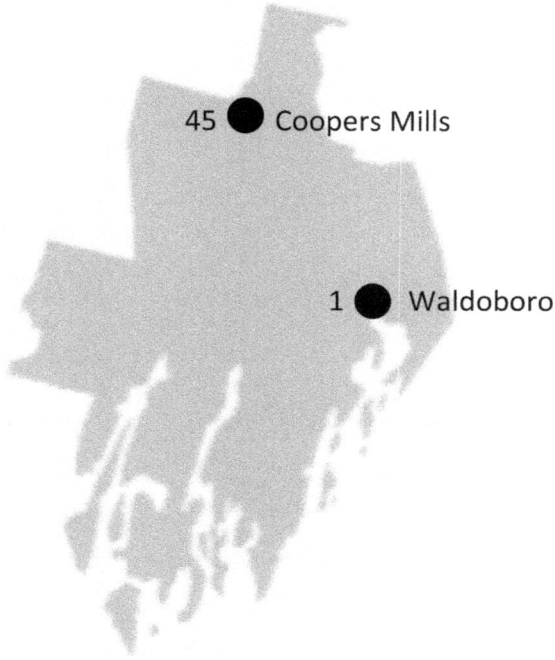

Sighting #	Year
1	1855
45	1994

Daniel S. Green

OXFORD COUNTY

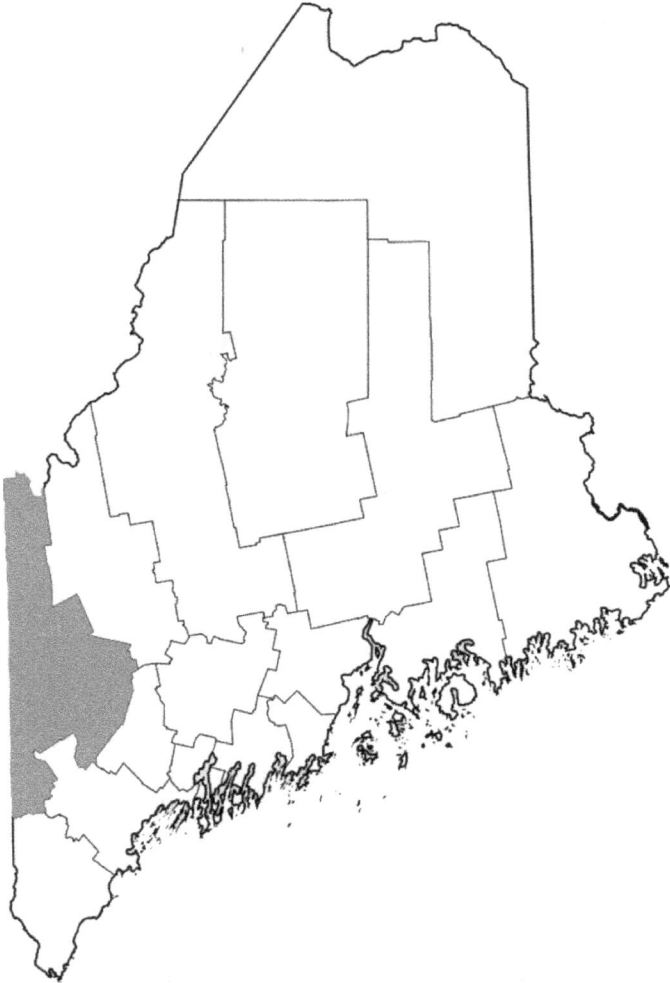

Oxford County boasts part of the Appalachian Trail, the Androscoggin River, Grafton Notch State Park, and White Mountain National Forest within its borders. The Oxford Hills are eight towns in the Western Foothills amidst ten lakes and ponds, including Lake Pennesseewassee.

Sighting #	Year
27	1977
42	1992
44	1993
49	1998
55	2001
60	2006

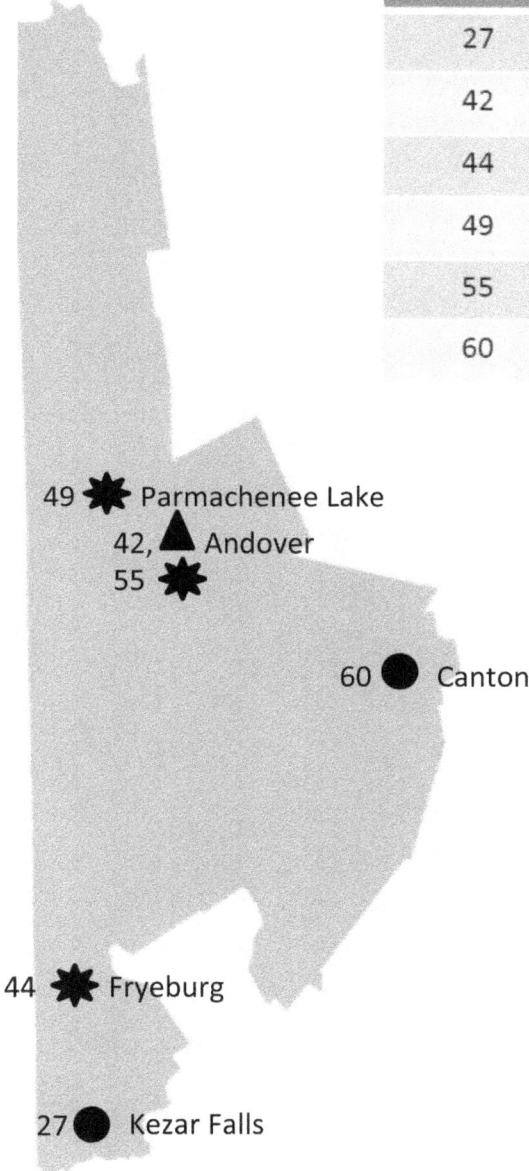

49 Parmachenee Lake
42, Andover
55
60 Canton
44 Fryeburg
27 Kezar Falls

PENOBSCOT COUNTY

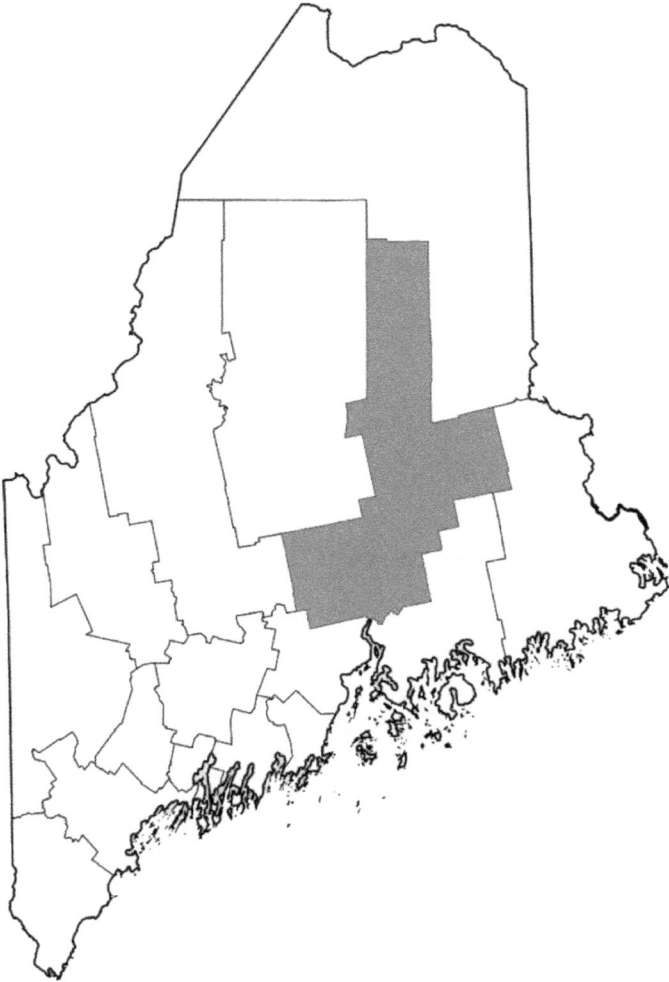

Penobscot County's greater Bangor area is the largest metropolitan area north of Portland. Maine's longest river, the
Penobscot River, runs through the county. Orono is the home of
the University of Maine.

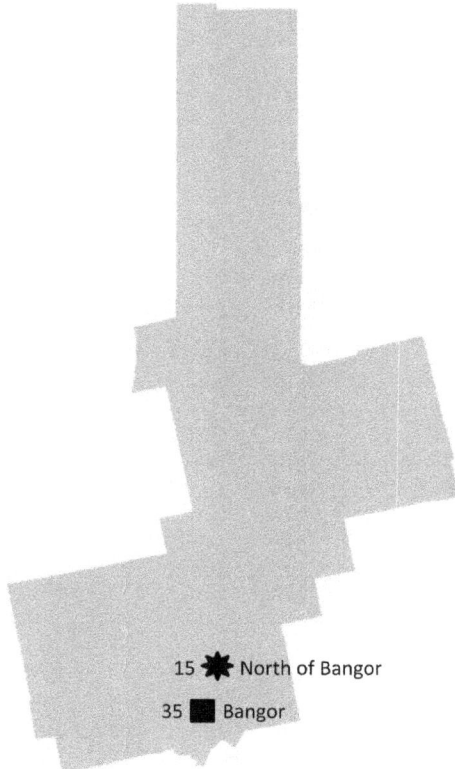

15 �֍ North of Bangor

35 ■ Bangor

5 ■ 1895, presumably north of Bangor

36 ✷ "Northern Maine woods,"

Sighting #	Year
5	1895
15	1973
35	1987
36	1988

Daniel S. Green

PISCATAQUIS COUNTY

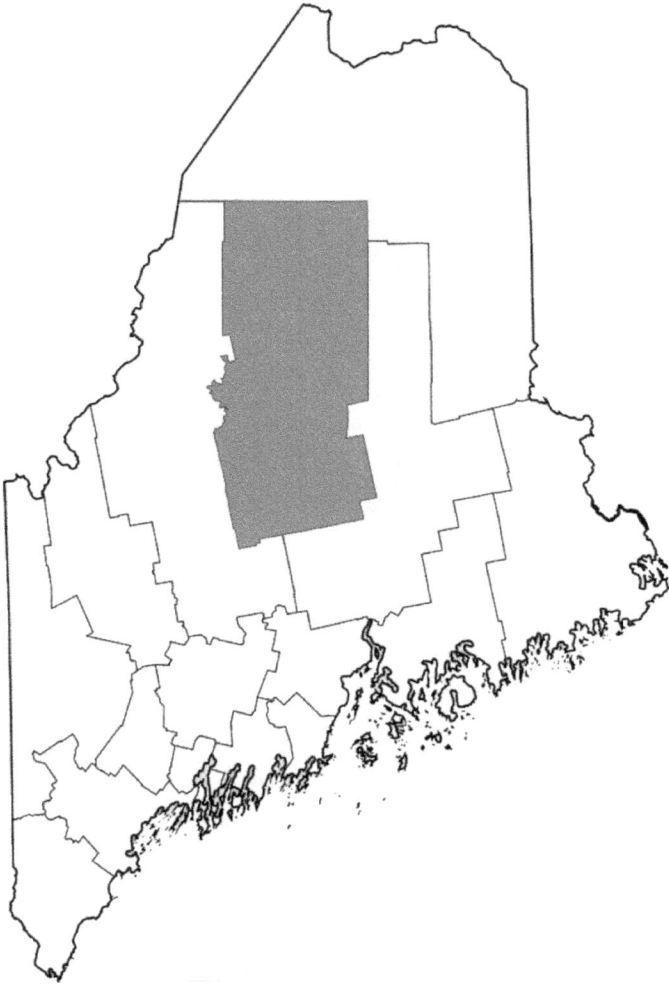

Piscataquis County is roughly the size of Connecticut. Hikers travel to Brownville to try Gulf Hagas, a series of gorges and waterfalls referred to as the Grand Canyon of Maine. Mount Katahdin of Baxter State Park stands a mile above the surrounding forests and is the tallest peak in the state.

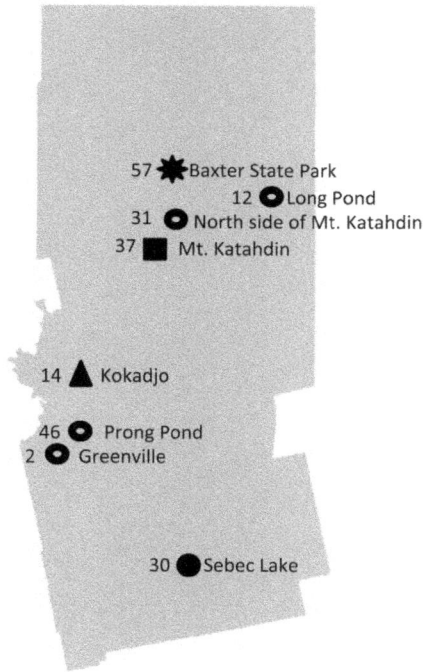

57 ✳ Baxter State Park
12 ◐ Long Pond
31 ◉ North side of Mt. Katahdin
37 ■ Mt. Katahdin

14 ▲ Kokadjo

46 ◉ Prong Pond
2 ◉ Greenville

30 ● Sebec Lake

67 ■ Precise location and time of year not given,
Presumably fall given encounter occurred while hunting

Sighting #	Year
2	1856
12	~1970
14	1973
30	1980
31	1982
37	1988
46	1995
57	2003
67	2008

SAGADAHOC COUNTY

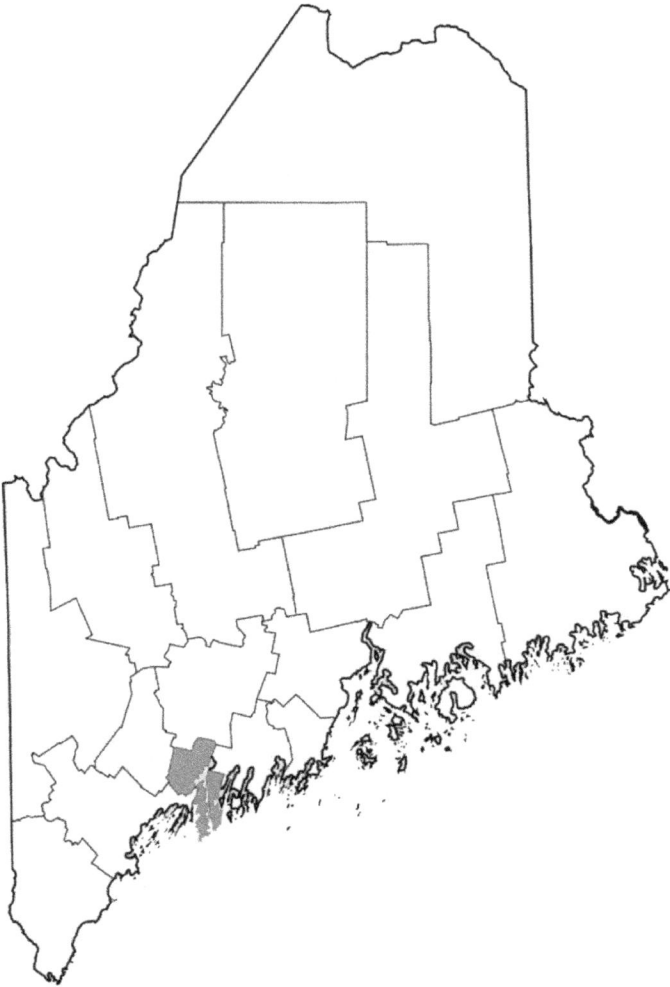

Bath is the County Seat of Sagadahoc County. Sagadahoc is an Abenaki term meaning "mouth of big river," referring to the Kennebec River, explored in 1605 by Samuel de Champlain. Bath Iron Works, founded in 1884, is a major American shipyard.

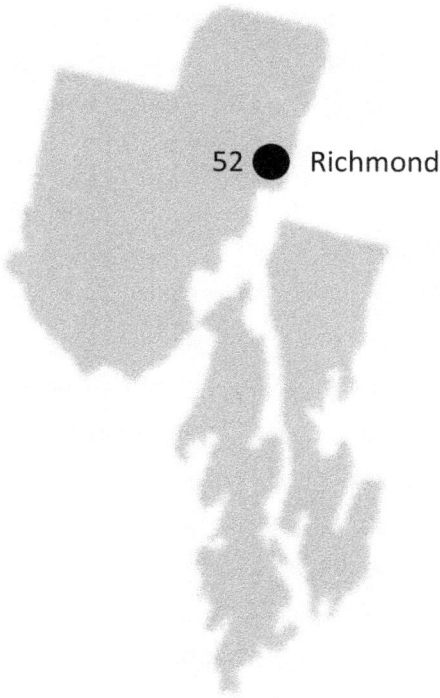

52 ● Richmond

Sighting #	Year
52	2000

SOMERSET COUNTY

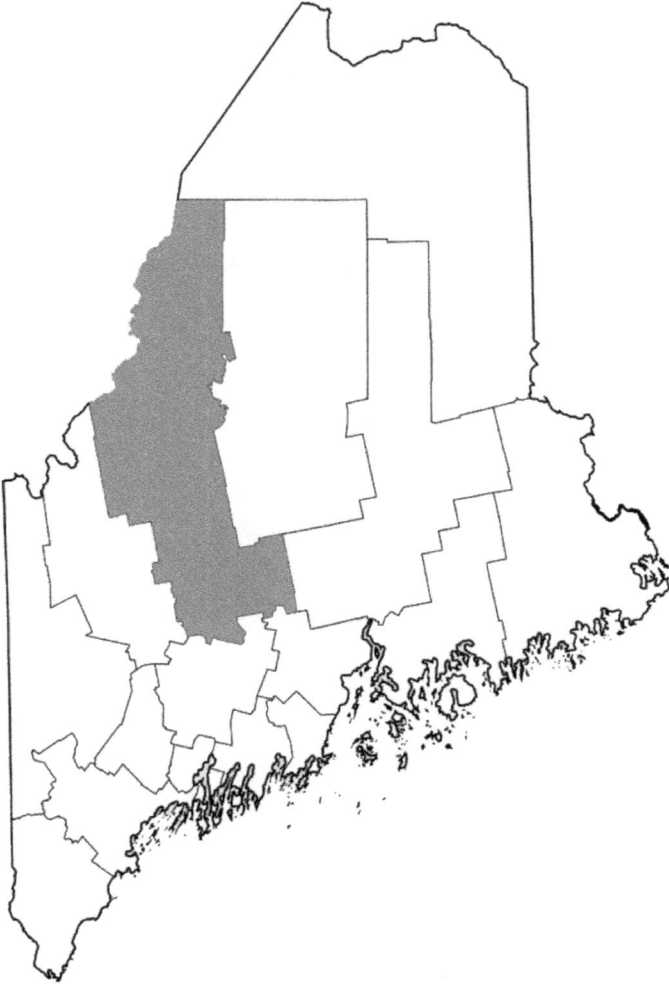

The Jackman-Moose River Region is one of the state's premier four-season areas. Jackman is known as the Switzerland of Maine, the first to get snow and the last to lose it. Benedict Arnold traveled through this region en route to Canada during the American Revolution. Rockwood is located on the west end of Moosehead Lake and is a gateway to the north country.

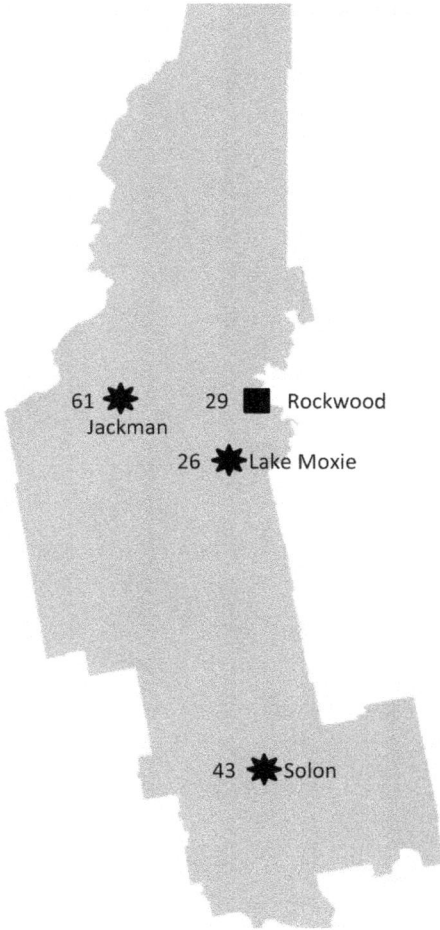

Sighting #	Year
25	1975
26	1977
29	Before 1978
43	1993
54	2001
61	2006
71	2010

61 ✹ Jackman
29 ■ Rockwood
26 ✹ Lake Moxie
43 ✹ Solon

25 ● Exact location unknown
54 ✹ Exact location unknown
71 ▲ Exact location and time of year unknown

WALDO COUNTY

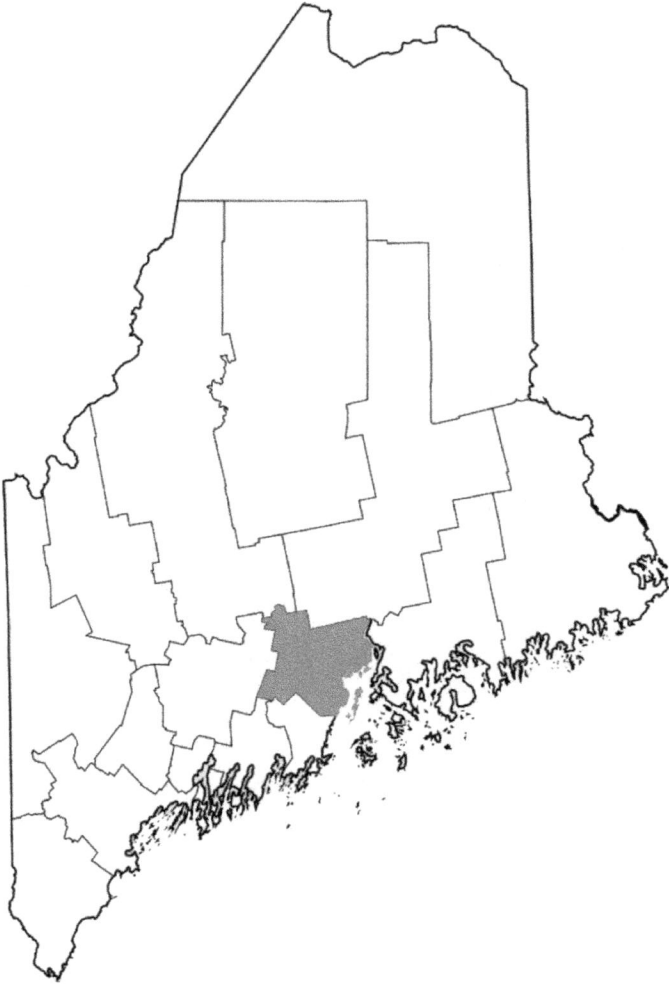

Waldo County is situated in mid-coast Maine, looking out over Penobscot Bay. The Penobscot Narrows Bridge & Observatory is the highest observatory in the Western Hemisphere. The original Fort Knox, in Prospect, was started in 1844, and was the first fort in Maine built of granite. Belfast, located at the mouth of the Passagassawakeag River is the county seat.

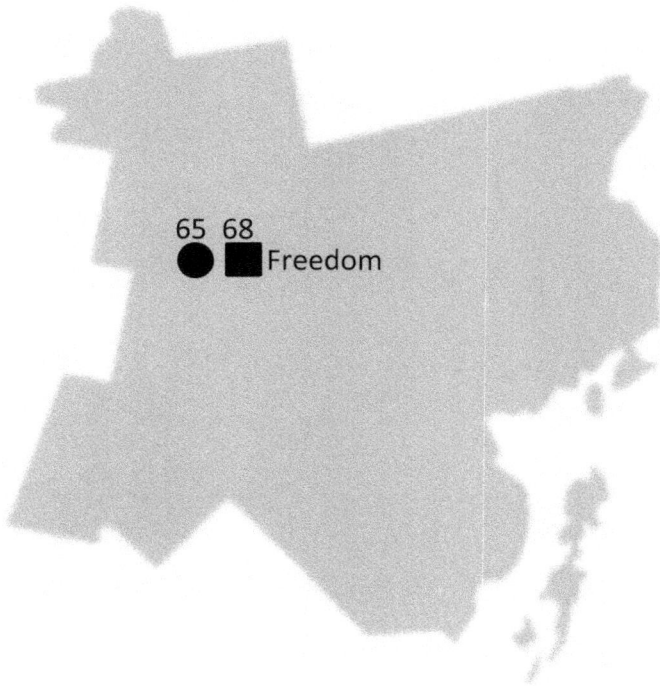

Sighting #	Year
65	2007
68	2009

WASHINGTON COUNTY

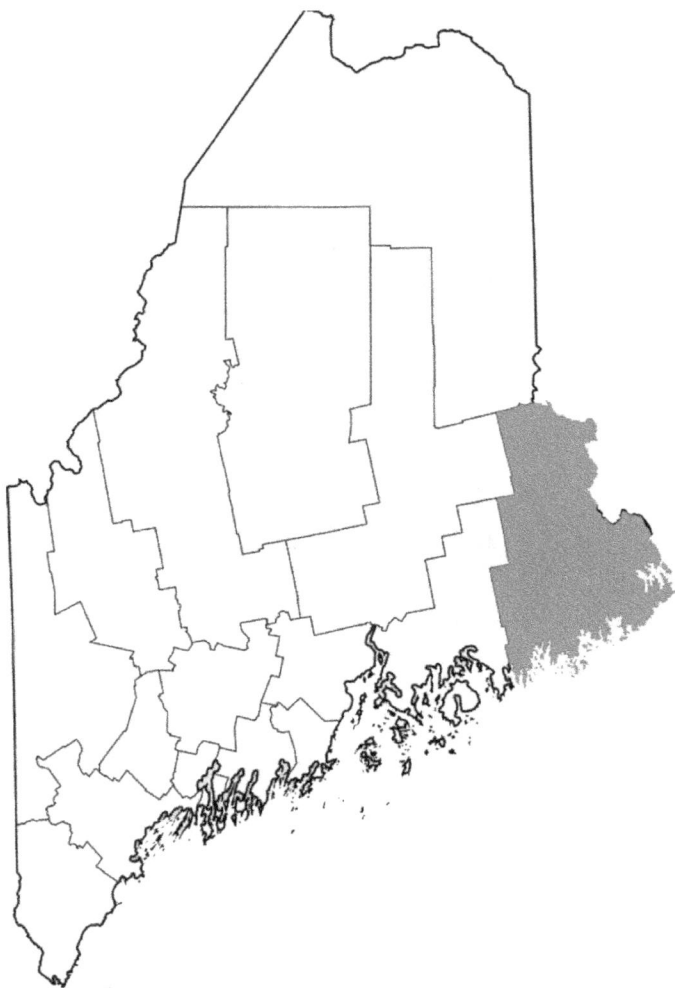

Washington County's Passamaquoddy Bay, an inlet of the Bay of Fundy, witnesses the highest rise and fall of tides in the continental United States. Moosehorn National Wildlife Refuge is an important habitat along the Atlantic Flyway for many bird species, while Cobscook Bay State Park near Whiting offers 800 acres of woods for exploring and roaming.

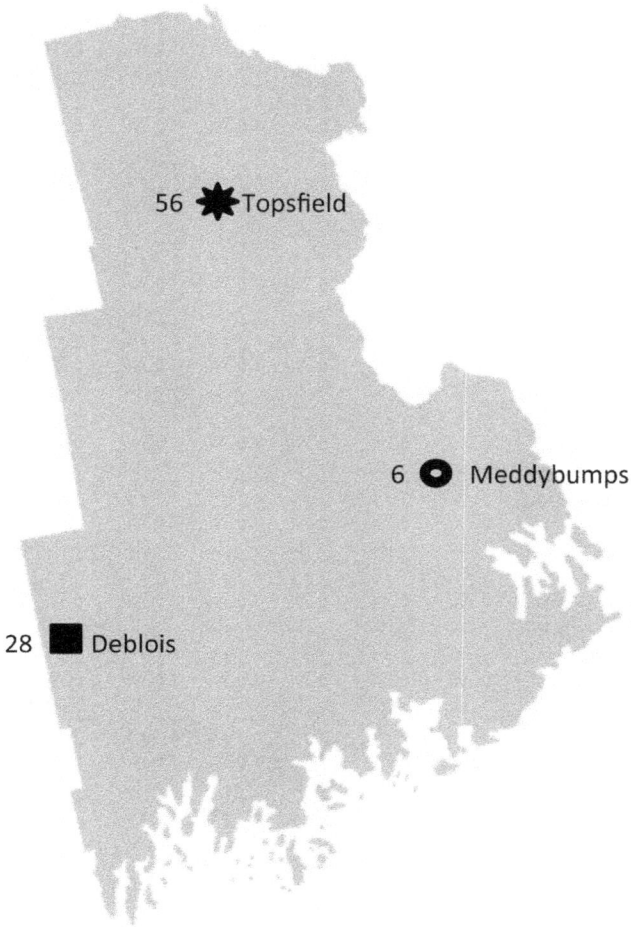

56 ✳ Topsfield

6 ◉ Meddybumps

28 ◼ Deblois

Sighting #	Year
6	1942
28	1977
56	2002

YORK COUNTY

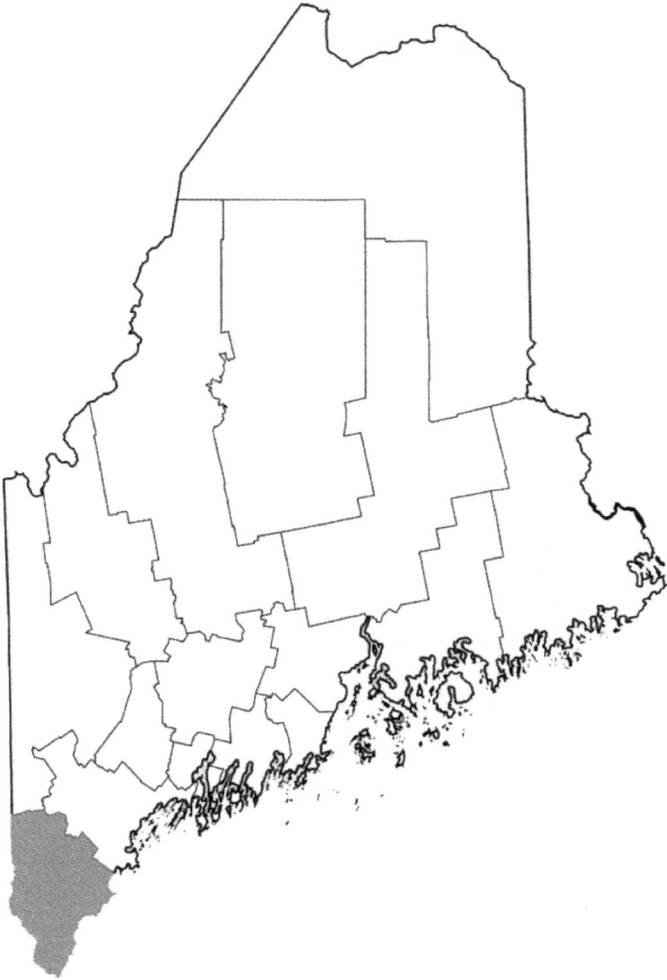

Alfred, first settled in 1764, is the county seat of York County. Old Orchard Beach is here, named for an orchard planted by Thomas Rogers in 1638. The Nature Conservancy manages the 1,600-acre Kennebunk Wildlife Management Area, a habitat for several species of plants and animals of special concern to Maine.

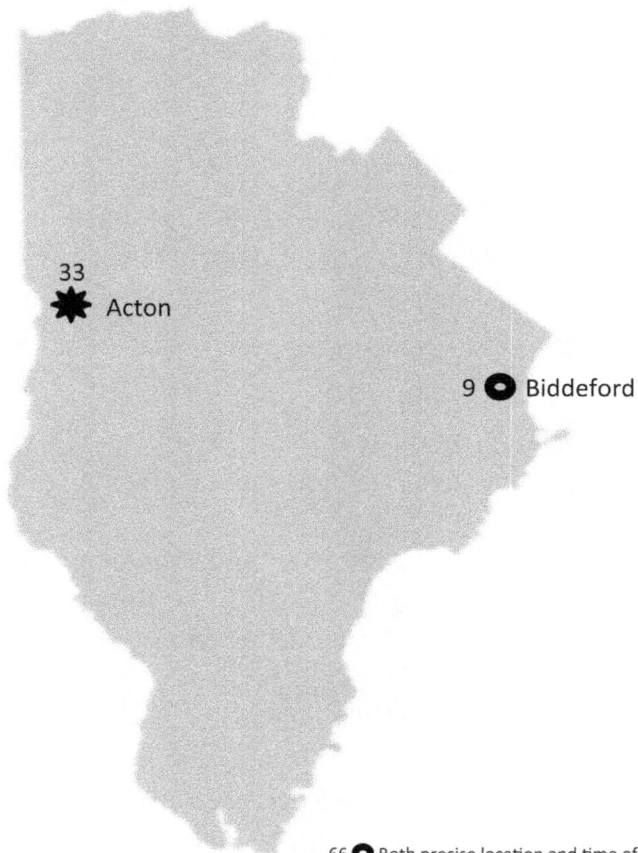

33
Acton

9 ⬤ Biddeford

66 ⬤ Both precise location and time of year not given

69 ⬤ Location unknown

Sighting #	Year
9	1951
33	1985
66	2007
69	2009

PERSONALITIES AND EXPEDITIONS

CHRIS JULIAN

"Nobody ever directly looked in Maine, solely in Maine. That's my deal,"—Chris Julian[1]

Chris Julian hails from the Maine Highlands and began researching Bigfoot in Maine in 1993. He started the now defunct blog *Bigfoot Research of the Northeast* in 2001 and *Maine Bigfoot Research* in 2006, "solving the mystery, one step at a time." As he wrote in the blog description, "The idea is to have a place where Maine people can report sightings and they can be investigated by people who live in Maine and know the area and are available to get to the (site) a.s.a.p. without losing valuable time!"

Julian's interest in Bigfoot started when his father took him to the movie, *Sasquatch, the Legend of Bigfoot*, which he cites as the point where the bug bit him. "My biggest dream was to go to the Pacific Northwest, go to western Canada and check this out. You read the books, and especially with Maine, it's been slim pickings as far as sightings go that were recorded. It was enough where it got me thinking, we could have one here. As you look and hear stories, a lot of the times the stories you hear, someone is not going to come right out and say, 'Bigfoot,' 'Sasquatch,' they've got funny names for it, especially here in Maine. I'm sure it happens everywhere. But there've been key phrases that I've asked people about, just names that have been passed along to me, and it's like opening up a floodgate. You almost just have to have the right key."

Julian joined Bob Coyne and Mike Killen as the first guest on the evening of November 27, 2007, for the premier episode of the now defunct Blog Talk Radio show *Bigfoot Quest*. He related an experience he called "unnerving." On Saturday, November 10, he had spent the day in the Rangeley, Maine, area with a reporter from the *Sun Journal*. It was a 350-mile-round-trip day and he got home Sunday at 1:00 a.m. and started unpacking his equipment (camera tripods, flares, parabolic microphone, lanterns, sound equipment to "call blast" Bigfoot sounds into the forest, and motion-sensitive trail cameras).[2]

Julian was standing with his fiancée on his porch when he heard what sounded to him like a woman screaming, "the loudest most terrifying scream, and then all in the same breath it turned into a howl," coming across his lawn from the woods. Julian reported he had no close neighbors. His initial reaction was that his head was spinning, the sound was so loud that he could not compute what he was listening to. He retrieved his recording equipment that he had started to unpack. While he did so his fiancée heard two more howls and saw movement of a large shape beyond the tree line and eye shine reflecting in the porch lights. Hearing nothing else, he left his recording equipment running.

The next morning he listened to his audiotape and DVR. The audiotape picked up nothing but the DVR had picked up a howl, not as close or even the same type of howl, but roughly similar minus the scream. For the next three nights Julian recorded sounds that in their totality did not make him feel comfortable. The initial noise induced a reaction to run and was one of the most unsettling noises he had ever heard.

Julian sent Bob Coyne four WAV files of his recordings that Coyne worked on and subsequently featured on the first airing of *Bigfoot Quest*. One of the files that Bob didn't play on the show, due to the amount of background clicks and scratches, reportedly contained two screams and wood-knocking sounds. A comment from the chatroom offered the opinion the howl didn't sound deep enough to be a Bigfoot and may perhaps have been a bobcat in heat. A subsequent caller suggested the howls sounded similar to coyotes. Julian mentioned he had lived in Maine his whole life and had heard coyotes many times before and, if he had thought the sounds were from coyotes, nobody else would have heard his recordings. Julian also had a picture of a healthy tree about four to five inches in diameter snapped right in half (no other trees near it were damaged) that occurred at the spot in the tree line on his property where the howls had come from.

Julian collected over 55 Maine Bigfoot sightings, marking each with a pin on a map board. "There's a lot here. I don't think they are all happening at once, I think they've been happening all along." "It bugs me to no end when somebody in Maine reports a sighting to someone in California. Report it right here, where it can actually be investigated."

Julian racked up about fifteen years looking for Bigfoot in Maine when he moved out of state around 2008. Citing a general aggravation with the effort, he stated he would pursue a more low-key approach. The Georgia Bigfoot Hoax[3] was pointed to as an example of distracting stunts and hoaxes that made the effort harder for those trying to honestly research the subject.[4]

MICHAEL MERCHANT

> "I believe people are reporting a phenomenon, a biological phenomenon. I don't believe they're all lying, I don't believe they're all confused. I think there are valid reports out there."[5]

One of the premier researchers and media personalities on Bigfoot in Maine is Michael Merchant. From his Discovery Bio: "An experienced wildlife tech, Michael leads animal biologists on hikes and frequently backpacks and kayaks." The 2010 edition of Discovery Channel's *Out of the Wild* was filmed in Venezuela and Michael was one of nine volunteers to participate in the show. The cast was required to travel 70 miles out of Venezuela back to civilization after a three-day crash course in both jungle and area-specific survival.

Michael has a degree in biology from the University of Maine, and cites work experience with the Maine State Department of Agriculture, Conservation and Forestry. His posted videos regarding Bigfoot including recorded Skype interviews with Mainers who have had encounters with what they believe to be Bigfoot, receive thousands of views. Beginning in early 2011, using the moniker SnowWalkerPrime, he has uploaded many videos on the Bigfoot subject.

One of his earliest posted videos concerning Bigfoot expressed the view that it would not be surprising that there are Bigfoot sightings in Maine given that the state is so heavily forested. He goes on to note that there are relatively few Bigfoot reports in the state. He mentions the Leeds' Line Road sighting of 2010 in an area he is familiar with (having conducted some marsh bird surveys in the area, providing him with intimate knowledge of the large swampy areas of that location). Merchant also mentions the account of a bowhunter in Auburn who had shot a deer and was paralleled out of the woods by something he thought was Bigfoot and the hunter's conviction that he would never go into the woods again unarmed. This early video also begins his stated philosophy on the creature espousing a no-kill position but more than happy to taser it in order to get the right samples and data including tagging the creature for further study, essentially a "catch and release program for Bigfoot."

"I believe it is a biological puzzle. A mystery. A real flesh and blood critter that doesn't levitate, isn't capable of telepathy, doesn't live in another dimension. It would have to be a mega-primate. One of the few remaining megafauna that survived the Quaternary extinction event 11,000 years ago. If that's true, given its high level of intelligence, it very well may have adopted a religion centered on avoiding contact with man. I can even understand why we're not finding bodies of this critter in the forest. It's rare. Why are we not getting photographs of it? It's extremely rare."[6]

346   Daniel S. Green

Michael's take on critical standards with Bigfoot research: "I think we need to raise the bar about the blobsquatches. I think it needs to be recognizable, relatively clear, the back story needs to be believable, you know, the puzzle pieces have to go together."[7] This view was further exemplified by his comments on the "Hovey photo"[8] from the Bigfoot Tonight Show (March, 2012), "It could be anything. It could be Bigfoot—I don't think it is. But, it might be. But, how would you know? There's no other information, there's no other data, there's just the one photo."

Merchant also has espoused a healthy skepticism, unwilling to accept photos, films, and accounts minus corroboration, supporting details, and unbiased analysis. "I am skeptical not so much of the existence of an undiscovered existence of a primate on this planet, as I am of people's motives. We need more than blobsquatches, in my opinion, to establish the existence of this thing. We need to be able to somehow duplicate the results, not to have a sighting with Sasquatch be a once in a lifetime occurrence, but something that you can follow the beast and understand the man-animal. If it's a biological creature we should be able to get our hands on it, we should be able to take photographs of it, it will leave behind trace evidence, physical tracks. When it goes through the forest, it will affect the forest. This is not a ghost; we're not looking for Casper, it's not a unicorn."[9]

Merchant has entered into many major stories of Bigfoot with his opinions. On the 1967 Patterson Gimlin film and the film subject's walk, Merchant talks about the calf of the film subject rising almost parallel to the ground in contravention to a normal human walking gait. Merchant presented a pair of contrasting views for this, one from the believer's camp, that a mid-tarsal break is responsible for the unusual walk. He then suggests the contrarian point that "clown shoes" could also lead to the walk. It could be a guy in a suit, and an approach from a scientific, logical point of view would suggest it is a guy in a suit. Merchant finishes with saying to his eye the gait is relaxed, that something very realistic to him about the footage—follow-on attempts to duplicate the walk from the footage come off with a clumsy look and feel.

Merchant: "Pretty much all the Bigfoot sightings that have been reported in the state of Maine are of smaller individuals, six foot, maybe six foot five, to my knowledge there's no 8-, 10-, 12-foot sightings in the state of Maine. I would be really surprised, if Bigfoot was a real critter, that we don't have it in the state of Maine, by all accounts we should have it here."[10]

Merchant was a participant in the first season of the 2014 Spike TV show, *$10 Million Bigfoot Bounty* hosted by Dean Cain.[11] Writing of the show before its premiere, Merchant commented, "If you want to be entertained with a high value production embracing awesome locals, and many surprises, then watch the show."[12] Merchant and his Maine teammate Kat McKechnie made it to the final two teams (each episode of the show saw the elimination of one of the nine teams), before ultimately losing to researchers Stacey Brown, Jr., and Dave Lauer.

The Blog Talk Radio show, *Chattahoochee Bigfoot Radio,* featured Michael Merchant and Ms. Kathy Villareal on October 6, 2013. They discussed an unusual incident that they had recently experienced.

> KATHY VILLAREAL: I had a strange encounter when he (Michael Merchant) took me out. He's taken me out camping before. We live in a wooded area, so I've seen plenty of moose and I've seen at least five bear with him.
> MICHAEL MERCHANT: Not a single one which has attacked you.
> VILLAREAL: He's always saying that he's saving me from these wild animals.
> MERCHANT: I protect her from the bears.
> VILLAREAL: That's what he claims (laughing). But, this was different. He actually was taking me for a whole Bigfoot trip. We hiked out there and it was quite a hike. At one point in time, Michael left, because we were going to be meeting up some friends, so he took off. And Jason took off. And I was left

alone for a while at the campsite. But then I got really tired there. So when the guys came back I wanted to go for a walk and get away from everything. So, what I did, I was walking up a stream and I was walking really slow because I wanted to take pictures of the bear, so I had my camera with me. And I was walking up this stream very slowly, taking pictures, and quiet. On my right, I saw basically a black figure run, just out of the corner of my eye, really fast so that really startled me. So I looked to my right and nothing was there. So I thought I just made that up, I didn't really see it, and that was just strange. My heart was just racing at that point in time. So I just thought well that's really silly. I forgot about it and kept on walking.

Eventually, about half an hour later I was walking and there was a whistle. Like a whistle that somebody does to get your attention. Michael does a similar whistle to me when we're out in the woods. I knew that wasn't a bird. So, I heard this whistle, and I knew the guys were back at camp, and I knew I was alone and I knew where we were was extremely remote, there was nobody else besides us there. So that really kind of spooked me. I just stopped. I realized it was something intelligent. I stopped in my tracks, I turned around and I started walking back. And that was the hardest thing I've ever done is to turn my back to something. I had no idea what was in front of me in the woods because I couldn't see it. I thought, you know what, I was a little bit foolish for walking away from camp.

I started walking back slowly, I didn't want to run because I didn't want anything to chase me. Luckily, Michael had been worried about me and he had been following me, he had been tracking me. So I came upon him. First off he spooked me because he was in his ghillie suit so I didn't see him right off the bat. He had a gun on him, too, and he kind of showed me that and I was like don't show me that, cover that up. Because I felt like, that's not even going to help, that's how I felt. It was one thing when I saw that black figure, and it was all black, there was no white skin on the face, nothing like that. It was definitely like a male because I could see the core of the body. I didn't even know what I was thinking. I knew when I saw something that fast, I knew the power of what I'd seen. It was really kind of spooky. I just told him don't show it (Michael's gun), because I felt like we were observed. I knew we were observed. I knew from the fact that something apparently paralleled me, and then went ahead and got in front, I guess. So that's my first experience with something that was strange.

And in addition to that, I forgot to tell you, there was also some really big footprints out there. I took a picture of one, and there were three more. I don't consider myself an expert or anything, I was just looking at them, I was going to talk about them with Michael later. Of course all that went out of my head at this point.

So I saw something, I heard a whistle, there were footprints. That was the first real experience of anything really unusual in the woods since he's been taking me.

After that, on other trips that we went out, for instance, I was with another person there, and another time with my son, and what I would hear out there, I didn't see anything, but what I'd hear was almost like really garbled male voices, but not English, not any kind of language. . . . I couldn't make it out, so that was very strange. I don't know what that is. But it comes from that section of the woods where I had seen the figure. But I'm not bold enough to go walking into the woods to go investigate it by myself (laughing).[13]

BILL BROCK

Bill Brock grew up in Ohio and West Virginia before moving to Maine in 1999. He caught the Bigfoot bug early, watching the Patterson-Gimlin film and pouring through all the reading material he could find. The outdoors enthusiast eventually moved to Durham to be closer to the epicenter of the Durham Gorilla sightings. Bill Brock's YouTube page says: "Bigfoot Hunter, Survivalist, Hunter, Fisherman, Trapper, Jack of all trades. Bill Brock is also the Northeast Leader for *TheSasquatchHunters.com*." As Brock told the *Sun Journal*, "I just like chasing cryptids, I didn't care if I'm chasing a leprechaun or Bigfoot."

During an outing purposefully looking for Bigfoot, Bill Brock and Michael Merchant recovered a 14″-long cast from a footprint found north of Moosehead Lake in early July 2012. The cast broke due to a lack of sufficient casting material. Bill Brock shared the cast for display at the International Cryptozoology Museum later that month.

Brock and Merchant found 14-inch footprints with 4′ 1/5″ strides at Chain of Ponds in Northwest Maine and they made several casts from this trackway. Other activity in the area of the tracks included scratch marks on trees and unidentified animal scat. The casts were loaned to the International Cryptozoology Museum for display. The website *Cryptomundo* cited an unnamed Maine Guide and certified tracker who leveled the following opinion on the tracks and a video Brock made of their discovery:

> I've looked at the casts, and there are some that certainly look like moose tracks to me. Especially where it appears as though they tracked this up a hill.

Moose will often step back foot into front while going uphill, as do deer. However, I cannot say how
the curvature to the track would occur, other than
the deepest depression would be in the center, where
the front and rear hoof overlap. I will say, however,
that my first thought upon seeing the photo with the
human foot for comparison is that it looked like the
rear paw of a bear to me.[14]

Brock offered, in part, this response to *Cryptomundo's* Chain
of Ponds story:

I had a Maine Biologist look at the tracks at the site
as well as after the casts were made. The Biologist
said it was not a Moose, as far as a Bear I also see a
bear track in the cast. So I don't know what we have
here. The only issue with me calling this a Bear is
that it has a legit big toe . . . and the other casts match
up to this one very well.[15]

On 15 December, 2012 Bill Brock encountered a series of intriguing tracks behind the Jones Cemetery near the Brunswick/
Durham line (setting for the Durham Gorilla flap in 1973). The
tracks were found paralleling a creek and measured around 14″
long and 6″ wide.[16]

Brock starred in a monster-hunting television show called
"Monsters Underground," which debuted on the Discovery Chan-
nel in September 2014. The first season comprised six episodes
filming the underground search for cryptids by the team of Brock;
Chris Onik, a doctor from Portland; Brock's Bigfoot-hunting friend
Craig McGhee; and Jeremy "The Beast" Bates, a professional boxer.

"Nobody else is doing it, I know that—to film underground is
not a joke," Brock said. "There were times we'd have a 50-foot drop
right next to us and nothing to hold onto and it was slick as glass.
It was a dangerous show to be on."[17]

MAINE GHOST HUNTERS (MGH)

The Maine Ghost Hunters teamed with Michael Merchant (representing Team Tazer Bigfoot) from August 3 to August 5, 2012, for a Bigfoot-hunting expedition. According to the initial news release, the location selected was an "undisclosed and remote wooded area in the deep woods of Northern Maine. True Bigfoot country, by all rights." After the obligatory stop at Dysart's,[18] the consolidated party of Maine Ghost Hunters and members of Team Tazer Bigfoot (Michael Merchant and Bill Brock) entered the North Maine Woods at Fish River Checkpoint.

Camp set-up was done by 3:30 in the afternoon and the group formed two parties to scout locations for their night investigation. Merchant and Brock each lead a group and provided pointers on field data collection, and reading tracks and signs. Areas of sustainability providing food, water, and shelter were key search criteria the group focused on.

Meeting back at camp, discussion about their respective scouting efforts ensued leading to finalizing the strategy for the expedition's first night. With excitement mounting as the evening hours wore on, two avian members of the group—roosters brought by Team Tazer, would be set out as lures to bring Bigfoot in. During the first night the party made wood knocks and a few calls but the eventual denouement was no takers on the free rooster bait.

The second day incorporated track casting at a water hole including prints of a Canadian lynx. A trail camera was set up at this water hole. During the second night, after beset by mosquitoes, and no encounters on the books, the party members got into their vehicles for a search of miles of lonely woods roads. This activity terminated about midnight and everyone returned to camp for a rest before an early morning pack-up and start of the (for some) 300-mile journey home.

Before heading out, the party retrieved the trail camera from the water hole and hooked up the camera to a Mac back at the campsite. Of four videos captured, none starred Bigfoot. This field trip was an interesting pairing featuring Merchant's thoughts of Bigfoot as a biological entity while the Maine Ghost Hunters were rooted in paranormal explanations and experiences.

"Because he's not a species of animal that's been proven to be in existence . . . we could say it's in the realm of the paranormal," Tony Lewis of Maine Ghost Hunters said. Loren Coleman commented on their expedition, "I think the possibility for Bigfoot in northern Maine is slight," he said in assessing the chances of success for the expedition. "If there are reports up there, and there are some sightings, that's probably some interlopers from the boreal forests, the mountain range forests of Quebec and New Brunswick. There don't seem to be a viable living population in northern Maine. . . . You get those kind of outbursts of sightings and then they disappear so it doesn't seem as if there's any routine, breeding population of Bigfoot in Maine right now."[19]

BFRO-HOSTED EXPEDITIONS

Nick Maione, business owner and trained biologist, led the Bigfoot Field Research Organization's first Maine expedition in June 2008. Fourteen people signed up for the three-night event, reportedly paying $300 for the experience. The selected location was Mt. Blue State Park, which was kept a secret during the expedition's initial planning to fend off potential pranksters. Amidst clouds of black flies the group did find some curious stick structures, and maybe had an unsettling experience. "I kind of got this weird feeling to get out of the car," he said. "As I walked toward this snapped branch, this growl came from the woods. I backed away. It sounded pretty vicious."[20]

D. L. SOUCY

D. L. Soucy is an author and photographer from the Mid Coast Maine region. He is a leading force in the *Maine Bigfoot Research Group*. Along with cryptozoology, Soucy has written on such subjects as folklore, history, and preparedness. Soucy has created a number of videos on his time in the woods looking for Bigfoot sign, accompanied by voice-overs of his thoughts on Bigfoot in general and specifics of what he found during his outings. Some thoughts from his videos follow below:

"The Hunt for Bigfoot video published on September 12, 2012."

Hello everybody and welcome to hunting for Bigfoot. This is going to be a first part of a multi-part series of videos that I'm creating dealing with my hunt for this elusive creature and to determine how much of a probability there is that this thing is a real creature and not a figment of imagination. And I do this through various means of studying folklore and anecdotal evidence, Bigfoot sighting reports, and stories both of fiction and factual nature coming from the past. I've read stories from various newspapers and books that relate to this creature that historically has been referred to many, many times as the Hairy Man of the Woods. These stories pertain to a creature that fits the description what we today call Bigfoot or Sasquatch here in North America . . . actually have been around for hundreds of years, and wasn't until recently that people became known of his identity as Bigfoot and Sasquatch.[21]

"31 March 2013"
I'm happy to say, it looks like the Durham Gorilla, if that's what we want to call him, is back.

One of the attributes that we've heard of in many, many cases is a fetid odor coming from these Bigfoot creatures. And I'm wondering if maybe the odor that we're smelling is actually this skunk cabbage. When you crush the leaves you get an oil out of it that smells really, really bad.

Here's the prints, the first one that I found was about 16" long and about 6"-wide. Rather squarish in shape but did have the toe outline.

I've been reading up quite a bit. Which I think leads me to the reason why I was able to find what I

did so quickly. Once you learn the attributes of whatever it is you're hunting, it becomes much easier to
find them. And I don't think, from what I've seen in
the Bigfoot community, there aren't that many Bigfoot
hunters out there, that are actually physically hunting
Bigfoot. I think they're for the most part going out on a
nice afternoon or a nice evening and hoping to come
across something. All too frequently they turn the
things they find into something that it isn't. Again, we
have to figure out [how] to hunt them. And food sources
is probably one of the best ways to do that. We have to
find out what he's eating. And again, that brings the
skunk cabbage that I discussed earlier into the picture.
I think one of the things we need to look for is to look
for areas that we know are rich in these sorts of
plants. We wouldn't normally eat them, but they are
edible. In the old days the Indians ate them, so it's
reasonable to assume that Bigfoot eat them too.

They seem to be quite adept at reasoning situa-
tions if in fact they do exist. We really need to start
delving into our resources, to come up with a more
comprehensive means of finding this creature.

I think there's only two options here. Number one
is to come across a dead one in the woods and Number
2 is to capture a live one. And the only two ways we're
going to do that is to learn more about its habits and
its habitat. And that's what I'm trying to do here.[22]

Soucy writes of his approach to Bigfoot, "I address the subject
from a folklorist's viewpoint and try to come to an answer as to
what this cryptid is, and whether or not it really exists."

MAINE BIGFOOT IN POPULAR CULTURE

Strange Maine was published in 1986 and edited by Charles G. Waugh, Martin H. Greenberg, and Frank D. McSherry, Jr. It included the story "Longtooth" by Edgar Pangborn, originally published in 1970 by *Magazine of Fantasy and Science Fiction*. "Longtooth" also made it into the Maine-themed anthology, *Otherwordly Maine* (2008), edited by Noreen Doyle.

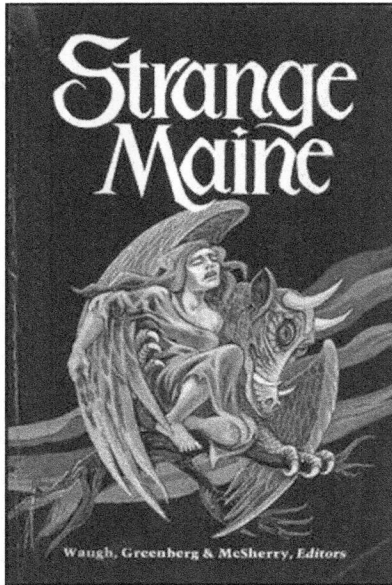

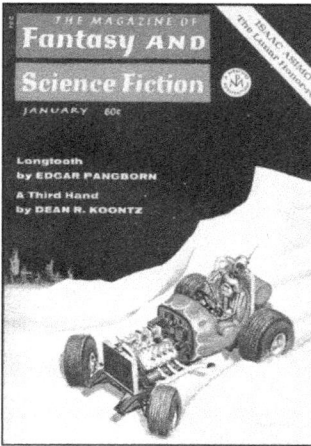

"Lonɠtooth" was oriɠinally published in the January 1970
Maɠazine of Fantasy and Science Fiction.

(riɠht) An Italian collection of Panɠborn stories
featurinɠ "Lonɠtooth" on the cover.

"Longtooth" tells the story of a Bigfoot-like creature haunting
a remote farm in Darkfield, Maine, where a young woman lives
with her older husband. Ben Dane has recently returned to his
home in Darkfield and visits the farm on a snowy night. The two
men discuss unusual experiences the old man has had at the farm
recently. This culminates with the sound of broken glass from the
woman's bedroom. When the men arrive they find the window has
been broken inward and the young wife is gone.

Pangborn was often cited as a compassionate writer, who could
infuse his stories with humor and empathy. "Longtooth" has a
melancholy tone and is unmistakably a horror story.

CROOKSTON BIGFOOT

Loren Coleman established the International Cryptozoology Museum in 2003, in Portland, Maine. The museum made *Yankee
Magazine's* "Maine Best Attractions 2010" list.[1] The International
Cryptozoology Museum's Mission Statement reads in part, "to
educate, inform, and share cryptozoological evidence, artifacts,

Loren Coleman with the Crookston Bigfoot
at the International Cryptozoology Museum.
(Courtesy Loren Coleman/ICM)

replicas, and popular cultural items with the general public, media, students, scholars, and cryptozoologists from around the world."

The International Cryptozoological Museum houses and preserves a collection of cryptozoological material, some of which Coleman first accumulated in 1960. One of the museum's features is the 8-foot-tall, 300-pound "Crookston Bigfoot," created by Wisconsin artist Curtis Christensen. The Crookston Bigfoot was added to the International Cryptozoology Museum's collection in 2004.

The genesis of the Crookston Bigfoot began in 1990, when Wisconsin taxidermist and artist Curtis Christensen decided to provide an outlet for his interest in Bigfoot by creating a life-size replica using the fur of musk ox and grizzly bear.[2] When it was complete the statue stood in his barn for several years.

In 1995, Crookston, Minnesota, attempted to proclaim itself "the Bigfoot capital of the world," following a spate of Bigfoot sightings. According to Don Holbrook, then executive director of the Crookston Economic Development Authority, Crookston wanted to create tourist interest in the town by establishing a Bigfoot museum. Though the museum would not become a reality, the Crooston Bigfoot statue found a home in the entry of RBJ's Restaurant on U.S. Highway 2 in Crookston. Kim Samuelson, RBJ's owner, purchased the Bigfoot statue, reportedly for $5,000.

Loren Coleman discovered that after almost ten years at RBJ's Restaurant the statue became the property of a tool company in Minneapolis that used it successfully at trade shows where it garnered much attention. Coleman obtained the statue and it undertook a one-week journey to Portland, Maine, where it resides as one of the many attractions at the International Cryptozoology Museum.

THE MAINE TARZAN

Some Wild Men are wholly contrived, whose tenures under the forest canopy have very definite and planned start and end dates. On August 14, 1913, Joseph Knowles, a *Boston Post* illustrator, went into the woods near Spencer Lake in Somerset County, armed only

Joseph Knowles, the Maine Tarzan before and
after his plunge into the Maine wilderness.

with the determination to prove that man could survive in the
wilderness of Maine while living in a primitive state. He was
essentially naked, and carried with him no weapons or food when
he trekked away from modern conveniences under a drizzly, over-
cast sky. He returned to civilization on October 18—two months
later, wearing moccasins laced with sinew and dressed in the skins
of a bear.

His plunge into the Maine wilds was done in the spirit of self-
reliance and old-fashioned toughness. Knowles was said to have
lost 30 pounds during his adventure. He was to be challenged on
his exploits however—the *Boston American* printed exposés claiming the bear skin was bought and said to contain bullet holes, and
it was held that Knowles knew the area and he was accused of

holing up in camps and eating well. Maine was the first of his wilderness forays, he did it again along the Oregon-California border and lastly in upstate New York. During the latter excursion Knowles was dubbed "Dawn Man" and accompanied by "Dawn Woman," Elaine Hammerstein, a Broadway performer.[3]

The main point of Knowles' adventure: Amid the anxiety of modernization and rapid changes inherent with progress, the wilderness was still important to American culture. During his solitary outing in Maine's wilds, Knowles sent out regular dispatches contrived by employing charcoal to write his missives on bark. The event was a stunt, developed by the *Boston Post*, which was then in a circulation war with the *Boston American*. Author Jim Motavalli states the *Boston Post's* circulation increased 30,000 daily copies thanks to updates on Knowles' progress.[4] Knowles' *Alone in the Wilderness*, telling the story of his exploits in Maine's backcountry, was published in 1913 and sold 3000,000 copies. "I'm still sorry I didn't manage to catch a bear cub, up there in Maine," he told the reporter from the *American Mercury* in 1936 (six years before he died). "It would have been a knockout—parading out of that timber a-leading a bear like a Scottie."[5]

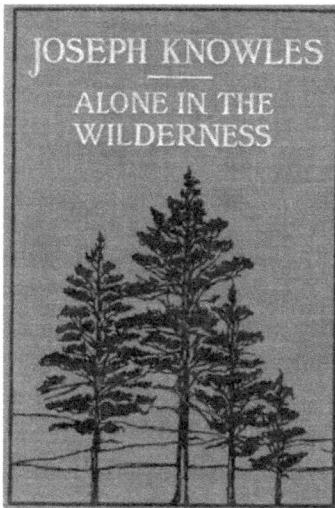

Alone in the Wilderness was first published in 1913 and is Knowles' own account of his experience with living in the Maine wilds.

Novelty Bigfoot patch.

The Border Mountain Militia patch inspired by the
Carolyn Chute book, *The School Heart's Content Road*.

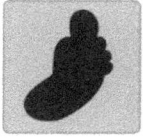

Made in Maine, USA

BIGFOOT INC.

Indoor Slippers for Outdoor Footwear

"Keep them by the door...
Keep dirt off the floor"

BIGFOOT SLIPPERS

Michael F. Thom is the designer behind the Bigfoot Shoe-in Slipper overshoe/overboot footwear that keeps garden mud, dirt and other debris off interior floors.[6] Billed as a durable, non-slip boot and shoe-in-slipper, this footwear is designed to quickly slip on and off. Sometimes affectionately referred to as the marriage-savers; a way to protect floor surfaces by covering wet and muddy soles. Made of 100% recycled wool. Bigfoot Inc. is headquartered in Belfast.

BIGFOOT STALKS THE COAST OF MAINE

Thomas A. Easton holds a Bachelor of Arts degree in biology from Colby College and a Ph.D. in theoretical biology from the University of Chicago. He was the book columnist for *Analog Science Fiction and Fact* magazine from 1979-2009, a contributing editor for *Biology Digest*, and a textbook editor with Scott, Forsman, and Co. Easton is a professor at Thomas College in Waterville, Maine. He was formerly on the faculties of the University of Maine and Unity College.

Bigfoot Stalks the Coast of Maine and Other Twisted Downeast Tales (2000) collects fourteen of Easton's stories previously published in magazines like *Argos* and *Amazing*. Easton sets his science fiction stories in an unnamed Maine town implied to be somewhere on Waldo County's coast. The mayor of this town, with his ever-present pipe at hand, has the uncanny knack for taking the very strange all in stride and figures prominently in many of Easton's stories. The use of this town as a recurring setting predates TV's *Haven* by around 30 years. The first story in this collection is

BIGFOOT

STALKS THE COAST OF MAINE
and other twisted downeast tales

THOMAS A. EASTON

"Mood Wendigo," published first in *Analog Science Fiction & Science Fact* Magazine in 1980. A Maine high school biology teacher, two of her students, and the mayor camp on a local hilltop to discover the truth about the wendigo.

> It screamed in the night, and anyone who sought the
> screamer disappeared without a trace. If they ever
> returned, they were mad, too blown of mind even to
> say what had happened to them.

The quartet confronted the mystery resulting in many more questions than they had when they started.

The second short story to note from *Bigfoot Stalks the Coast of Maine* is "Alas, Poor Yorick" originally published in *The Magazine of Fantasy and Science Fiction* in August, 1981. The story begins enigmatically, offering a hint of the soon-to-unfold discovery that some legends do indeed come to life:

> The world is full of ghosts, of myths and legends and
> bogeymen. Most are mere stories, vaporings of idle

minds, but some are real. Real enough to touch and talk to.[7]

A visiting anthropologist from Boston stumbles upon one of the town secrets and the resulting explanation by the mayor makes up the bulk of the story, exploring a theory held by some for explaining Bigfoot's identity.

JERRY CARDONE

Jerry Cardone gained the nickname "Dinosaur Man," a moniker coming from the myriad homemade pop culture figures and creations adorning his Houlton property that includes a large wooden gate with the words, "7 Wonders of God Creatures." Among the visionary art are Santa Claus, aliens, flying saucers, and dinosaurs. Cardone pointed out a piece of his art to a field reviewer for *RoadsideAmerica.com*, a towering Bigfoot bearing a garish smile of rubber teeth and covered all over in moose skins. "Jerry says that he doesn't believe in Bigfoot, and he doesn't believe in Santa or dinosaurs either, although they're among his most frequent subjects."[8]

Jerry Cardone

Cardone did garner some critics who complained about the accumulation of stuff on his property. Cardone's artwork has also attracted positive attention both locally and nationally. Photos of his work taken by Portland photographer Tonee Harbert were part of a University of Southern Maine exhibit titled "Off the Grid: Maine Vernacular Environments."[9]

BIG AND HAIRY

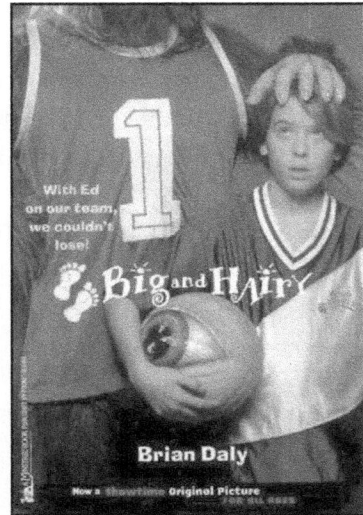

Big and Hairy was published by Pocket Books in 1994 by Portland, Maine, native Brian Daly. Daly, who did his own hard-court action for Deering High School, authored a story about a new kid (his family has moved to Spruce Island, Maine where his father has landed a job designing ceramic lawn ornaments) trying to fit in and making new friends. These goals are accomplished when twelve-year-old Picasso Dewlap befriends a young Bigfoot who just happens to have incredible skills on the basketball court. Dewlap successfully enlists his new friend onto the losing Spruce Island Middle School basketball team, nicknamed the Lawn Ornaments.

Daly wrote the *Big and Hairy* screenplay for *Showtime's* "Original Pictures for all Ages" which debuted on Sunday, November 22, 1998.

MONSTER IN THE WOODS

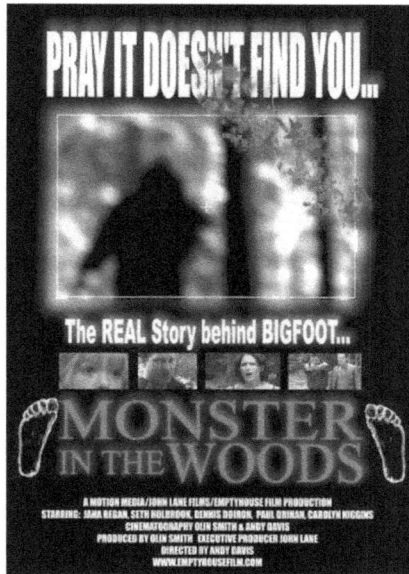

Poster art for *Monster in the Woods* (2007).

Monster in the Woods is a feature length film that takes place in Maine, focusing on two cryptozoologists (played by Jana Regan and Seth Holbrook) investigating Bigfoot sightings. Events begin to move quickly—and chaotically—when a reward for Bigfoot's capture is offered and the cryptozoologists race to save the beast from gun-toting locals. The movie is based upon a screenplay by Al Lamanda and Andy Davis; the latter also directed. Davis told *Fangoria* that he was heavily influenced by the 1972 cult Bigfoot movie *Legend of Boggy Creek*.[10] The world premier for *Monster in the Woods* was held at Merrill Auditorium in Portland, Maine, on January 12, 2008.[11] Maine humorist Tim Sample introduced cryptozoologist Loren Coleman who presented the film to the Merrill crowd.[12]

FINDING BIGFOOT SHOW COMING TO MAINE?

WQCB located at 106.5 on the FM dial is a country music station located in Brewer, Maine, and serving the eastern portion of the state. Morning host J. R. Mitchell watched an episode of *Finding Bigfoot* that included a map of the Appalachian Trail showing reported sightings in Maine. In a posting to the Q106.5 website he wrote, "So I Googled Finding Bigfoot Coming to Maine, and found a post on the Craigslist from Matt Moneymaker, posted the very same day, here's the post below, complete with HIS typos . . . not mine . . ." The post attributed to Matt Moneymaker follows:

> Hi all, Matt Moneymaker here from the B.F.R.O. and this is one of the many ways we utilize in order to get the message/information out. Craigslist has been good to us.
>
> We have received many reports of possible sightings in the Richmond Maine specifically from the Kennebec River up to Rt. 201, especially Beedle road area.
>
> The reports have become questionable enough to allow our team to examine the area with your help of course.
>
> Currently we are working on a Bigfoot and Beer—Town Hall Meeting—I will keep you informed, in the meantime if any residents of Richmond or passers by have suspected sasquatch please inform us asp! Happy Squatching![13]

Emily Burnham, reporter for the *Bangor Daily News*, learned of Mitchell's post and contacted Moneymaker to confirm if *Finding Bigfoot* was indeed coming to Maine. Moneymaker told her the Craigslist post was a hoax and that *Finding Bigfoot* did not have plans to come to Maine. Burnham concluded her article with a hopeful denouement:

> As for "Finding Bigfoot" coming to Maine, Moneymaker said he'd love to bring his crew up to the Pine

Tree State—but he needs more witnesses to come forward with their Bigfoot sightings. Have you seen Bigfoot in Maine? Let us know, and maybe we actually CAN bring the show up north."[14]

HAVEN—WHO, WHAT, WHERE, WENDIGO?

Haven is a TV series that premiered on SyFy in July 2010. The show depicts a fictional Maine town called Haven and is loosely based on Stephen King's novel *The Colorado Kid*. The show is based on the handling of emergent crises when a character's "Troubles" manifest. Troubles are a supernatural affliction that impact certain characters at certain moments and have been a recurring "plague" besetting the town for centuries. Manifestation of a citizen's Troubles appear to be idiosyncratic—one person's strong emotions affect the weather, another Haven resident has the power to literally hold the town together.

Dave Teagues (John Dunsworth), a resident of Haven, Maine, and a co-owner of the *Haven Herald*: "It's a human spirit, but stronger, faster, tracks prey like a lion, it survives on human flesh."

Teagues' reference is the titular mystery of the tenth episode of *Haven's* second season, "Who What, Where, Wendigo." The Troubles in this episode play out through three sisters who have become Wendigo. A serial killer is lured into the woods and becomes a food source for one of the girls, she being the only one amongst her siblings to partake of human flesh. The other two sisters, despite their strong Trouble-provoked cravings, resist eating people and satisfy their hunger on animals.

This *Haven* episode perpetuates characteristics from the Native American tales of Wendigo—namely the propensity of the Wendigo to eat human flesh. [Spoiler: The story concludes with the girls taken to a slaughterhouse where they may remain and feed until the Troubles are over.]

HISTORICAL LITERATURE—
MUSINGS ON INDIAN DEVILS AND OTHER STORIES

The importance of period literature to the subject of Bigfoot is underscored by a *Skeptical Inquirer* article by Hugh H. Trotti exploring a theory that fiction may have helped start the idea of the North American Bigfoot. Trotti noted Daniel Boone's claim of killing a ten-foot, hairy giant that Boone labeled as a "Yahoo." Yahoos happen to be large, hairy, and man-like creatures Jonathan Swift writes of in *Gulliver's Travels*[1] in the chapter "A Voyage to the Country of the Houyhnhnms." Had Boone merely recounted Gulliver's encounter as his own, thereby creating a tall tale and by extension a tenacious North American legend?[2]

Dr. David M. Zueffle wrote a response to Trotti, also published in *Skeptical Inquirer*, in which he provides support for Trotti's speculation.

> Consider a research note by Leonard Roberts from the journal *Western Folklore* (Roberts 1957). Roberts reported that he had encountered "four or five versions" of what he called "a curious and strange legend" he collected from an isolated region of the Kentucky mountains. The stories centered around large, hairy wildpeople who lived in the woods. In one version of the story, told by a Mr. Lee Macgard of Harlan County, Kentucky, the cave-dwelling wildperson is specifically called a "Yeahoh."[3] Could "Yeahoh" be a corruption of Swift's "Yahoo," and

could this be evidence that Boone's tall tales survived over one hundred years of retellings in the Kentucky highlands?[4]

The similarity to the Yoho legend from Maine is unmistakable. Eliot Wigginton, a folklorist and oral historian, was well known for having developed *Foxfire* magazine, which featured articles based on interviews exploring oral history, traditions, and local culture. In *Foxfire 4*, a story collected from Maine sounds similar to the tales from Appalachia.

> A Maine lobsterman once told me of a "cove about two mile below where I live called Yoho Cove" named for a wild man who lived there years ago and always hollered "Yoho, yoho." When a party of natives went raspberrying by the cove one time, the Yoho ran out of the woods, grabbed a girl from the group and made off with her. A couple of years later this party returned to the cove, and the girl rushed toward them, carrying a year-old baby and screaming for help. The berry pickers lifted the girl into the canoe and pushed off from the shore, when the Yoho ran down to the water, seized the baby, tore it in half, and with a hideous yell threw one part at the retreating canoe. "So it's been called Yoho Cove ever since."[5]

Folklorist Richard Mercer Dorson conducted an oral interview and recording of the Yoho tale in 1956.

> DORSON: This is July the 13th, Friday the 13th (1956), a very good day to record folklore, and I am again in Machias with Curt Morse the famous story-teller, in the hospitable home of his daughter, Mrs. Eva Hall, and Curt is going to tell some of his yarns. Now there's one about how Yoho Cove got its name. How does that go, Curt?

CURT MORSE: Cove about two mile below where I live called Yoho Cove and the old fellas years ago allus said there was some kind of wild man lived there, and all they could understand he holler, "Yoho, yoho" all the time, especially at night. So he kinda slacked off and there was some of the natives down around the shore, don't 'cha know, and took kinda of a dugout canoe I call it, dug out of tree, went across there rasperrin'. Well they got about ready to come home and they heard this Yoho hollerin'—they all him a Yoho. So before they reached the boat this fella, this man, ran out and grabbed this girl and took her back in the woods with him and left the rest screechin'. So they went home and a little while afterwards why they . . . it kinda died out, don't 'cha know? They missed the girl a lot. Well they thought she was dead and about two years afterwards or about a year and a half afterwards they had kinda forgot about it and they was over there raspberryin' or blueberrin' again and they heard this screechin' and they looked up and this girl there, their relation, was runnin' and screechin; for help. So she had a baby with her chasin' along—a year old—some little year old baby, somthin' like that. And she—they got her in the canoe anyway, started off from the shore. And the Yoho come down on the shore and caught the baby, or took the baby, tore it apart, tore it to pieces, throwed one part at the canoe as it was leavin', and took the other part back in the woods. So it's been called Yoho Cove ever since. That's all of it that I know about. It's always been called Yoho Cove.

DORSON: Mrs. Eva Hall, you've heard that, you say?

EVA HALL: I've heard it since I was a little girl, I've heard that story.

DORSON: Who'd you hear it from?

EVA HALL: Oh, I don't know, uncles, father, people who lived around there. We always heard that story. Whether it was true or not, I guess it was handed down from way back years ago when people were quite uncivilized.

CURT MORSE: Probably about a hundred and fifty years ago.[6]

As will be seen in this chapter, Wild Men and Bigfoot-like entities are rooted in Maine folklore and are present in several examples of period fiction set within Maine. The greatest concentration for these creatures' appearances within this literature falls between the nineteenth to early twentieth centuries.

COMMENTS ON INDIAN DEVIL
INDIAN DEVIL AS LARGE CAT

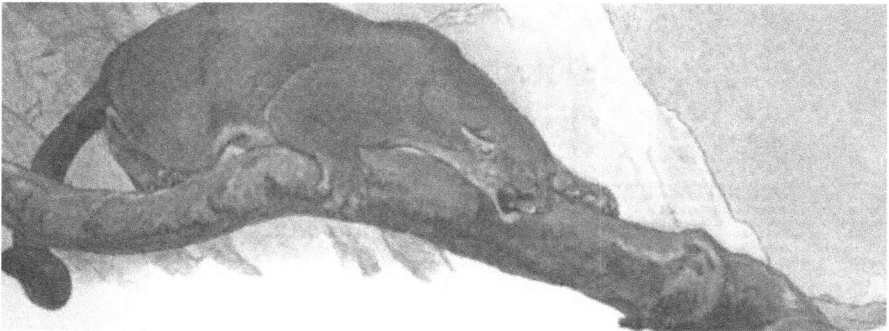

The preponderance of references to an "Indian Devil" in Maine refers to a creature specifically or very similar to the eastern panther. A less-used application of Indian Devil may refer to a wolverine. There are many instances in modern Bigfoot research of the term Indian Devil used as a synonym for Bigfoot, showcasing its historical relevance and association to man-like beasts by Native Americans and early settlers. This section will show that the term Indian Devil in period literature, newspaper articles, and stories often, though not exclusively, refers to a big cat.

In *Sasquatch: The Apes Among Us,* John Green includes mention of a letter from an Air Force sergeant from Connecticut who stated that as a boy he spent many summers at a camp in eastern Maine near Topsfield. The letter related its author knew people in the area "who almost take the existence of this animal for granted just as a New Yorker may believe in the fact that taxi-cabs exist." "Indian Devil" was what the locals had called it for as long as anyone could remember. Its description matches up with other terms used to describe Bigfoot-like creatures across the country. The sergeant specified the Indian Devil the locals spoke of lived in the woods, was "seven-foot tall, reddish-brown, fur-covered, almost-human-looking creature."[7]

Green also includes C. A. Stephens' *Camping Out* with its "Pomoola, the Injun Devil" (presented later in this chapter). Green cites *Camping Out* as evidence of the tradition of the "hairy, half-human 'Injun Devil', existed in New England in the mid 1800's."[8]

Author David Emblidge recounts reading C. A. Stephens stories as a boy:

> Our home in West Virginia enjoyed three generations of *The Youth's Companion*, and its weekly arrival was an event: Homer Greene, James Wilard Schultz—there were many stars among its contributors, but none so bright as C. A. Stephens. His stories of Maine—the Old Squire's farm, the winter frolics, the Big Woods—were unforgettable. In some of his early fiction (I never knew where fiction began and biography ended) four boys, Kit, Wash, Wade and Raed, cast their vote for an education less formal than the one they might receive in college. They planned a camping trip to Mount Katahdin, hoping that there they might find some mineral deposit which would finance their education by travel. The resident Indian devil of the mountain, Pamola, did his best to stop or hinder them; but right prevailed, and persistence paid off. They found a vein of graphite, solid

it for fifteen thousand dollars, and were off to Labrador, Iceland, and the Amazon. Each trip inevitably resulted in a book, and I had them all."[9]

C. A. Stephens employed the term Indian Devil as both a manlike being and as a large cat, the latter usage sometimes accompanied by the scientific name *Felis concolor*. "Owing to our severe climate panthers were never very numerous in northern New England—not nearly so numerous as panther stories, in which the 'panther' is usually a Canadian lynx. Even at present we occasionally hear of a catamount or an 'Indian devil'"[10]

Following are several examples of Indian Devil referencing a big cat.

Henry David Thoreau, *The Maine Woods* (1864):

> When we returned, he observed a much larger track near the same place, where some animal's foot had sunk into a small hollow in the rock, partly filled with grass and earth, and he exclaimed with surprise, "What that?" "Well, what is it?" I asked. Stooping and laying his hand in it, he answered with a mysterious air, and in a half whisper, "Devil [that is, Indian Devil, or cougar] lodges about here—very bad animal—pull 'em locks all to pieces."
>
> "How long since it was made?" I asked.
>
> "Today or yesterday," said he. But when I asked him afterward if he was sure it was the devil's track, he said he did not know. I had been told that the scream of a cougar was heard about Ktaadn recently, and we were not far from that mountain.[11]

"The Indian Devil," *The Hartford Courant* (1890):
Though few know and still less believe it, it is a fact that there is in the northern part of Maine, an animal that is known by the Indians as the "Indian Devil."

He is the genuine North American panther and is the only one ever seen or heard in the state. His haunts are in that region where the counties of Penobscot, Piscataquis, and Aroostook meet, but he has most frequently been heard about Millinocket Lake in Piscataquis county and every little while a story appears in print that the screeches of a panther are heard thereabouts.[12]

Isabel Hornibrook, *Camp and Trail: A Story of the Maine Woods* (1897):

And many eyes dilated as he told of blood-curdling adventures with the "lunk soos," or "Indian devil," the dreadful catamount or panther, which was once the terror of Maine woodsmen.[13]

"Boys, that catamount is the only animal that an Indian is skeered of. Ask a red man to hunt a moose, a bear, or a wolf, an' he's ready to follow it through forest an' swamp till he downs it or drops. But ask him to chase a panther, an' he'll shake his head an' say, "He all one big debil!" He calls the best in his own lingo, 'lunk soos', which means 'Injun devil'; an' so we woodsmen call it too."[14]

"Injun Devil," *Boston Daily Globe* (1898):

In the old days, the days of half a century ago, when sportsmen were few and big mast pines were more in evidence than now, strange and startling stories were told of the untamable savageness of this fearful animal. The superstitious among the woodsmen of upper Maine and Canada credited the fierce cat with the power of looking right through the spruce walls of a logging camp.[15]

"Real 'Indian Devil'," *Lewiston Saturday Journal* (1900):
The panther is also known as the catamount and "Indian Devil" in this part of the country. In the West they are called mountain lions, and everywhere they are feared with an unholy dread for they are without doubt the most ferocious animals on the American continent. They were the only things in the forests that the Indians feared, hence the name, "Indian Devil."[16]

"An 'Injun Devil' Yarn," *Daily Kennebec Journal* (1907):
The following yarn now going the rounds will be interesting reading to the people of the locality mentioned:

West Gardiner, Me., July 21, 1906—The "Injun" devil is frightening the West Gardiner berry pickers. They have seen him lying on his side in fields and pastures a strange dun brown thing with lolling chops and tasseled ears. They have caught glimpses of him flying through the thickets at dusk, and he has been faintly seen on distant hills against the twilight, a ghostlike creature scenting the evening woods. His height is five feet, and the tracks that he has made, as determined by local observers, measures 10 feet apart. Some call him "Lucifee," some call him the Indian devil.

The range of running is around Cobossecontee, Manchester, West Gardiner and Purgatory Mills. He has injured nobody as yet, but the berry pickers are afraid to go to the pastures. No attacks have been made by him upon domestic animals and hence it is supposed that he lives on rabbits and other food most desired by Indian devils.[17]

"'Injun Devil' Was a Big Maltese Cat" was a *Lewiston Evening Journal* headline from the January 10, 1912, edition. In this one

Flattened His Body for a Fatal Spring.

The illustration accompanying Ella H. Stratton's story, "A Real Heroine," from the *Stevens Point Daily Journal*. The picture shows an "Injun Devil" sizing up his quarry in the Aroostook Valley.

story is another example of how the term Indian Devil is used to refer to a large cat, however, this one more mundane than a lynx or an eastern panther. The story is steeped in mystery and mood, of locals being terrified of "blood curdling cries" occurring nightly and the eyes that burned like coals staring out from the black night as witnessed by Mr. McDonald.[18]

> Bangor, Me., Jan. 10—Spencer Boyd and Fred Lyman of Greenbush came down to Bangor from Schoodic with reassuring news. Early on Sunday morning with their trusty rifles they slew the Injun Devil which has been infesting the neighborhood of Ebemee pond [sic] for the past few months and striking terror to the hearts of timid people who have been hearing

his blood-curdling cries almost nightly thereabout. According to Messrs. Boyd and Lyman the animal who has caused all the commotion turns out not to have been an Injun Devil at all but only an ordinary pussy cat gone wrong. Taken altogether their story is an extraordinary one.[19]

Early in November Herbert McDonald, who camped near Ebemee, heard the most awful howling one can well imagine along in the middle of the night. He wasn't in the least alarmed, but he threw some more wood on the fire and sat up to see if the howls appeared to come nearer. They did and, along about 2 o'clock, Herbert saw eyes like coals of fire—that's just what they looked like, coals of fire—staring at him out of the darkness. It made him feel creepy. Just as it would almost anyone. Herbert didn't waste much time after the eyes appeared but snatched up his 30-30 Winchester and blazed away. There was a sharp scream and then all was still. As soon as daylight came Herbert investigated and found clouts of blood on the leaves about where the eyes had been. But that was all.[20]

The story states the commotion caused in the area had resulted from "the largest maltese cat—just a common house cat, but an enormous one" which Mr. Boyd was able to bring down with his rifle.

INDIAN DEVIL AS . . . WOLVERINE?

As the preceding examples show, Indian Devil was a commonly used name for an eastern panther. Wolverines could also be the animal referred to by this moniker as explored by author Tom Seymour.

Wolverines are a completely different critter. Little is known of the Maine wolverine, a near-mythical creature that hasn't even been thought about for many, many years. Still, tales from the past surface every once in awhile. My neighbor, a man of 80 years, whose grandfather lived a long life as well, has fascinated me with his second hand accounts of what the old-timers called the "Injun devil."

And, according to my neighbor, the Injun devil still roams the most remote sections of the far north. He knows, because he can identify them by their powerful odor. It's wishful thinking on his part, though, to imagine a wolverine living in Maine. At this point in time, no section of our state, no matter how wild, is completely remote. And wolverines need perfect wilderness in order to survive. They cannot co-exist with man.[21]

INDIAN DEVIL AS OTHER THAN CAT

Mrs. Lucy Thompson, a full-blooded Klamath Indian, wrote in her book *To the American Indian*, "In this book I will endeavor to tell all in a plain and truthful way without the least coloring of the facts, and will add many of our fairy tales and mother's stories to their children."[22]

"Our Indian devils are Indians who for some reason or cause leave the tribe and go far away into the lonely mountains, and into the depths of the forests, where they live near the streams and places almost inaccessible. In their loneliness they roam through the forests and over the mountains like some wild animals of prey. They forget the language of their mothers and become something like wild beasts, fleeing from the sight of human beings."[23]

"In olden times, the women, especially were always careful to keep together on their camping trips when they were gathering the acorn crop, grass seeds, pine nuts, etc., for fear of the Indian devils. These Indian devils would sometimes watch the camps of the Indians very closely and follow them about as they moved from place to place, watching for an opportunity to seize one of the young women and carry her off to make her his wife. If a young woman strayed away too far by herself, she was often made a captive by one of these devils. The women of the tribe had great fear of them as they had great horrors of becoming the wife of a wild man."[24]

"Sometimes the women would be captivated by the Indian devils and would be gone away from their tribe for years, when they would return and tell of their wild life and experiences. They would become the mother of children and the children would inherit the wild habits of their father, as they would always be whistling, making strange noises, romping wildly about and always on the go, roaming everywhere in the wilds. These women were never happy when they came back to their people, as after a time they would long to go back to their devil husbands and children."[25]

"When the Indians would go on their hunting and camping trips into the mountains, as soon as they heard an owl screech or hoot, they would stop and listen, and try to distinguish if it was an Indian devil imitating the owl or the cry of a wild animal. The Indians would stop at once, kindle a fire and hallo; this was given as a warning to the devils that they were awake and ready to fight them if necessary.[26]

Indian Devil is in evidence in other parts of North America as a term for strange creatures, and is directly associated with Wendigo (Weetigo) in the following story. "Ghostly Figure Haunts Indians of North Wilds" was the United Press story title, from the November 4, 1935, *Berkeley Daily Gazette*, dateline Kaskesiu, Saskatchewan:

> Appearance of a weird, hooded figure which stalks and attacks settlers during the night has aroused fears among Indians and superstitious whites that a "Weetigo," mythical Indian devil, again is roaming this northern country.
>
> The legendary evil spirit is described as a ghost-like figure of a man, wearing a sack over the head with slits cut out for eyes.
>
> Several settlers and Indians have reported to police that the figure had sneaked near their lonely homes and camps in the north area around the National Park and frightened them with weird antics.[27]

Loren Coleman is a noted and respected cryptozoologist, curator of the International Crytptozoology Museum in Portland, Maine, and former contributor to the *Cryptomundo* blog where this note on Indian Devils by Coleman comes from: "I've interviewed some individuals, oldtimers actually, in the Rangeley, Maine, area who told of seeing Injun Devils, and describing them as bipedal hairy hominoids, like Bigfoot, with glowing eyes apparently reflected from campfires—near lumbering sites near the Rangeley lakes around the turn of the 19th to 20th century."[28]

HISTORICAL LITERATURE

Jack's Injun Devil
R. B. Nason

We were coming out of the woods from a fishing trip, and Jack had been beguiling us with a yarn about the trout in Chairback pond [sic], which, he said, had four legs and no fins or gills. We had set him down as another of the tall story tellers of the Maine woods, when one of our party, who was something of a naturalist, identified Jack's four-legged trout as a Spotted Newt or Salamander, and so established the truth of the main parts of the story.

We were in this somewhat reconciled and chastened frame of mind when the talk turned upon "Injun Devils."

"You can't tell me nuthin' about Injun Devils," said Jack shaking his head with an air of profound conviction.

"Did you ever see one?" we chorused.

"You bet yer life I hev, and it warn't more 'n ten miles from here, neither."

"This isn't another four-legged trout, 's it, Jack?" we asked.

"Well, you needn't believe it unless ye want to, but 'twan't no four-legged trout nor nuthin' er the kind. 'Twas an Injun Devil, I tell ye, and them's straight facts. Bill Cobb was with me and he'll tell ye the same thing."

"What did it look like?" I asked.

"Didn't look like nuthin' I ever saw before nor sence, an' I've seen everything that travels these woods. 'Twas big 's a man, an' covered all over with hair jus' like a darn dog."

The naturalist suggested a bear.

"Bear nuthin'," snorted Jack. "Don't ye suppose I know what a bear looks like? I've hunted bears ever since I was knee high to a spavined to'd, an' seen more of 'um than you could shake a stick at."

"Well, tell us about it, anyway," all of us urged.

Jack slowly "loaded up" his pipe, and commenced:

"Well, me an' Bill was fishin' Orson, an' 'long toward night we come to a camp where Babcock had been gittin' out knees an' spool-bars the winter before. We went in an' built up a fire an' made some tea an' cooked some fish for supper. By that time it'd got pretty dark, but pretty soon the moon come up full an' made the clearin' round the camp's light as day. You could see's far as you'd a-mine to any-wheres in the clearin' on both sides of the brook, so ye needn't say I couldn't see it jus' as I am tellin' ye.

"Well, long after supper Bill he found a kag with some mullarsis in it, an' I went to work an' made some candy."

"You didn't have any canned goods with you, did you?" I asked.

"Well, I can't say we hadn't been drinkin' *some*, but nuthin' to 'mount to nuthin'," he continued. "We didn't start with only a pint anyway between the both of us, an' that warn't enough to keep the flies off. No, sir, we warn't drunk, neither of us. I can handle a long-necker myself alone, an' then I ain't what ye'd call drunk neither. I never was what ye could call real loaded more n' two 'r three times, an' there was more 'n a pint to two of us, I want ye to know."

"Well, go on, anyway," we put in, heading him off. "You had only a pint for fly medicine, and couldn't have been very much affected, that is true; and you had made some molasses candy."

"Yessir, an' when I got it cooked enough I went down to the brook to cool it off. There was a big hemlock slantin' across the brook just above where I was, lodged pretty low down on the other side in some old fir tops, an' while I squat there on a rock coolin' the candy off, I heard them bushes crack, an' I looked up quick, an' I want ye to know there he was! He had his forward legs up on the log an' was lookin' me straight in the face. His face looked jus' like a cussed monkey grinnin' at ye, an' he had teeth as long as my finger, an' eyes shinin' like two coals o' fire. I want ye to know

I was that scairt I couldn't wiggle for more 'n a minute, an' he was a-lookin' right at me, an' then I dropped the fryin'-pan jus' as he come up on to that log like a cat an' stood up on his hind legs as straight 's a man. I could see him as plain as I can see you this minute. I want ye to know I never was that scairt in my born days before nor sence. I let a yell out o' me an' started for the camp an' he after me. I hadn't gone more 'n twenty feet 'fore I stubbed my toe over a root an' went flatter 'n a pancake. You better b'lieve I did some tall scratchin' an' yellin', but I hadn't more 'n time to wink twice 'fore he lit on top o' me. Jus' that minute Bill come runnin' out o' the camp, an' that must o' scared him off."

Jack stopped and pulled at his pipe, his face set and stern. After a moment he continued:

"Well, sir, I got up out o' that an' I was so weak I couldn't scarcely stand. We didn't wait for him to come back, neither. We struck a bee-line out over the tote-road for home, an' we didn't stop 'till we got out o' the woods out to the main road. We went an' wok' up Doll Crandlemire an' went in there. My shirt was ripped clean off my back, an' there was tracks on my shoulders where he lit on top o' me as big over as the palm of your two hands. Doll Crandlemire'll tell ye the same thing, too, for he saw um that night. An' I've got the marks right there with me to this day."

"No, I haven't been in there sence an' I don't never callate to. You needn't b'lieve there ain't no Injun Devils unless ye want to, but, by Judas, I *know* there is!"

* * *

Some time after Jack told us this story I learned that a man had escaped that summer from a camp in the neighborhood during a terrible attack of delirium tremens, and was never seen again—alive, at any rate. But the body of a naked man was found in the woods in the early winter half buried in the mold and leaves.[29]

C. A. STEPHENS

C. A. STEPHENS

Charles Asbury Stevens, known for his youth-oriented travel and adventure stories and for reminiscences of his youth, was born in Norway, Maine, October 21, 1845, and passed away in that town on September 22, 1931. He graduated from Bowdoin College in 1869 and received his M.D. from Boston University in 1887. A teacher at Norway Liberal Institute for five years, he started writing juvenile literature in 1870 and for over 40 years was connected with the *Youth's Companion* magazine, being one of their most popular writers. Editor Daniel Sharp Ford paid Stephens $2 apiece for his first two short stories and it is estimated that Stephens wrote over 3,000 short stories and over 100 chapter serials as well as numerous books.[30] Among the latter were the popular Old Farm Stories, Camping Out Series and Knockabout Series.

Stephens' writing was infused with his knowledge of the Norway, Maine, region and its associated lore. Stephens made use of the Wild Man motif several times in his stories, leaving Chad Arment, author of *The Historical Bigfoot*, to ask if Mr. Stevens hadn't written these stories based upon a foundation set in local folklore. Speaking specifically of *Camping Out*, Arment says, "How much of this tale Stephens may have heard from actual witnesses or local legends is unknown; we certainly can't be adamant at this point."[31] And Green in *Sasquatch: The Apes Among Us*, writes, "Even if the book was actually fiction, it probably indicates that the tradition of a hairy, half-human 'Injun Devil', existed in New England in the mid 1800's."[32] C. A. Stephens employed the Wild Man motif in at least three of his stories.

Was it an Indian Devil?
C. A. Stephens

Can a myth come into existence spontaneously? Can a story, utterly truthless, obtain widespread belief through hundreds of years, and thus become a tradition?

At the close of a five years' residence among the hunters, lumbermen and river drivers of the Northern Maine forests, in connection with the lumbering business of my uncle's firm, I find myself puzzling at these questions, as I recall the persistent and ever-recurring tales and accounts which everywhere come to my ears, of that strange being, or animal, which the Indians used to call "Pomoola," and which the white woodsmen have translated "Indian Devil."

As nearly as can be gathered from the legends of the Pequankets, Androscoggins and Penobscots, Pomoola was a beast, a man and a devil, in whimsical trinity (a savage relative of the old Greek Cerberus, perhaps), possessing certain supernatural and fiendish powers, and thus capable of doing infinite mischief if offended. With this fanciful character the savages seem to have invested something they did actually see and meet occasionally in those regions.

Katahdin and the neighboring mountains was a locality especially infested by it. Indeed, the summit of Katahdin has been sacred to Pomoola from earliest times. No Indian, in his senses, would venture to ascend it. The Schoodic Lake region was another of its haunts.

Numberless are the stories of the strange tricks and antics of this curious devil of the woods and lakes.

I have heard woodsmen, plain sensible bodies, who were neither "long-sighted" nor "biscoes," and who were clothed in their right minds, undoubtedly, tell the story, straightforward and honest enough to silence the most skeptical; not of what they had heard, either, but of what they had actually seen.

It would seem well-nigh impossible that these accounts should originate in nothing. If verbal testimony could substantiate them, enough could be collected to establish, in law, the existence of a whole legion of Pomoolas. Now setting all superstition aside, will not this array of testimony warrant the following question: May there not have existed, and perhaps be at present existing, in the great forest region to the northward, some creature, "wild man," or something of the sort, not down in the books, which has been the kernel of the "Indian Devil?"

I have my own little story to tell (for I did see something rather queer one day); and right here seems to be the place for it, after remarking that some have thought the "Indian Devil" was merely the Felis Concolor or catamount. This is entirely a mistake, arising, perhaps, from the well-known dread the savages had for the cougar.

During the winter of '67, the headquarters and general depot of our lumbering gangs was at the head of the Chesuncook, a long narrowish lake to the northwest of Katahdin. The logs went down this lake with the spring freshets, into the west branch of the Penobscot. And during the frozen season it offered an easy open thoroughfare up into the forest. From this point the various "camps" within a radius of twenty and twenty five miles took their provisions. One of the most distant of these had been established at the head of the Caucomgomac, a smaller lake, some twenty miles to the northward. And during the month of February, a "row in camp," from the incompetency of the "boss," had made it necessary to pay off and discharge a part of the hands; and I was sent up from Bangor for that purpose, having for guide and escort (I was but seventeen then) an old hunter named Hughy Clives.

The trip up to the "Head of the Chesuncook" occupied a week; and after resting a day at the supply camp on the Caucomgomac.

The trail ran through an almost unbroken wilderness; and the snow lay from four to five feet deep. But on our snowshoes we didn't mind the depth; the main thing was to keep out of the brush. And hence it seemed a great relief to come out just as the sun was setting, upon the foot of the Caucomgomac, stretching away, in a broad frozen plain, ten miles to the northward.

As the camp was no more than a mile in from the "head," Clives had anticipated getting up there early in the evening. To him the distance was as nothing; but to me it now seemed infinite. It was my first whole day on snowshoes, and my legs had not taken kindly to the length and novelty of the performance. And despite all my attempts to disguise the fact, I knew that Clives saw that I was about used up.

"Think you can hold out, youngster?" said the old fellow, looking round to me with a hard grin. "How's your wind? Think you can straddle the lake? Or will you have to dig a hole in the snow here, and burrow like a rabbit? Nice chance there under those little hemlocks."

I still preferred going on to burrowing. We struck out upon the lake, and had soon left its black spruce-lined shore a mile to the rearward.

Meanwhile the twilight had deepened. The sky was clear, and in the west deep red; but the startling phenomenon peculiar to frozen waters, and said to indicate a storm, the moaning and groaning of the air beneath the ice, began with frightful distinctness. Never had I heard such sounds before. It seemed as if a thousand demons in agony were gurgling and drowning beneath us. Now a wild sigh would struggle up from the depths and echo to a distant moan, when, anon, a deep fearful groan would sweep along, making the thick ice shiver and vibrate like a drumhead.

But Clives only laughed at my apprehensions that the lake must be breaking up, and was saying that this was a common enough thing to hear on these waters, when a short gruff bark from "Vete" (Hughy's hound) caused to look suddenly around. Some thirty or forty rods to the rearward, and barely distinguishable in the rapidly-falling dusk, a dark object was coming on behind us. We both took it for a man, with the first glance.

"Strange though," muttered Clives, hastily examining the cap on his rifles. "Must have come up pretty quick, too. Looked round not five minutes ago; saw nothing then."

Strange indeed. Unpromising, too. I had quite a large sum in currency about me, for the payment of the gang. The circumstance of being followed in the dark was not pleasant. But then, we were two to his one; some one out like ourselves, perhaps. "Somebody out hunting, I suppose," said I.

"Possible," said Clives, walking on, with one eye turned back over his shoulder. "Barely possible. Don't often fall in with a man so, in here though."

"At any rate he means to come up with us," said I; for although we were now walking quite rapidly, the distance between us was visibly lessening.

"That's plain enough." Replied Hughy; "and to tell the truth, youngster," looking uneasily at me, "on your account

I should like to know what's wanted before he comes much nearer."

"You'll have to find out in a hurry, then," said I, with another glance behind. "He's coming up at a great rate, surely."

"Hallo! Hallo there!" shouted Hughy, cracking his rifle. The stranger was now no more than a hundred yards distant; but for all reply (that we heard), raised his arms and brandished them wildly about his head.

"Who are you? What do you want?" shouted Clives again. And again those wild gesticulations.

"Well, by Jehu!" exclaimed the old man. "What signs and motions! Must be a dumb man. Stopped, though, hasn't he? Why in the world don't he answer?"

"Perhaps he did," said I. "The ice makes such a noise we may not have heard him."

"Humph! Perhaps!" But he don't act right, and I've a mind to let a ball fly after him."

"Set Vete on him," suggested I. "Here Vete, my boy, go for him?"

But Vete, a fine dare-devil of a hound, only winced and stuck the closer, almost tripping us up in his efforts to keep near.

It grew darker momentarily; and as nothing could be got out of the stranger, at this distance, save those inexplicable gestures, and since, to tell the truth, we didn't care about going back to cultivate his acquaintance, we at length started on again.

"Can't make him out at all," muttered Clives. "Shouldn't be surprised to hear a ball go by any minute. But I don't quite like to fire first. Perhaps he'll go back, now he finds we've got our eye on him."

And for a while we fancied he had gone back; but a moment later discerned him flitting along after us, sometimes coming up within fifty yards, then halting till invisible in the darkness.

Several miles were thus passed over. What with the sighing and gurgling under the ice, and that mysterious object flitting along behind us, the situation was a rather singular one, to say the least.

Presently the moon began to shine up over the forest-clad mountains to the eastward; and a little broad bright disk peered up over the distant ridge. Things were thus revealed in a much clearer light. In the darkness our pursuer had at times assumed gigantic proportions. It now seemed the figure of a medium-sized man.

"But he's bare-headed," said I, as this fact became apparent in the moonlight.

"No gun, either," said Hughy. "And that must be a mighty snug-fitting suit of clothes he's got on. Hallo! Hallo there! You man in tights! What's wanted?"

Thus called to, the curious object halted, gestured as before, then dropping on all fours, ran off to the left of us, coursing along with amazing swiftness, and describing a broad semi-circle, came around to a point directly in front of us, some fifty rods ahead. We stood still in our amazement.

Arriving at this point the creature began to dig in the snow, throwing it up in silvery wreaths, and in a few moments had buried itself from our view, save a black crest peeping up over the edge of the drift.

"Good heavens!" I at length ejaculated. "What running! That was never a man!"

"No, that was never a man," repeated Hughy.

"But what is it?" cried I. "What can it be?"

"You saw it run?" said Hughy interrogatively.

"Saw it run! I guess I saw it run! I never saw anything run like it! Must have been a spook."

"And you saw it beckon and make signs?" continued Hughy, reflectively.

"Of course I did—a dozen times. What in the world are you driving at?"

"Well youngster," the old man went on, "that's an Indian Devil. This is the third one I've seen; or the third time I've seen it; for maybe it's the same one. The last time was eight years ago, down on the Junior Lake. I might have known what this was at first; but I didn't think till I saw it run."

"But what's to be done?" said I. "Going to fire at it?"

"Fire at it! No, not for the world."

"Won't it meddle with us?"

"Not if we let it alone, and go on about our business."

"But it has got square in our path."

"We must go round it, then."

"Isn't that being a little cowardly?"

"Young man, you don't know what you're talking about."

In short, Clives would hear to nothing like offensive operations against the creature, devil, or whatever it was. The old fellow was pretty stoutly tinged with the superstitions of the region in which his life had passed. There was no help for it. Turning aside, we made a wide detour around the hole in which the creature was crouching, quite similar to the one it had made to get in front of us, and then walked quickly on again. But our tricky pursuer was not yet done with us. Before we had got a quarter of a mile away, we again beheld him careering around us as before, and again concealing himself in the snow.

To be playing at such a game of bopeep, on a wintry lake, after being on my feet all day, was anything but amusing to me. And not having the fear of Pomoola sufficiently before my eyes, perhaps, I should certainly have risked a shot at him. Indeed, I'm of the opinion that a well-directed rifle-ball would not only have cleared up the "Indian Devil" mystery then and there, but would have added a new and important specimen to some zoological collection. It was undoubtedly a beast, "wild man," or something like that. The track (for it did leave a track) was that of a large longish foot, pressed down deeply into the snow. But to have fired at it would have been doing violence to Hughy's prejudices. We again "made our manners to the devil" by sheering round his hole, and repeated the same programme twice more before reaching the head of the lake. After entering the woods again, we caught several glimpses of it dodging among the trees; but lost sight of it for good about half a mile below the camp.

And thus ended my adventure with an "Indian Devil." I know I did see something queer. And I respectfully add this to a legion of very similar stories which one may hear any evening at a logging camp.[33]

The Young Moose Hunters
C. A. Stephens

We sat beneath the shed and watched the sparks darting up, and the slower wreath of black smoke rising toward the stars, momentarily clouding their silver sparkle.

Just then the cry of some animal was heard from the mountain side above us. It was not loud nor startling, but a lonely cry of discontent or hunger. Such sounds impress one strangely in the forest at night. We listened to hear it again, and soon it resounded anew; rather more distinctly this time, or else it was because we were hearkening with intent ears.

"Do you know what that is?" Scott asked.

I could not even guess. It is often very difficult to identify animal cries heard in the woods at night time. The forest echoes change the character of the note. This sounded somewhat like a man shouting rather disconsolately at a distance. We continued to hear it, at intervals of ten of fifteen minutes. But it did not alarm us much.[34]

Something was round the camp, too,—some creature. Perhaps it had followed us in from gumming, though we did not hear it till as late as eight o'clock in the evening. If we

MOOSEHEAD.

Art from *Harper's Magazine* evoking the activities
offered in Maine for sportsmen and women.

had only had a gun we would have made it scamper! Fred did not care for it. He lay down and went to sleep. But it disturbed me thoroughly. I could not go to sleep. No more could Scott; and I think Farr did not rest any too well. I noticed he kept one hand on the butcher-knife. We could hear twigs break, off a little way. I fancied I could see its eyes. Repeatedly I heard its step. It kept prowling about till near morning. We had little idea what it was; the night was too dark to discern any thing off from the fire.[35]

That night Fred had the second watch, and at about twelve he waked us.

"Just come out here a minute," he said.

We roused up and went out.

Fred was standing on the log platform; and we got up beside him.

"Hark!" he said.

We listened. Some moments passed. Then, distinct on the cold air, there came a singularly prolonged and piercing cry from seemingly a long way off.

"I've heard that more than a dozen times," said Fred.

"Any idea what it is?" Farr asked him.

"No; never heard anything like it before in my life."

It was repeated again and again, at intervals of five or ten minutes.

"I don't believe that it is an animal," Farr said.

"Isn't it the 'Cannucks' trying to frighten us?" Scott said.

That question made us laugh. It was a rather improbable supposition.

We went back to our sleep.

Fred said next day that the sound had continued for an hour or over after we had gone to sleep.

And the next night Farr waked us at a few minutes after two to hear the same cries again.

They seemed even more distinct this time. But we could gain no idea as to what produced them.

The second night after, Scott told us that at a quarter before five o'clock he had heard it twice, but very faint and far off. I do not think that even the second time we heard it that it was within three miles of the camp.

"Ah, I tell you, fellows, there are things in these woods that folks do not know of," Fred would say occasionally. This was a pet idea of his; and, indeed, we never did know what made that noise; we could not even guess with any certainty.

The fourth night after moving up to our fortified camp it was very dark and cloudy; and, at a few minutes after eight, it came on to snow,—a driving storm. I had the first watch, but was glad to get down from the post of duty and take refuge inside the shed.

"I guess the 'Cannucks' won't stir out to-night," Fred said.

It was agreed to watch inside our camp-curtain. But at about half-past one there was a noise outside, on the log-fence, as of some one trying to climb it. Farr was on guard. He instantly cocked his gun, listened an instant, then peeped out very cautiously; for he knew, that, if there were enemies inside the fence, they would fire into the camp at the slightest indication of our wakefulness.

The storm was driving so thickly, and the darkness was so great, that he could see nothing. But he stood ready for instant defence for fifteen or twenty minutes; then he quietly waked the rest of us, and in whispers informed us of what he suspected. We all took our guns, and listened a long while. At length Fred crept out under the curtain, with his revolver in one hand, and the butcher-knife betwixt his teeth!

He was out ten minutes or more; and, on coming in, reported that he could find nothing either within or without the fence. But Farr was positive that he had heard a considerable noise. Afterwards we thought that it might have been a wild cat, or a bear that had smelled our meat.

But the alarm had so excited us that we none of us went to sleep again till near five o'clock.[36]

The next night the "hungry man" came at a little past twelve. Fred had the watch. He waked Farr. They told us next morning that they had set out to fire at him, and either kill him or scare him off.

"If it's the Devil, we ought to kill him!" Fred said; and this shows what a bad hold the thing had taken on our minds.

Scott was more sensible.

"That's a human being," said he, "as much as we are. To shoot him would be murder."

Farr said, that, if it was a human being, it was the queerest specimen that ever he saw; for his part, he believed it to be a woods-witch, and if we did not look out it would bewitch us.

Scott ridiculed this talk.

"I'll be one of three to go down there and catch him," said he. "It's some poor woodsman who has got lost, and perhaps turned light-headed."

Fred admitted that he had heard of these cases, where men had got lost in these forests, and become crazy from wandering about. But he declared that he did not care to be one of the three to catch him. He should be very loath to lay hands on that *creature*, he said: should be afraid he might vanish, leaving a smell of brimstone behind him.

"Oh, what stuff that is!" Scott exclaimed in derision.

Thus we talked of it.

Of one thing we were pretty confident, namely, that he had no weapons.

The next night Scott tried to induce us to go and help catch the man.

"If he's crazy, wandering about here, we ought to do something about it," he argued. "By and by he will freeze to death, as the weather gets colder."

But he could not bring Fred or Farr to see it in that light at all.

"I guess he will manage to keep warm," Fred would say.

"Looks to me like a chap that would not have any difficulty in finding a hot brick 'most any time."

"Humph!" Scott would exclaim. "What's the use to be a fool, Fred!"

Evidently the many stories that Fred had heard from the lumbermen had not been without some effect on his mind. He declared that he was not afraid of the man, but he did not mean to interfere with him.

Scott, on the other hand, argued that it was our duty to find out what ailed this person, and assist him. That very evening he roasted a piece of moose-meat in the oven, and, taking it down to the grain-shed, hung it up by a bit of rope directly in the doorway. It was his watch (the first watch) that evening, and he watched till one o'clock (two watches) to see what came of it.

Next morning he told us, that, at a little before twelve, the man had come; and that, on espying the roast meat hanging there, he had seized upon it with strange, wild exclamations of what Scott took for delight.

Fred told him that he had better not go to holding communications with the Devil.

But Scott now gave us no peace; and during the next two days, first I, and then Farr, agreed to help him catch the unknown; and at length Fred consented to help.

For my own part, I had by this time very little fear that it was a supernatural being; but I did dread to touch the poor filthy creature.

Accordingly, that night, at ten o'clock, we all four went down to the ox-camp, and hid ourselves there in ambush, as before. And this time we did not have long to wait. We had not been there more than fifteen or twenty minutes before we heard him coming through the brush, on the other side of the stream.

"He's more prompt since he has got a taste of your moose-meat, Scott," Farr said.

The strange being came up to the door of the grainshed, looked about it a while, then went inside. We held ourselves in readiness.

"Disappointed that he didn't find one of your moose-steaks waiting for him there," Farr whispered.

Presently the wretched creature came out with a piece of pork, and sat down on the stump.

Said Fred, "I had rather tackle a catamount than go near him."

"What foolishness!" Scott whispered back. "The bare fact of his eating that pork shows that he is human fast enough."

"Don't know about that," retorted Fred. "Perhaps he needs it to grease down his brimstone with!"

"Well, come on," Farr whispered. "If we must, we must. Now for him!"

We had laid down our guns; and, at the word, made a rush at the unconscious pork-eater. But I must needs confess, that we did it with no great alacrity. I think that each one of us was very willing that some of the rest should be the first to lay hold of him. We had but a few yards to go, and were upon him before he had even time to turn. Had we seized him pluckily on the instant, we should have held him beyond doubt; but we all held back a little.

Up leaped the unknown.

"Whooh!" he snorted. "Moon-tykes! Moon-tykes!"

Scott seized hold of him; so did I, and so did Farr. But the man whirled, kicked, and struck with such effect, that he threw us off and ran.

But now that our blood was up, and we were fairly into it, we gave chase hard after him,—Fred ahead. Down the bank, on to the ice, and across the stream, went the "hungry man," screaming "Moon-tykes! Moon-tykes!" at every leap. Half a dozen times going across the river, we had our hands on him—almost. The opposite bank was three or four feet high, and set thick with alders. Among these the man

" MOON-TYKES ! "

Fred, Scott, and Farr give chase to the Wild Man.

leaped; but, before he could force his way through them, Fred grabbed him, and threw him back upon the ice. We all lay hold of him, by guess, but it was slippery as glass there. Round and about we went, and all came down together—wallop! I, for one, had both hands fastened onto that old coat, and held on. But the coat did not hold the wearer! It gave way like brown paper. The pork-eater jumped out of it, and regained his legs. Fred seized one ankle; and the wretch ran, dragging Fred, stomach down, on the ice. His *bare feet* stuck, while our boots slipped. Fred said that the man kicked him in the head, and for that reason he let go his ankle. At any rate, he got away, and ran off up the stream for twenty rods or more, and thence into the woods—*naked as when he was born*!

"I guess he will freeze to death now!" said Fred as we listened to his departing footsteps.

Scott was disposed to blame the rest of us for not holding him.

"We had better let him alone than used him in this bungling way," he said.

Farr laughed as if it were a good joke.

We hung his coat up on the alders, so that if he ventured back after it he might take it.

But next morning there hung the coat! I went down to take a look at it by daylight.

Of all the coats I ever set eyes on, that was the shockingest one! It was a mere bunch of rags, filthy and malodorous to the last degree! I thought that it might originally have been of black tricot; but, indeed, it was hard telling what it was originally. The pockets had been torn out, or worn out, long previously. There wasn't a single button on it. In front it looked as if it had been tied together with strings.

We watched the following night from ten till after one. The "hungry man" did not come; and next morning there hung his coat still. We never saw so much as a hair of him afterwards.

Farr said, that, as he had nothing to wear, he was probably too modest to pay us another visit.

Scott regretted the way the affair had turned: he talked of little else for several days.

Three nights after, the old coast either blew away, or else the owner did actually come after it. And the man may even have come back to the grain-shed after more pork, for the moon did not now rise till toward morning, and cloudy weather had set in. As to who or what he really was, we never knew further than I have related.

At present writing I am inclined to believe that it was a person more or less light-headed, very possibly one of the "Cannuck" gang we had known of, whom the others had unfeelingly turned adrift to shirk for himself. The existence of several of these roving gangs is a well-ascertained fact. Sometimes they have plundered the fields, and stolen horses from the pioneer towns and plantations. The Moose-river settlement were seriously troubled by a party of nine of these woods-thieves only recently. Six or seven horses were taken; and the gang was dispersed and driven off only after a sharp and bloody fight with the citizens.[37]

Camping Out
C. A. Stephens
(From Chapter 10)

"Old Cluey" and his Strange Tale of the Pomoola

During the evening we explained to Cluey the object of our expedition. He had heard the "lead story."

"Were you ever on Mount Katahdin?" asked Raed.

"I never clomb to the tip-top," said Cluey; "but I've tramped all around it after moose and caribou."

"We would like to have you go with us," said Raed. "We will give you two dollars per day to go with us as hunter and help carry the luggage."

Cluey said he would take till morning to consider the matter.

"Do you ever see any thing of the Indian devil here?" Wash asked.

"The Ingin dav'l!" exclaimed Cluey. "Where'd ever ye hear any thing about the Ingin dav'l, yonker?"

"Oh! I've heard that he lives in this vicinity," said Wash, laughing.

Old Cluey shook his head.

"We've often heard stories of the Indian devil," said I. "Can't you give us one?"

"Do you know what the Ingins say about the tip-top of Mount Katahdin, yonker?" demanded Cluey, suddenly turning to me.

"Oh, yes! They say that Pomoola will destroy every one who seeks to reach the summit. But that does not apply to white men."

"Ow do you know that ar?"

"Dr. Jackson ascended it without any inconvenience in 1838; so did Mr. Bowditch; and so also did Mr. Holmes and party."

"Be ye sartin that they went to the tip-top? Thar ar' three or four different peaks."

"They reached the highest point on the Katahdin ridge, if we may believe their statement; and there is no reason why we should not," I added, seeing that Cluey still looked a little doubtful.

"What made you shake your head when I asked you about the Indian devil?" queried Wash. "Do you believe in it?"

"Wal," said Cluey, fairly cornered, I don't myself; but thar's plenty as do. An' they do tell some cur'us yarns about it,—mighty cur'us."

"But did ever you see any thing of the sort yourself,—any thing that looked diabolical?" persisted Wash, determined to pin at least one of these singular tales.

"Wal," began the old man, after hesitating considerably, "I did see suthin ruther cur'us wunst. I did, no mistake."

"Ah, you did!" exclaimed Wash. "Tell us about it, please."

"Yes, tell us about that," we all put in.

Thus, brought to brook, Cluey related the following incident. Wash has straightened out his English, and fixed it up fit for perusal. The story is Cluey's in Wash's words, as recorded at the time in his note-book, from which I extract it. Wash entitled it,—

OLD CLUEY'S INDIAN-DEVIL STORY

"'Twas years ago," said old Cluey Robbins. "I was nothing but a youngster then. My brother Zeke and I used to hunt in company with an old woodsman named Hughy Watson. This was either our first or second trip with him up to the lakes. After a tramp of five days through the woods from Norridgewock, our native town, we had come out on the shore of a wild-looking sheet of water, now called 'Ragged Pond.' Its notched, scraggy, and craggy shores might well have suggested the name. Near us a noisy brook came rattling down into it; and, not more than a quarter of a mile farther on, the outlet comes out in a parallel direction with equal noise and foam. Some idea may be obtained, from this circumstance, of the rough surface of the country about us.

"'It ought to be a clever place for mink,' said Hughy; 'and we may find a family of beaver up this brook. I never was on these waters before.'

"We made up an open camp, Indian fashion, under some large spruces; and just at dusk we had the good fortune to see and shoot a caribou. It was cloudy, and came on very dark. I never, before nor since, heard such a serenade from owls as our fire drew around us. Screeches and the most dismal hoots blended in horrible concert. Round and round us they glided in noiseless circles. There were scores of them. It was utterly impossible to sleep; and the frequent

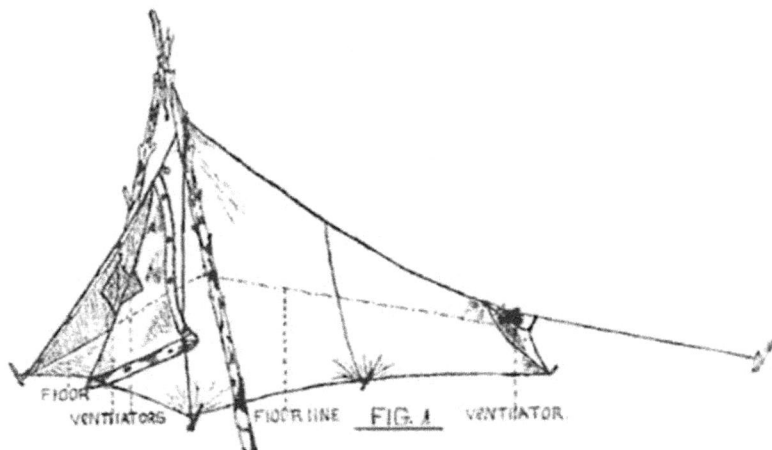

"The pocket edition" is 6 x 8 x 4 ft., is easily set up, and weighs only 33½ lbs.

discharge of our guns failed to disperse them. But in the morning the merry notes of the king-fisher told us there were plenty of trout in the stream we were on; and, where there are trout, there are always mink: so we fell to lining the banks with 'figure-four' traps, which occupied us during the whole of the following day. There were no indications that the stream had ever been trapped before; and we anticipated a full pack of fur.

"'This is what I call freedom,' said Hughy as we sat round our fire that night. 'Every thing just as old Mother Nature made it; and she made it pretty rough and wild too,' continued the old fellow, gazing off at the black spruce-clad peaks of Katahdin far to the eastward, where the 'hunter's moon' was looming up over that desolate ridge. 'Like enough we are the first white folks ever in here. The lumber men wouldn't come into such a region as this. We crossed their old trail ten miles below.'

"Likely enough we were; at least, we had no reason to complain of the trapping-ground we had thus stumbled upon. We began to reap a fine harvest of fur ere the first three days had passed; and for boys of sixteen, like Zeke and I, no better entertainment could have been got up.

"But, as days passed, we began to notice that Hughy seemed uneasy and watchful.

"'What can ail the old man?' asked Zeke as we were making the round of the traps one day. 'He don't act at all as he did the first few days we were in here. Haven't you noticed it?'

"Yes. I had noticed it; and we agreed to rally the old chap a little when we got back. Well, after supper that night, seeing Hughy looking sulky and absent, I asked all at once,

"'What is it, Hughy? Aren't things going on right here?'

"The old man turned and looked at us a moment, as if not certain what he should answer. Then he said,

"'I never like to be laughed at, especially by boys. I thought, at first, we'd struck a fine stream; and perhaps it's all fancy; for I haven't seen or heard a single thing wrong yet. But I've been feeling for several days just as if there was something, either man or beast, hanging round us here. It may be a catamount; or it may be some mean thief of a river-driver, sneaking about for a chance to steal our fur; or some Indian who hunts here, and would be glad to be rid of us. Can't tell. And perhaps it's all my notion; but I can't get rid of it. I remember, once when I was up at the Telos Lake, I felt just so several days; and finally one night I hid in a clump of hemlocks a little ways from my camp, and didn't go to it at all. Along in the night I heard a noise about it, and saw what I took for men there. I didn't speak, or fire on them. Things were upset round the next morning; but I had moved my fur the day before. And, another time, I was up beyond Katahdin; and, several days before I had seen any signs, I began to feel that something was watching me. A night or two after, I waked up, and saw a catamount glaring at me from a tree-top. I suppose he had been prowling round, but had kept out of sight. And I think we shall find that there's something unusual lurking round us now.'

"Old Hughy's presentiments served to keep us wakeful and vigilant; but several days passed without the least sign of any one's being near us, and we were beginning to forget it, when one evening I saw what certainly justified Hughy's suspicions. I had left the fire to bring some water from the brook, which was within a few rods of us. I had stooped to dip it up, when, as I rose, I caught a glimpse of what I took to be a man, standing at a little distance. In an instant it vanished behind a shrubby fir. I felt quite positive; yet it was so dark, and whatever I had seen was out of sight so quick, that I knew I was very liable to have been mistaken. Checking my first impulse to run to the camp and give an alarm, I decided to say nothing at present, but watch.

"The evening passed. By none o'clock Hughy and Zeke were both asleep. I lay down, but kept awake.

"Hour after hour went by. At length, the moon rose. It was one of those still, late autumn nights when frogs are silent, and birds and insects are gone; when only the larger beasts of prey are abroad. There were no owls that night. The leaves had fallen, and covered the ground with a dry and rustling carpet.

"After a while I began to distinguish footsteps among them at a distance. They were faint and stealthy; and I was somewhat in doubt whether it were not my fancy, till the sharp snap of a twig convinced me. It might easily have been a 'lucivee,' or a 'fisher,' or a bear; but somehow I at once connected it with what I had seen in the evening.

"I listened breathlessly.

"The steps were coming nearer. But it was very dark under the thick spruce-boughs. Suddenly the steps ceased; and for a few moments all was still. Then I saw a dark shadow pass a narrow vista where the moonlight fell through the black tree-tops. It had the shape of a man. The steps went on as if the creature, or whatever it was, were passing around us, keeping at about the same distance. Gradually it came around to the point where I had first heard it. There was another pause; and again I saw it cross the moonlit line, to continue its walk around our camp. I wasn't much scared; but its movements gave me a strange

sort of feeling. I remember thinking it was no use to wake
Zeke, or Hughy, who was snoring away at a great rate. So,
cocking my gun, I crept noiselessly down the path we had
beaten to the brook, to get nearer the place where I had
seen the shadow in the moonlight. Creeping up within two
or three rods, I crouched at the root of a fallen tree, and
waited. The footsteps were again approaching in their circuit. There was the same pause as before and again the form
stepped into the moonlight a moment, and was again in the
shadow. But the moon was pouring down brightly; and I
distinctly saw its shape,—the figure of a man, looking brown
and naked, save where a hairy outline showed against the
light. A feeling of sickness or of horror came over me. The
idea of using my gun did not even present itself. I crept back
as silently as I came down. I heard the steps come round
again; then they grew fainter and fainter as the walker
moved off into the forest.

"It was getting toward morning. I sat down to think the
matter over. Presently Hughy woke.

"'You up?' said he. Whereupon I told him what I had
seen. He listened without a word, till I was describing how
it looked as I last saw it; when he exclaimed,

"It's an Indian devil! It's old Pomoola! That's just as I've
heard the Oldtown Indians describe it a hundred times; but
I always thought it was all a lie. They always left a place as
soon as they'd seen one of these things; and I reckin we'd
better.'

"But we didn't leave; and our good luck with the traps
continued, despite Hughy's hints at Indian superstitions.
We were pretty cautious, however, and kept together a good
deal. It was not that we were particularly afraid of it as a
beast; but its singular movements had given us a sort of
dread of it.

"Nothing further was seen for some time. We had begun to fish in the lake for trout. It was alive with them too,
splendid fellows. We frequently caught them as heavy as

ten pounds; and one day Zeke caught a togue which must have weighed twenty or twenty-five pounds. He fairly drew our canoe after him when he was hooked, and it took all or skill to land him.

"I remember we were up near the head of the lake that afternoon. Our camp was at the foot, or lower end. It was getting dusk as we paddled back along. There were several islands in the lake, nearly all of them craggy or high. Just as we were passing the lower one we heard a curious noise,— a sort of 'Waugh, Waugh!' and, looking round to the island, we saw a strange, manlike creature standing upright on a rock overlooking the water. We were not more than eight rods off, and it was not so dark but that we could see it plainly enough. As we stopped paddling, it uttered the same sound again,—a noise between a grunt and a bark.

"I knew at once it was the same creature I had seen before, and told them so. It must have swum half a mile to get to the island. If we hadn't been fools we should have gone up, and found out then and there what it was, and so solved the mystery; for the island was small, and we should have had it completely penned up, and at our mercy. But we were boys then, with our heads full of Hughy's big stories; and as for Hughy himself, all the fur in Maine wouldn't have hired him to go a stroke nearer. Zeke hallooed at it: whereupon it raised its fore-paws, or arms, and swung them about like a drunken man, making the same noise as before. It was growing dark; and we came off and left it.

"The next day we went down round the island; but it wasn't there. It had gone away during the night.

"It was now November; and one morning we woke to find the ground white and a smart snow coming. Towards night it cleared up cold and wintry. Our open camp wasn't very comfortable that night. We waked up shivering. Hughy was wincing under twinges of his old foe the 'rhematiz.'

"'We must get out of this, boys.' Said he. 'Winter's coming.'

"During the day we took up our traps, and prepared for our long tramp southward. We packed our fur in bundles; for we had to back it out for the first forty miles. It was to be our last night there; and we sat about our fire talking over home-matters, and thinking of what might have happened since we left. All at once, Hughy remembered our canoe.

"'We may come here again,' said he; "and its some work to make one. You go down, Cluey, and pull it up out of the lake, and hide it in that little clump of cedars close to the water. It'll keep sound there two or three years.'

"So I ran down to the lake. It wasn't more than a hundred rods. Drawing the canoe out of the water, I stowed it away, bottom up, among the cedars at the foot of a low crag which overhung the lake.

"I was just coming away, when I heard behind me the same queer sound we had heard at the island, and, looking up, saw the beast-man again, standing at the top of the crag. He wasn't more than a hundred feet off: so I had a pretty good view of him as he stood out against the clear sunset sky. It was the same form and shape as before, fully as tall as a man; and I could now see his face. Perhaps it was partly fear; but I did think it had a *devilish* look. There was a tuft of thick hair on the head, which lent a frightful expression to the face.

"If this was what the Indians used to see, I don't wonder they thought it was the Devil. I had my gun, and slowly raised it as if to take aim. The creature raised its arm in the same way. But I had no thoughts of firing; I didn't dare to: and, when I lowered my gun, the creature dropped its arm with another 'Waugh, Waugh!'

"I know I was frightened; yet I saw it plainly enough, and could have sworn to its identity anywhere.

"I don't know how long we stood staring at each other: but I saw it was growing darker; and, stepping backward till I was out of sight behind a cedar, I went into camp about as fast as my legs would carry me.

A Logger's Camp

"Zeke was for going down all together, and shooting at it; but Hughy wouldn't hear of it. He was pretty strongly tinged with the old Indian whims concerning Pomoola, the demon of the mountain near us.

"'We'd no business with it,' he said; 'and he'd have nothing to do with it whatever, unless he was obliged to.'

"The next day we started for the settlements. That was the last we saw of it. Of course, Zeke and I told our story after getting home; and I presume it never increased our reputation for veracity among our neighbors. Hughy showed an old hunter's wisdom by keeping still about it. When persons who had heard us asked him, he merely said that we did see something rather queer; and that was all they could get out of him. Zeke and I pitched into him once for not substantiating our account better.

"'No use, no use at all,' said the old man; 'and I ain't going to get laughed at for nothing.'

"I've thought about it a great deal since; but I never could satisfy myself what it was we saw. I've heard of wild men, of children carried off and reared by wild beasts; and the Indians were always telling of Pomoola: but I never could settle it in my mind. I know there are a great many things in the Northern wilderness which the 'scientific men' would laugh at a person for seeing or trying to describe.

"But here's my story. You can take it for what it is worth; and so must the reader. But we record it as a very fair specimen of hundreds of similar 'yarns,' common among the lumber-men and Indians, concerning the fabulous being or demon of the Katahdin region. My opinion is that it is all pure bosh, not only this story of Cluey's but the whole batch of them."

I heartily concur with Wash; though it does seem strange that there should be so many stories with no foundation whatever in fact.[38]

The Adventures of Six Young Men
in the Wilds of Maine and Canada
C. A. Stephens

The Woods-Demon. Returning after our hunt that afternoon, we passed a deserted lumberman's camp, half hidden in the black, rank firs which had sprung up in the clearing about it.

"That ere's the old haunted camp," Nugent remarked.

"O, haunted is it?" said Rike. "What's it haunted with, bear-ghosts?"

Nugent affirmed that the place was commonly reported by woodsmen to be haunted by the spirit of a strange unhuman creature which had years before been killed there by a logger crew in which Nugent himself had worked. That evening he told the story. I give it in substance, as setting forth a strange, yet possible, physiological fact.

THE HAUNTED LOGGING CAMP—NUGENT'S STORY
There were seventeen "choppers" and four teamsters, besides the foreman and cook, in our crew that winter at the camp over here—twenty-three men in all, partly French, and more than half of them Catholics.

In February, towards the last of the month, after they had been in the woods thirteen weeks, there came on at night one of those fearful northeast snow-storms such as are known only in British America and Siberia. Overhead the wind roared and shook the tree tops, and the snow, fine as meal, was sifted blindingly down through the frozen boughs.

So full was the air of snow that the voices of the men seemed muffled as they came in from their work. In an hour and a half a foot of snow had fallen, and the old "loggers" predicted a fall of four feet by morning.

But, gathered before the fire in their warm log camp, saluted by the savor of a bountiful supper, the hardy fellows cared little for the terrors of the storm outside.

Supper was nearly ready, and Lotte, the cook, had taken up the three-gallon teapot to fill, when he discovered that the supply of tea was out. If there was any tea drank that night, it would be necessary for some one to go down to the wangin, on the river bank, half a mile below.

The wangin, or storehouse, was a strong structure of heavy logs, wherein was kept the winter supply of flour, pork, sugar, molasses, etc., which had been poled up the river in bateaux before the ice had formed.

Leading down to it from the camp there was a road cut through the timber. The men, however, generally avoided going there after dark. There were wolves about, and the loup cerviers of that section are remarkably large and fierce.

Bidding them wait patiently a little while, Lotte began to muffle himself for the tramp, when one of the French boys, a youngster of twenty, named Marc Lizotte, offered to go in his place. Lotte very thankfully handed him the key and a basket, and Marc set off at a run through the snow.

He had been gone scarcely ten minutes, when terrified shouts startled the camp.

The men rushed to the door. Hardly had they opened it, when Lizotte leaped in amongst them, shouting,

"Frap la porte! Frap la porte! Pour Dieu!"

One or two of the men declared that they heard a hoarse cough at the same moment, close at hand; but it might have been the wind.

They saw nothing, and all cried out, "What is it?"

"I don't know," panted Marc, "Je crois que c'est le diable lui-meme!"

A laugh followed this honest avowal.

The boy said that he had come within a few rods of the wangin, when, distinctly above the roar of the storm, he heard heavy blows, as of some one pounding on the door with an axe or club. He stopped short, and, peering sharply through the darkness, made out the form of a man, as he at first thought, breaking in at the door.

He watched a moment; then called out to know what he was doing there. Instantly the creature turned, uttered a most unearthly cry, and sprang after him with uplifted club. Marc ran for his life.

He said that the "tison d'enfer," as he called it, had chased him to the very door of the camp, and at one time was almost at his heels.

Many of the men ridiculed the whole story. "Marc got scared," they said. Others thought that it might have been a thief, or some wandering hunter, possibly an Indian, who had tried to break into the wangin.

Three of four were for going down, but the "boss" said that nobody could break through the log door, even with an

axe; and as it was storming fiercely, none cared to make
the trip in the snow and darkness.

Early next morning, three of us on snow-shoes set off
to get tea and look for traces of "Marc's bugbear." We found
them easily enough.

The wangin door was broken open, the great bolt
wretched from its socket, and the huge log door-post split
and splintered. The snow had drifted in.

It was not without many cautious glances that we ventured to enter.

The room was empty. So far as we could discover, nothing of value had been stolen. From one of the open barrels
of pork a chunk had been taken out, gnawed, and the rem-
nant thrown into the snow. That was all.

The snow had covered the tracks, both Marc's and the
creature's. We should have believed it the work of some wild
beast, but from the fact that no wild beast could have broken in that door.

This happened on Tuesday. The next Sunday two of the
choppers, named Leverett and Corbain, went out to hunt a
caribou, the tracks of which we had lately seen in the snow
near where we had been felling timber. About three in the
afternoon they came back and told a curious story.

They had come upon the caribou not an hour after starting
out; but it had run off through the snow, and they had to
follow it an hour or two before getting near enough to shoot
at it with the old musket they had carried from the camp.

Finally they fired, and wounded the animal so severely
that after running a little way it fell in the snow. They had
no difficulty in killing it.

Leverett then began to skin and cut up the carcass, with
a view to take the best portions of the meat back to camp.
While thus engaged, they heard a trampling of the snow,
and looking up, saw a "geant"—so Corbain called it—coming towards them, snuffling horribly and brandishing a
great bludgeon.

Leverett dropped his knife (they had not stopped to re-load the old musket) and both he and Corbain ran away as fast as they could.

They described the creature as a monster, a giant bare-headed and bare-legged, with matted hair, but wearing, they thought, the skin of a bear about its shoulders and body. Its arms were bare, brown and hairy, so also was its face, and the hair on its head was thickly matted and stood up in a frightful tuft.

We could scarcely believe so queer a story; several of the men openly ridiculed it. Eight or nine took their axes and followed Leverett and Corbain back to where they had shot the caribou.

They easily found the spot; but the carcass of the cari-bou had been dragged away. They followed the trail in the snow for a mile or two, and then, as night was coming on, turned back.

Of course there was plenty of talk and surmise. The French boys would hardly stir out of camp alone, and the Province men teased them continually.

A few nights after, the wangin was broken into again. The "boss" had repaired the damage done the door by the former attack, and strengthened it by two new bars of green ash, a foot in width and four inches thick. This time the hinges were torn out of the post, and the door wrenched outward, leaving the bars intact.

But, as before, only a few chunks of pork were missing. It was plain to anybody that it must have taken prodigious strength to pull out the hinge irons, fastened and clinched as they were into the door post.

The men began to look serious, and to talk less about it.

That day four of us Province men agreed to go to the wangin, and watch for the man, creature, or whatever it was. The boss was willing.

We quit work early, got our supper, and went down to the wangin a little after sunset. We took our axes and the old gun, which we used to shoot game with, loaded with slugs.

Going inside, we pulled the door together, and drove the hinge-irons back into the post, and then barricaded it on the inside with barrels of flour. There was a number of bales of hay in the wangin. We cut open one of these, and piling the hay against the front side, lay down on it, first pulling the moss out from between the logs, so that we could look out.

It was a still, clear night with a new moon, which soon set behind the tree-tops, leaving it dim, yet not very dark. The names of the three men with me were Hanley, Carnise and Foley.

The evening passed tediously enough, for it was pretty cold lying there with no fire. Nothing stirred about the wangin till after eleven o'clock, when I heard steps in the snow at some distance.

I whispered to my mates, and we listened intently. The steps came nearer. We peered out between the cracks, but it was too dark to discern objects off in the woods.

A moment later Foley muttered under his breath, "There it is!" and then I saw a dark, indistinct figure, not twenty yards from the wangin. For some moments it stood there motionless, then slowly came up to the door.

It seemed to be the form of a very large man, but even in the darkness we perceived that he was half naked, and of strange mien.

We noiselessly got on our feet, grasping our axes, Carnise holding the gun. I confess to have been a little scared.

There was a sound as of nails on the door, then a heavy push against it, which made it crack and the barrels rattle.

"Who's there?" shouted Hanley.

For a reply there was a kind of snort, followed by a tremendous blow from some heavy instrument. At the same instant there came a blaze of fire and a stunning report.

The old gun had gone off in Carnise's hands. He was holding it cocked and it went off itself—he said.

As soon as we recovered from our flurry, I looked out through the cracks, but could see nothing of our disturber; and after looking and waiting an hour or more, we ventured out and went up to the camp, for it had grown unbearably cold.

Our story re-excited the French boys very much.

Two nights after, one of them saw, or fancied he saw, the creature stealing up behind him as he came in from the ox-camp, a few rods back of our shanty. Then their terrors took voice.

They were going to leave and get out of that place. Parbleu! 'twas the devil himself.

The foreman, Lamson, was used to the French.

"It may be the devil," he said, and started four of them off to the settlement after a priest. He knew that the presence of a priest was the only way to keep the gang from bolting. When once these French Catholics are scared, nothing can quiet them but a priest.

The night after the priest was sent for, we were roused up, about one o'clock in the morning, by a tremendous uproar at the ox-camp. There was a great racket, and the oxen bellowing terribly.

"Some of them are loose and goring the others," Lamson exclaimed, and half a dozen of us turned out to quiet the hubbub.

Foley was ahead. The great door of the ox-camp was open. This circumstance of itself ought to have warned us, but we were too stupid with sleep to reflect.

As Foley entered the doorway, something struck him on the head and knocked him down, senseless. A blow was also aimed at Corbain, but he jumped aside, and we all ran back to the shanty for our lives.

Lamson caught up the gun. The rest took their axes and rushed out.

We found Foley lying in the snow. He was beginning to stir a little. There was a fearful cut on his head, but whatever had struck him was gone.

The ox-camp contained nothing but the oxen.

One of them was bleeding from his nose profusely, and kept shaking his head as if half stunned by blows. That same night the wangin was again broken into and more pork taken.

Nothing now but the expected arrival of the priest kept the "Frenchers" from stampeding.

The Province men kept quiet. They did not know what to say or think. Lamson spent the forenoon, gun in hand, looking about in the woods for signs and tracks. There had no snow fallen for over a week. Our own tracks and those of the teams had trodden the snow hard about the camp; but out in the woods Lamson found in many places the print of a large moccasined foot. It left no heel print. The boots and moccasins worn by the men had raised heels.

That afternoon the boss called Carnise and myself off from work an hour an hour or more before sunset.

"I want you two men to go down to the wangin with me," he said.

We went to the shanty, and then set off for the wangin, Lamson taking the gun.

"I'll know whether this critter is flesh and blood or not," he said, as we went down. "I'm going to set a spring-gun for him. It doesn't act like a man; but if it isn't a man, what can it be? At any rate, he's a murderous wretch. Devil or not, I'm going to try a charge of slugs on him."

We helped Lamson set the gun inside the wangin, with a line running from the trigger across the doorway, and the muzzle pointed at about breast height of a person entering the door. The door itself we left just as it had been found that morning, wrenched outward and forced back on the outside.

Lamson told the men at supper what we had done, and bade them keep away from the wangin. Foley was getting better of his wound.

So much had the setting of the gun and the attendant circumstances excited me, that I did not go to sleep that night, but lay thinking and listening.

The night wore on. It was not till after two in the morning that the gun was sprung. Its dull, heavy report startled me, and set me trembling all over.

All the men were expecting it, and seemed to be but half asleep. In a moment everybody was astir. Lamson went to the door.

"We've fetched him!" he explained; and, on going out, we could plainly hear cries, as of one in pain and rage.

The French boys refused to stir out of the shanty; but the most of the Province men set off with Lamson, who had lighted a lantern.

The cries burst out at intervals as we hurried down the wangin path.

It was with strange feelings that we approached the place. By the light of the lantern we saw the brown, hairy body of a man, half covered with a bear hide, lying just outside the door. There had been a desperate death-struggle.

The Wild Man

The five slugs with which Lamson had charged the gun had pierced him through and through. But he died only after many struggles and outcries.

I had never imagined anything like the ferocious expression of the face. The skin of the body was very brown, and much covered with hair. The legs, especially, were almost shaggy. The feet were tied up each in the hide of a lynx, and the bear hide hung from his shoulders, fastened with a thick thong around the neck. There was no other clothing.

Both arms and legs were rough, dirty, almost horny. The face was repulsive, and the hair matted. Though lean, the body must have weighed fully two hundred pounds.

There was every indication of great physical strength,—instances of which we had certainly seen. Near by lay a large, greasy bludgeon, between three and four feet long, which must have weighed ten or twelve pounds.

The idea which we settled on was that it was a backwoodsman, half-breed Indian, perhaps (it was hard telling), who had years before gone crazy. By chance he had been lost in the woods, and had gone on leading a wandering life, eating raw flesh for food, till the man had well nigh changed to a fierce animal. Instances are known to lumbermen where persons getting lost in the woods became deranged.

That night the men whom the foreman had sent down to the settlement came back, and a priest came up with them.

The corpse of the madman we had killed was put into a roughly constructed coffin and buried in the snow.

Lamson asked the priest, whose name was Villate, whether he wished to conduct a funeral service over the body. He declined to do so.

Lamson then asked him what he thought of what we had done, and whether he deemed it a murder. And to this question from the French boys, he went back, taking two men as guides, and these men did not return.

But the general opinion of the men was, that it was not a murder, and that such a being as this man had better be put out of the way than suffered to go at large in the woods.

Such was Nugent's story.[39]

Charles Clark Munn's popular book, *Uncle Terry, A Story of the Maine Coast*, written in 1900, is thought to have influenced the story of the Marr family of lighthouse keepers at Hendricks Light finding a baby washed up on their shore.[40] Could Munn's *The Hermit*, likewise have contributed to the athletic myth of the Wild Man?

The Hermit
(Extract from Chapter 2)
Charles Clark Munn

When the camp-fire was reached, Martin had arrived and was cutting boughs with a hunting-knife, while Jean and Levi were just entering the tent. A fire had been started, a blackened pail had been hung from a stick over the flame, and preparations for a night in the woods were well under way. Into this little group leaped the terrified doctor, breathless, with face scratched and bleeding.

"Pack up, quick!" he exclaimed in a husky voice; "we've got to get out of here at once! There's a wild man back up in the woods, and I wouldn't stay here for a million dollars!"

Martin and Levi exchanged quick glances, and a halt came in the camping work. For a moment the two looked at one another, and then, as if recalling that curious footprint they had seen twenty miles away, they glanced furtively up the bush-choked log road. One instant only Martin hesitated, and then he recovered himself.

"Doctor," he said, "I expected you would get well scared the first time you went into the woods alone, and I see you have. What you saw, most likely, was a blackened stump half hid in the bushes, or possibly it might have been a bear. If so, he is a mile away by now, more scared than you are. Here, take a drink, brace up, and help us to make camp. It's almost dark."

But Dr. Sol was obdurate. "I tell you, Martin Frisbie," he replied sternly, heeding not the proffered flask, "I wouldn't stay here a night for love nor money. We are watched, and by the most savage-looking creature I ever set eyes on." Then, with many additions, as might be expected, he told the story of his fright.

Martin and Levi exchanged knowing looks once more, but made no comment until the tale was told, and then Martin spoke.

"Levi," he said, "what do you say; is it go on, or stay?"

"We've got to stay!" came the resolute answer; "thar ain't a campin' spot within five miles either up or down the Moosehorn, and it's too late to cut one out!" And once more he began work.

As for Martin, he was inwardly nervous but outwardly calm. He had not quite recovered from the previous night's experience and the queer footprints, however, and yet it did not occur to him that that had any connection with the cause of the doctor's fright. And yet, it might have.

Then another thought came, and it added to his fears. They had started early and paddled a good twenty miles up an almost currentless stream; on either bank lay an impassable wilderness, much of it swampy. No hunter or trapper stealing along ahead had been sighted that day, and if this wild man the doctor had seen was he who was prowling around their tent the night before, how had he reached this spot?

But Martin had already decided upon his own course, and though startled somewhat by the doctor's fright, he now pulled himself together once more and attempted to calm his frightened comrade.

"It may have been some hairy-faced, old trapper that you saw, doctor," he said finally, "and they are harmless. If it was, he will show up by and by, and hang around till we offer him a drink. I've met them many times here in the wilderness before, and a little good rum secures their friendship for life, so don't worry." And Martin resumed his cutting of boughs.

When supper was over and night had quite shut them in about the camp-fire, conversation was resumed and Levi told a story with the seeming intent of allaying the doctor's fears.

"Thar's a good many hunters 'n' trappers livin' up here in the woods somewhar," he said, "'n' thar's no tellin' when one on 'em 'll show up. I was campin' one fall, way up on the Allagash, me 'n' Pete Roncou,—that's a cousin o' Jean's here,— 'n' we had some traps set for otter. Well, thar was a storm coming up that night, an' though we'd been thar a week, we hadn't heerd a single loon, an' knew 'twas so late they'd all gone South, an' all 'twonce we heard one clus in shore. 'Pete,' said I, 'that's queer; we ain't heered a loon sence we came, 'n' now thar's one squallin'.' Well, a loon, long way off, sounds good deal like a human, but clus to, not a mite. In a minute it squaked agin, 'n' this time it sounded jist like 'h-e-e-e-l-p!'"

"'I tell ye, it's a human,' said Pete, an' with that we each on us grabbed a brand 'n' started. I don't just know what made me, but I grabbed my gun, too; but Pete, he never thought o' his'n. Well, we heerd it again, 'n' this time 'twas a reg'ler human holler for help. We hallooed back 'n' it answered 'n' we kept on goin' 'n' wavin' the sticks to keep 'em goin', an' finally they went out, an' arter they'd gone out, we kept on hollerin', but didn't git any back. We hed hard work to git back ourselves, tho', fer we hed to guess at it."

He paused, as if the sequel were not worth telling.

"Well," put in the doctor, "what then? Did you ever find what it was?"

"Not 'zactly, but we sorter guessed from the moccasin tracks 'twas an Injun that had sneaked into our camp while we was loon huntin', fer Pete's gun an' all our shells was missin'. That was more'n ten years ago, 'n' Pete ain't heerd the last on't yit."

"Are there many Indians wild in these woods now?" queried the doctor, glancing up to where the zone of firelight outlined the entrance to the old tote-road; "I thought they were all civilized."

ONE MOMENT ONLY HE SAW THE GRAY, HAIRY VISAGE — Page 24

"So they are," replied Martin, not waiting for Levi, "and that's why some of them adopt white men's methods of getting what they want."

"But the face I saw belonged to a white man," interjected the doctor, who had not recovered from it, "and it wore a most demoniac look, with grizzly hair all around and a mat of it on top."

"That may be," returned Martin, "and so would any old trapper look when you saw him. They never shave or get a hair cut from one year's end to another, and all look alike— ragged, hairy, and dirty. I've met them often, and, as I told you, they are all harmless and love rum. If you saw one,— which I doubt,—he is like all the rest, and by now is fast asleep up back of here in the bushes."

With that Martin arose, for it was time to turn in, glanced first at the starlit sky and then up at the opening in the forest back of the tent. At that moment Levi chanced to throw a handful of fir boughs on the dying fire, and as the flames flashed in response and the zone of light widened, Martin caught the full view of a hideous human face peeping out from behind a stunted spruce.

One moment only he saw the gray hairy visage; the next it had disappeared.[41]

OLD KING SPRUCE

". . . In the night when he awakens, all a-shiver in
   his bunk,
And with ear against the logging hears the steady,
   muffled thunk
Of the hairy fists of monsters beating there in grisly
   play—
Horrid things that stroll o' night-times—never, never
   seen by day.
                                    The Ha'nts[42]

Holman F. Day

Holman Day was a native of Vassalboro and author of over 300 short stories, 25 novels, as well as many plays and several volumes of poetry. Day joined the *Lewiston Evening Journal* in the 1880s and began writing his popular "Up in Maine," column in the early 1890s.[43] Day loved to write of the Maine woods and held the lumberjack in high esteem.

Old King Spruce
Holman Day

Chapter IX. The Ha'nt of the Umcolcus
Abe was on his knees, stretching up his neck and twitching his head from side to side with the air of an agitated fowl.

"We'll make it a rule after this to have only common songs—like Larry Gorman's," continued the boss, with a quizzical glance at the woodsman poet. "These high operas are too thrillin'."

But those who stared at Abe promptly saw that his attention was not fixed on matters within, but without.

"He heard something." Muttered one of the men. "He's got ears like a cat, anyway."

If the giant had heard something it was plain that he heard it again, for he dropped his knife and scrambled to his feet.

"Me go! Yes!" he roared, gutterally; and, obeying some mysterious summons, his precipitateness hinting that he recognized its authority, he ran out of the camp.

"Catch that fool!" yelled the boss; but the first of those who tumbled out into the dingle after him were not quick enough. The night and the swirling storm had swallowed him. A few zealous pursuers ran a little way, trying to follow his tracks, lost them, and came back for lanterns.

"It's not use, Mr. Wade," advised the boss. "He's got the strength of a mule and the legs of an ostrich. The men will only be takin' chances for nothin'. He's gone clean out of his head, and there's no tellin' when he'll stop."

And Wade regretfully gave orders to abandon the chase. He and the others stood for a time gazing about them into the storm, now sifting thicker and swirling more wildly. He was oppressed by the happening, as though he had seen some one leap to death. What else could a human being hope for in that waste?

"He's as tough as a bull moose and as used to bein' out-doors," remarked the boss, consolingly. "When he's had his run he'll smell his way back."

Teamster Tommy Eye was the most persistent pursuer. He came in stamping off the snow after all the others had reassembled in the camp to talk the case over.

"Did ye year it?" demanded Tommy.

"I did, and I run like a tiger so I could say that at last I'd seen one. But I didn't see it. I only heard it."

"What?" asked Wade, amazed.

"The ha'nt," said Tommy. "I've always wanted to see one. I was first out and I heard it."

"What did it sound like?" gasped one of the men, his superstition glowing in his eyes.

"It's bad luck forever to try to make a noise like a ha'nt," said Tommy, with decision. "Nor will I meddle with its business—no, s'r. 'Twould come for me next. Take a lucivee, an Injun devil, a bob-sled runner on grit, and the gabble of a loon, mix 'em together, and set 'em, skim off the cream of the noise, and it would be something like the loo-hoo of a ha'nt. It's awful on nerves. I reckon I'll take a pull at the old T. D." He rammed his pipe-bowl with a finger that trembled visibly.

"I've seen one," declared positively the man who had inquired in regard to the sound. "I've seen one, but I never heard one holler. I didn't know it was a ha'nt till I'd seen it half a dozen times."

"Good eye!" sneered Tommy. "What did it, have to come up and introduce itself and say 'Please, Mister MacIntosh, I'm a ha'nt?'"

"I've seen one, I say," insisted the man, sullenly. "I was teamin' for the Blaisdell brothers on their Telos operation, and I see it every day for most a week. It walked ahead of my team close to the bushes side of the road, and it was like a man, and it always turned off the same place and went into the woods."

"Do you call that a ha'nt—a man walkin' 'longside the road in daylight—some hump-backed old spruce-gum picker?" demanded Tommy.

"The last time I see it, I noticed that it didn't leave any tracks," declared the narrator. "It walked right along on the light snow and didn't leave any tracks. Funny I didn't notice that before, but I didn't."

"You startingly ain't what the dictionary would set down as a hawk-eyed critter," remarked Tommy, maliciously. "It must have been kind of discouragin', ha'ntin' you."

"It was a ha'nt," insisted the man, with the same doggedness. "I got off'n my team right then and there and got a bill of my time and left, and the man that took my place got sluiced by the snub-line bustin' and about three thousand feet of spruce mellered the eternal daylights out of him. Say what you're a mind to—I saw a thing that walked on light snow and didn't make tracks, and I left, and that feller got sluiced—everybody in these woods knows that a feller got killed on Telos two winters ago."

"Oh, there's ha'nts," agreed Tommy, earnestly. "Mebbe you saw one; only you got at your story kind of back-ended."[44]

Hobgoblins of the Wild North Woods
Winthrop Packard

Tozier, the lumber scaler, told me tales one night when the big stove roared and glowed dull red in the womgan. The spruces stood waist deep in the drifted snow outside and now and then cracked with the crushing grip of the frost on their mossy boles. No breath of wind stirred their tops and the white peak of Katahdin stood silver in the moonlight, pointing toward the myriad stars that almost blot out the blue of a northern Maine winter night sky.

Bitter cold it was in the open spaces as we tramped in that night, and the snow seemed to ring under our snow shoes, so hard were its crystals. Yet under the shade of the spruces the warmth of the day's sun seemed caught and held prisoner that the wild creatures of the wood might not suffer too much. I had started and asked, "What's that?" once as he drew near the camp, for shadowy forms with glowing eyes stared at us for a moment, then slipped away into the night, ghosts or bogies I verily thought; but Tozier only smiled and said that it was a buck deer and two does that were feeding on the tops of the trees that the choppers had felled that day. Then when we were almost at the womgan door he stopped and held up his hand. I listened, and down the slope from the base of the weird white mountain came sounds such as I have never heard in the woods before, a babbling wail as of a madman talking and mourning to himself. It might be near or far, one could not tell, for sounds carry a long way in the still, frosty air of the mountain night. There was something almost human in it, yet not sanely human, and there were notes, too, such as one would think only a beast could utter. It gave me a strange shiver of fear and pity to listen to that voice that mourned and snarled and complained to itself in utter loneliness. Then it ceased, and Tozier dropped his hand without a word, tramped up to the womgan door, unfastened his snow shoes and walked into camp. I followed speculating whether it might not be foxes or possibly a lucivee, and I asked no questions. Foxes in the rutting season make all sorts of weird noises, and with the lucivee's cry I was not familiar.

We ate supper with the gang in the cook house, then retired to the womgan again, where sleep the boss, the foreman, and any chance visitors. Tozier puffed at his pipe for a long while. I eyed him in silence. At length he spoke.

"What do you think it was?" he asked.

"A lucivee," I replied, promptly. Nothing is so serene as a little knowledge. But the man of the winter woods shook his head.

"Foxes," I guessed.

"I don't know but a fox might make noises like that," he said. "They make all kinds of queer talk in the mating season, but—" There was silence again. I noted the white rim that the frost had painted about the cracks of the closed door, where the below-zero chill from without met the warmth of the shadowy interior. The weird chill of the frozen wilderness seemed creeping in on us, and I bustled a bit ostentatiously and put another white birch log on the already roaring fire.

"Men," said Tozier, "who get lost in the woods go crazy after a little, and the best of us are liable to get lost up here, especially in winter. Scaling lumber as I do, I go from camp to camp all winter long. Now and then I get mixed and have to lie out a night, and I would not think of going without a compass and a map. The best of woodsmen will get turned round in a snowstorm, and more than once has gone out and never been heard from again. They lost their bearings, then their wits, then they die of exposure, unless—"

He listened again intently, but there was no sound from the mountain. I listened too with something of the creepy sensation of a man who expects to see a ghost. I had heard vague hints of the wild man of Katahdin, a giant in strength, who howled like a wolf and lived as did the beasts of the forest, but I had hitherto thought little of it. The lumbermen had their superstitions, no doubt, but they had been reticent about this one. My companion had seemed a hard-headed man of business, a man of figures and routes, of the lumber rule and much knowledge of townships and land owners. It would not be like him to repeat or believe bogie stories.

"There was old Bill Henderson," he said, "a man who had lived all his life in the woods, lumberman, trapper, timber-looker, guide; familiar with the mountains as any man could be. He started out one winter day to go eighteen miles across to another camp. He had his little dog with him, or that might have been him that we heard just now. A light snow fell that afternoon, but the people at the camp weren't expecting him back for a week or so, and they knew he was all right. They didn't know he was coming, over at the other camp, so they did not worry about him. Three days later some one came over from the other camp and reported that old Bill hadn't been there. We all set out right away to hunt for him, and on the third day we found him. Some one heard the little dog yapping, and we went over the ridge, and there, sure enough, was old Bill, running about on his snow shoes, jumping up in the air and grabbing at things. He would laugh in a crazy way, and the little dog was frolicking about as if he was having the best time in the world. Bill never took notice of us till we spoke to him, and then he turned and ran like a moose. We had to run him down to catch him, and never could have done it except that he was weak from hunger and exposure. He came into camp like a lamb then, and although he was out of his head for a day or two, mumbling and laughing, he finally came round all right. You can't get Bill Henderson to go out in the woods now for love or money. He's afraid of them.

"What was he doing when we caught him? Oh, he thought he was catching birds. They were the finest and most beautiful birds he ever saw and it made him laugh to think how easy it was to catch them. Crazy as a loon, you know. If we hadn't found him just as we did, it would have been the last of old Bill. He might have died or he might have turned wild man and lived the way this one does."

"Look here, Tozier," I said. "You don't believe there is really a wild man on Katahdin?"

Tozier shrugged his shoulders—a gesture he must have caught from the French inhabitants of the northern counties. "Why not?" he said. "You heard that crying up on the mountain side tonight. I have heard it before. There are men who say they have seen him, a great hairy fellow with big red eyes, half clad in rags and fur. Why should not a man who knows the woods well live in them for years even is he is crazy? There's lots of strange things in the woods that folks of the town don't understand."

Now, I don't believe in spooks and hobgoblins when I am away from their probable location—down in the electric-lighted city, for instance, but there is much in the deep woods that is a cure for skepticism. Up there the forest feeds fancy and the voices or the unseen speak to the dullest in the wilderness at night.

A little wandering wind came up and puffed frost through the crack of the door; shook it, then fled away with a sigh. I drew a little nearer the stove. Somehow I felt a chill.

"Are there any ghosts up here?" I asked. Tozier replied uneasily. "I guess that was the wind. If we were up on

Chesuncook I should not feel so sure. The French people who live about there say that old man Farrington comes down the lake on the ice. I suppose old man Farrington is the only man that was ever buried in so strange a fashion, and it is no wonder he walks. He died long before I was born, but this is the story as the old people tell it: Farrington came into the woods way north of Chesuncook when the lumbering operations first began up there, He cleared a good-sized farm and made money selling farm produce to the camps. His wife died, his children grew up and moved away, but the old man lived along there alone and was thought to have a good deal of money. Finally word came down the lake that he was missing. People went to his house and found him there dead and buried. He had gone about it in original fashion, too. He had taken up the floor of his house and dug a grave there. He put a box at the bottom for his money, then he piled the dirt on a swinging platform just above it, so rigged that a little would change the balance and dump the earth into the grave. Them, they suppose, when he felt that his end was near, he lay down in this grave in such a way that when he died and his head fell back it would pull a string, loose a toggle, and dump the earth in on top of him. That's the way they found him, at the bottom of a self-filling grave, the only one of the kind ever invented, but they found no money in the box. Therefore, they said that he was murdered after all and the thing rigged that way to avert suspicion. At any rate the French people say that he comes down the lake on the ice on the trail of his murderers. Lots of them claim to have seen him."

"Did you ever see him yourself?" I asked.

"I have seen something up there," he said, "and I would not care to see it again. Chesuncook is a queer place in winter. You see all sorts of things there sometimes. The tote road lies along it on the ice for miles and they mark it with birch tops stuck in the snow so that when it blows and drifts over level the teamsters can find the road. I used often to go up and down this on snow shoes. Sometimes I have seen those birch tops grow in the distance till they made a regular forest, and a man who did not know would think the woods had come down and camped on the lake. I have seen a little wooded land loom up like a mountain, and when you got near it it would go up in the air and vanish completely. Yet it was just ahead of you, and if you walked on you would find it again. This happens on still days, most often about sunset."

"Mirage," I said, with something of the serene confidence with which I had guessed that the weird sounds on the mountains were the cries of the lucivee. I had seen the same strange effects on still days in the arctic ice.

"Maybe it is," said Tozier, without enthusiasm for my theory. "I have seen a man coming down the lake look as big as a meeting house when he was three miles away, then go out of sight entirely. By-and-by he would be right in front of you as if he had sprung out of the ice. I have seen what I thought was a tote team till I got right up to it, and then it was nothing but a little bunch of dried leaves. Those things make a fellow feel uncomfortable when he is out along on the lake at dusk.

"One night I had taken supper at McLean's Camp and went on up the trail. The teamster's wife was sick down below and she was expecting him out that day. At dusk when I got well up the lake I thought I saw him coming toward me. Then he changed to what was a tall woman dressed in brown. As he got quite near he seemed to be a big man again, dressed in fur, a mighty big man and not the teamster at all. I began to feel queer about it and wish I was back in camp again. Then he came right at me, eight feet tall, all furry, and with big red eyes that stuck right out of his head. It was too much for me. Why, I saw him as plainly as I see you and he was almost as near. I turned and ran for the shore, which was near by, he following, but not very fast.

"So I cut a big hickory club and waited for him and waited for him. He stopped and glowered at me, and someway my mad got up. I was ashamed to think I had run at all, and I started back toward him with the club. As I approached he edged off on a slant, glaring at me all the time and making for a point that juts out into the lake, and I after him. But it is no use to chance one of those things. When he got to the point he simply turned into a rotten wood stub standing there, eight feet tall, with two broken branches that stuck out like arms. It gave me the queerest feeling I had got yet in the whole affair. I had been by that point forty times in my life and there had never been any stub there. I went up to it and hit it a whack with my club, and it gave a grunt just as a man would who had been struck in the chest. Perhaps that was mirage, too. But how do you account for the stump that was never there before and is not there now, and why should it grunt when I hit it?"

It was a strange tale, this one of the furry man with great red eyes that became a stump when attacked, where no stump had been before and no stump was afterward. I did not believe in the ghost of old man Farrington, of course. I did not believe in demon stumps or hobgoblins of the north woods—that is, when I was away from their supposed habitations, but I felt queer little thrills of dread as Tozier told his simple straightforward story in the gathering gloom of the womgan. The fire was low anyway and I reached for another stick to put on it.

What was that? The little wandering wind had come back again. I could hear it breathe as it stood outside the door, then it tapped again on the little womgan window. Then I could have sworn that something walked away in the direction of the dingle. I sprang to my feet, seized the rifle that stood by the door, and flung the door wide.

Night and the keen chill of the winter woods—that was all. But Tozier held my hand and pointed as I would have closed the door again. A shadow moved by the dingle: Was that the sound of deep breathing that I faintly heard? Then we both jumped, for there was the crash of a heavy blow, the ripping of broken broads, and the sound of someone or something that ate, with the champing of a hungry hog. We rushed toward the dingle with a shout. A great furry creature stood just in the doorway, erect, eating something which it held in both paws, or both hands.

It looked fiercely with glowing eyes at us for a moment but yelled as I leveled the rifle at it and fled away into the shadow of the spruces, now erect, now on all fours, and babbling and wailing to itself in the same tones that I had heard up on the mountain side hours before.

Tozier caught the rifle as I would have fired. "Good God, man!" he cried; "Would you kill a man?"

"It—it was a bear." I stammered.

In answer Tozier pointed silently to the tracks in the snow, visible in the dim light of the lantern which the cook, soonest roused of the crew, had brought. They were the tracks of great bare hairy feet with stubby nails, or claws. A bear would make tracks something like that, so might a wild man. But I can't make a lumberman believe that a bear would come out in severe winter weather. I never heard of one myself. The creature, whatever it was had broken open a barrel of pork, ripped the staves asunder with giant strength and was feasting on the contents when we arrived.

As I have said, I do not believe in ghosts and hobgoblins, at least not in town and in the daylight; but all the same I'd give a good deal to know what it was that whimpered and babbled on the mountainside that bitter winter night and ran yelling from the dingle pork barrel when I leveled the 43-70 at it.[45]

BIGFOOT IN MAINE RESOURCES

www.bfro.net/GDB/statelisting.asp?state=me
A popular and respected site devoted to aggregating Bigfoot sightings. The Bigfoot Field Researchers Organization (BFRO) online site is a community of active participants engaged in the search for Bigfoot. The BFRO also hosts excursions into the wilderness in search of Bigfoot. The founding of the BFRO dates back to 1995, making it one of the oldest organizations of its kind. The link provided is for the BFRO's Maine page of state sightings.

www.bigfooteencounters.com
Bigfoot Encounters has been in existence since September 1995 and was started by Bigfoot researcher Bobbie Short. As a repository for information on Bigfoot, the site has many articles, discussions, and a state-by-state listing of sightings, including over 20 for Maine. Check the Sightings' link (for a state-by-state breakdown of bigfoot sightings) as well as both the Newspaper & Magazine Articles and Stories, Sightings, Encounters, Letters links for additional encounters and stories in Maine.

bigfoottales.worldpress.com
This site is by Maine author and photographer D. L. Soucy. The site's inaugural posting on May 23, 2012, provides its *raison d'etre*: "This blog will primarily address the folklore behind these creatures,

the history of the strange world of Sasquatch and the Yeti-like animals that inhabit the dark recesses of our forests, as well as our minds." The site contains several postings focusing on Bigfoot in Maine and serves as a forum for Soucy to share notes and results from his fieldwork.

www.facebook.com/MaineBigfootSociety
This is the Facebook page for the Maine Bigfoot Society "to be used as a network for witnesses, believers and local bigfoot stories."

groups.yahoo.com/neo/groups/MaineBigfootResearch/Info
"This group will focus on Bigfoot activity in Maine. The idea is to have a place where Maine people can report sightings and they can be investigated by people who live in Maine and know the area and are available to get to the sight a.s.a.p. [sic] without losing valuable time!" This group replaced the defunct Maine Bigfoot Research Group.

strangemaine.blogspot.com
This blog was founded in 2005 by Michelle Souliere. "This site is a nexus for conversation about Maine's unique strangeness, people who love it, people who have experienced it, & people who are intrigued by it. History, mysteries, legends, current events, cryptozoology, & more." Souliere's important research has dug up many fascinating stories on strangeness in Maine including several stories and accounts of Bigfoot and Bigfoot-like creatures.

AUTHORS

Loren Coleman: A leading and respected cryptozoologist living in Portland, Maine, Coleman has been conducting fieldwork and research on cryptids for over 50 years. Coleman began the International Cryptozoology Museum in 2003 in Portland, Maine, and began contributing to the cryptozoological site Cryptomundo.com in 2005. His blog continues at CryptoZooNews.com. A few of Coleman's many published works include: *The Field Guide to*

Bigfoot and Other Mystery Primates (2006); *Bigfoot! The True Story of Apes in America* (2003); and *Mysterious America: The Revised Edition* (2001).

T. M. Gray: Gray lives in Bar Harbor, Maine and is a writer of horror fiction as well as several nonfiction books dealing with spirits and apparitions like her *Ghosts of Maine* (2008). She is also the author of the two-volume *New England Graveside Tales* (2010) which includes some stories and accounts of cryptid creatures, like Bigfoot, lake and sea monsters, and thunderbirds.

D. L. Soucy: Soucy's *Maine after Midnight* (2008) has a small section on Bigfoot from pages 146 through 148. Recent books like *Sasquatch: Searching for the Answer* (2013) and *Could Bigfoot be Human* (2014) deal specifically on the topic of unknown hairy hominids. Soucy has been conducting field research in the greater Brunswick area for several years.

Michelle Souliere: Souliere maintains the *strangemaine* blog and is proprietor of the Green Hand Bookshop in Portland, Maine. (The International Cryptozoology Museum and the Green Hand Bookshop both occupy the Trelawny Building.) Her *Strange Maine* published in 2010 has a section devoted to Bigfoot from pages 74-76. The uncovering of the Gorham Wild Man anecdote (see Wild Man chapter) is an example of her adroit research on Bigfoot in Maine.

C. A. Stephens: See chapter on Historical Literature. Charles Asbury Stevens (1844-1931) often drew upon reminisces of his youth spent in Norway, Maine when writing youth-oriented travel and adventure stories. From 1870-1930, it is estimated that Stephens wrote over 3,000 short stories, including the *Camping Out* and *Knockabout* series. As referred to earlier in this work, Stephens' use of the Wild Man in his stories has led some researchers to wonder if perhaps he was not leveraging local Norway-area legends as inspiration.

C. J. Stevens: Stevens is a Maine native and author of *The Super-natural Side of Maine* (2002), which contains a section on Bigfoot.

BOOKS

The large and growing library of Bigfoot-focused literature is re-plete with many worthwhile, professional works recommended to those interested in delving deeper into the subject. It seems likely that ten people would commend ten different lists of works for fur-ther suggested reading. The author's own list of three leading au-thors follows:

John Bindernagel has no doubts in the existence of Bigfoot. Bindernagel was raised in Ontario and educated at the University of Guelph where he graduated with a degree in wildlife ecology. He came to British Columbia in large measure for the opportunity to research Bigfoot, or Sasquatch. "I use [the term] Sasquatch in-stead of Bigfoot, because the latter has such a jestful connotation."[1]

Bindernagel has authored two books on Bigfoot: *The Discov-ery of the Sasquatch* (2010) and *North America's Great Ape: The Sasquatch—A Wildlife Biologist Looks at the Continent's Most Misunderstood Large Mammal* (2008). Both provide extensive content on anatomy, ecology, food habits, and behaviors for Bigfoot, as well as problems the author thinks hinder an open and critical evaluation of the subject.

John Green is the author of several books on Bigfoot and former editor and publisher of the *Agassiz-Harrison Advance* in British Columbia. Green began exploring reports of Bigfoot in the 1950s. His ongoing research efforts led to his amassing of thousands of reports and one of the largest and important databases of Bigfoot encounters ever developed. Green's own review of the evidence, including investigation of tracks at Bluff Creek in 1958, have led him to believe that Bigfoot is a real creature and he has been at the forefront of exploring the mystery since the inception of the modern Bigfoot era.

In contrast to some searching for Bigfoot, Green has made his database of sightings accessible to others interested in conducting their own analysis and research. "There is no second prize," Green notes, in the quest to find Bigfoot. "The man who actually collects one sees a pot of gold and fame out there. And if you pre-select a group of individuals who are willing to fly in the face of society, you tend to get some pretty opinionated individuals."[2]

Considered by many to be an indispensable book on the subject, Green's *Sasquatch: The Apes Among Us* was first published in 1978. Level-headed in approach, Green includes evidence from the earliest records up to the late 1970's and is credited with causing many people to reconsider the idea of North American primates unknown to science. Other works by Green are considered classics of Bigfoot literature: *On the Track of the Sasquatch* (1968), *Year of the Sasquatch* (1970), and *The Sasquatch File* (1973).

Jeffrey Meldrum is a Full Professor of Anatomy and Anthropology at Idaho State University. His *Sasquatch: Legend Meets Science* (2006) is considered requisite reading on the subject. Meldrum specializes in physical evidence, often and perhaps most famously manifested in the form of footprints. A popular conception holds that Bigfoot tracks are purposefully fabricated by hoaxers. This may be an oversimplification of a real and large body of evidence. Meldrum, upon a 1996 visit to Paul Freeman in southwestern Washington, was shown a 35-45 print trackway that was to forever change his opinion of the reality of Bigfoot. This would become a singular, pivotal moment for Meldrum and was his first opportunity to see purported Bigfoot tracks in the ground.

"They were amazing." Meldrum recalled. He had been forewarned before meeting Freeman by no less a Bigfoot luminary than Rene Dahinden. But the prints Freeman showed Meldrum turned out to be, in Meldrum's analysis, remarkably clear tracks in fine silt, providing skin-rich patterns persistent even though drizzly weather was obliterating the details. The animation evident—tension cracks, pressure ridges, "drag-ins" and "drag outs", and half tracks—was not indicative of a prankster strapping on rigid, faux feet and planting a line of tracks. "There was clear detail of toe position variation, in some instances the toes were quite extended and in others they were tightly flexed and dug-in to gain purchase in the soil, and in some cases they slid and made these grooves with extrusions up between the toes. Then finally the pads dug in and got a purchase in the wet soil." To Meldrum this could only be the real thing.[3]

Sasquatch: Legend Meets Science is a thoughtful, well-produced book providing entertaining and informing reading for skeptics and believers alike. "A responsible skeptic's obligations do not extend to poking fun at the people who are looking for evidence, considering the lack of evidence to be proof of no evidence, or making personal comments about people. That's not good science. In some cases, Dr. Meldrum, and other scientists like him, are being better skeptics that the skeptics."[4]

And for balance, the author includes and recommends, *Bigfoot Exposed* by David Daegling, a thought-provoking contrast to the previously mentioned, certainly pro-Bigfoot, books. Daegling grew up in northern California in the late '60s and early '70s when a lot of Bigfoot activity was going on including the Patterson Gimlin film of 1967. Daegling took notice and loving monsters anyway, collected Bigfoot stories. Daegling's thoughts on the Bigfoot enigma:

> "This book delves into the contradictions and the enigmas that make Bigfoot a problem seemingly without resolution. There is, I believe, a satisfactory explanation to the phenomenon that is neither simple nor necessarily intuitive. To arrive at this explanation, we have to delve into the mystery in considerable detail from a variety of angles."[5]

Using his training and experience as a biological anthropologist, Daegling's *Bigfoot Exposed* appears more a thoughtful debunking than callously dismissive work. The paleoanthropologist is admittedly a skeptic, citing the want of verifiable data. Leveraging his specialty in primate anatomy and biomechanics, Daegling offers a skeptically-grounded contribution to serious Bigfoot literature. *Bigfoot Exposed* chapters include "The Social History of Bigfoot," "The Patterson Film," and "The Eyewitness Problem." In the end, Daegling retains his skepticism, having given his refutation and counterargument to many facets of Bigfoot's enduring mystery.

EPILOGUE

". . . the search for the wild man renders its best
results when we investigate the territories that
extend from the other side of the mirror in which
we behold ourselves."[1]

BIBLIOGRAPHY

Adams, Glenn. "Wolves may have returned to Maine." *Lawrence Journal-World* (October 10, 1993).

Anonymous. "2-Year-Old Girl Missing in Woods." *Meriden Journal* (July 14, 1956).

—. "25 Years Ago Today." *Lewiston Evening Journal* (September 12, 1938).

—. "50 Years Ago Today." *Lewiston Evening Journal* (July 9, 1946).

—. "Acted Like a Wild Man." *Biddeford Weekly Journal* (July 19, 1912).

—. "Again the Wild Man." *Lewiston Evening Journal* (January 27, 1897).

—. "A Hairy Giant." *Lewiston Evening Journal* (August 30, 1895).

—. "A Hermit in Maine." *New York Times* (August 05, 1883).

—. "A Hunter's Encounter with the Devil." *Lewiston Evening Journal* (October 17, 1876).

—. "All-Day Hunt for Wild Man." *Lewiston Evening Journal* (August 21, 1913).

—. "A Man Thought to be the Wild Man." *Bar Harbor Record* (December 04, 1895).

—. "And it was a Jolly Party." *Lewiston Evening Journal* (October 2, 1913).

—. "An Eerie Character." *Lewiston Evening Journal* (August 4, 1914).

—. "An 'Injun Devil' Yarn." *Daily Kennebec Journal* (July 23, 1907).

—. "Anyone Lose an Ape at Durham?" *Lewiston Daily Sun* (July 27, 1973).

—. "Are the Sea Serpent and Maine Windigo Doomed?" *Boston Daily Globe* (January 11, 1920).

—. "Are They the Last Cave Men?" *The Lincoln Star* (July 29, 1934).

447

—. "Aroostook County Woman Believes Danville Wild Man is her Better Half." *Daily Kennebec Journal* (August 08, 1898).

—. "A Terror in the Woods." *San Francisco Call* (November 27, 1895).

—. "A Wild Man." *Bangor Daily Whig and Courier* (August 17, 1896).

—. "A Wild Man." *Lewiston Evening Journal* (November 30, 1896).

—. "A Wild Man." *The Toronto World* (August 25, 1913).

—. "A Wild Man of the Mountains." *New York Times* (October 18, 1879).

—. "A Wild Man's Prisoner." *Mansfield Daily Shield* (January 22, 1893).

—. "Bad Man with Red Hair." *Bath Independent* (June 18, 1885).

—. "Beast Reported Prowling in Durham Area." *Bangor Daily News* (July 27, 1973).

—. "Because his Mother." *Lewiston Evening Journal* (March 20, 1915).

—. "Bigfoot Sightings Bring VT National Attention," www.wcax.com/story/17402034/Bigfoot-sightings-bring-vt-national-attention

—. "Bigfoot Still Lurks in Crookston." *Grand Forks Herald* (January 11, 1999).

—. "Bull Caribou Spied in Moxie Lake Region." *Lewiston Daily Sun* (January 7, 1930).

—. "Bullets Await 'Ape Man'." *Daily News* [Frederick, MD] (April 03, 1926).

—. "Caribou Return to Maine Woods." *New York Times* (March 24, 1912).

—. "Chased by a Wild Man." *Montreal Daily Herald* (June 25, 1888).

—. "Closing of Goldthwaite's Much Regretted." *Biddeford Journal* (July 27, 1920).

—. "Complaints were made." *Bath Independent* (June 25, 1898).

—. "Current News and Notes." *Lewiston Evening Journal* (December 04, 1879).

—. "Deer and Caribou in Maine." *New York Times* (January 29, 1888).

—. "Devil's Tracks." *Reading Eagle* (May 11, 1902).

—. "Down-East Backwoodsman." *The New York Times* (August 07, 1892).

—. "Durham Men on Wild Man Hunt." *Lewiston Evening Journal* (August 20, 1913).

—. "Durham Wild Man Left Clothes." *Lewiston Evening Journal* (September 22, 1913).

—. "Durham Wild Man May Be Artist John Knowles." *Lewiston Daily Sun* (August 22, 1913).

—. "Durham Wild Man Not Yet Captured." *Lewiston Daily Sun* (August 21, 1913).

—. "Durham Wild Man Now at Pownal." *Lewiston Daily Sun* (September 15, 1913).

—. "Durham Wild Man on Rampage." *Lewiston Evening Journal* (August 22, 1913).

—. "Every Day Was April Fool's for Connecticut Publisher." *Record-Journal* (March 30, 1986).

—. "Farmer Snow's Sweet Corn." *Biddeford Journal* (July 31, 1920).

—. "Fifty Searchers Out." *Lewiston Evening Journal* (August 25, 1913).

—. "Finding Bigfoot Coming to Maine." maine.craigslist.org/rnr/4187198740.html

—. "Finding Bigfoot in Maine!" www.topix.com/forum/city/lincoln-me/TNUT3RGL25KBGANHT (November 25, 2007).

—. "Find Missing Thorndike Boy Safe." *Lewiston Evening Journal* (July 19, 1954).

—. "Fired at Wild Man." *Lewiston Daily Sun* (August 27, 1913).

—. "Footprints of the Devil." *Bath Independent* (July 13, 1901).

—. "Footprints on Maine Shores." *Lewiston Evening Journal* (December 4, 1897).

—. "General and Personal." *Philadelphia Record* (January 23, 1890).

—. "Ghostly Figure haunts Indians of North Wilds." *Berkeley Daily Gazette* (November 4, 1935).

—. "Gorham is the Latest Town." *Daily Kennebec Journal* (August 15, 1905).

—. "Gorilla is Back." *Lewiston Evening Journal* (August 13, 1973).

—. "Gorilla Spectators Cause Durham Property Damage." *Lewiston Daily Sun* (August 15, 1973).

—. "Heard Boy Crying." *Lewiston Evening Journal* (August 5, 1911).

—. "Her Memory is a Blank." *Lewiston Evening Journal* (August 04, 1911).

—. "Hiking: Myth Versus Reality of Sasquatch in Ossipee Range." www.Mid-AmericaBigfoot.com/forums/viewtopic.php?f=47&t=6732

—. "History from our Files." *Portland Press Herald* (November 17, 1947).

—. "Hobgoblins of the Wild North Woods." *Boston Evening Transcript* (December 31, 1902).

—. "Holman F. Day Dies in California." *Lewiston Evening Journal* (February 21, 1935).

—. "Hunt for Wild Man." *Lewiston Daily Sun* (August 23, 1913).

—. "Hunting a Wild Man." *Lewiston Evening Journal* (November 21, 1893).

—. "I Believe in Bigfoot or Sasquatch." www.experienceproject.com/stories/Believe-In-Bigfoot-Or-Sasquatch/1488085

—. "Injun Devil." *Boston Daily Globe* (March 12, 1898).

—. "'Injun Devil' Was a Big Maltese Cat." *Lewiston Evening Journal* (January 10, 1912).

—. "Is not Insane." *Lewiston Evening Journal* (November 20, 1907).

—. "It is All There." *Lewiston Evening Journal* (September 07, 1900).

—. "It was Not a Bear." *Biddeford Journal* (October 7, 1921).

—. "Killed Animal a Wolf." *Bangor Daily News* (March 29, 1994).

—. "Leaves from the Experience of an Old Hunter." *Lewiston Evening Journal* (January 26, 1884).

—. "Lee's Wild Man." *Boston Daily Globe* (November 06, 1896).

—. "Looking for the Duroculi." *Biddeford Journal* (October 6, 1921).

—. "Lost 24 Hours, Girl, 2, Found." *Pittsburgh Press*, Sunday (July 15, 1956).

—. "Lowell Man Flees Hollis." Nashua Telegraph (May 10, 1977).

—. "Maine Best Attractions 2010." www.yankeemagazine.com/article/travel/best-me-2010/attractions-me-2

—. "Maine Bigfoot: Maine Ridge Monster, Meddybumps Howler or Pomoola." (December 31, 2011) www.examiner.com/new-age-in-bangor/maine-Bigfoot-maine-ridge-monster-meddybemps-howler-or-pomoola

—. "Maine IBS Sighting Reports." MABRC Forums www.mid-americaBigfoot.com/forums

—. "Maine's Giant Moose." *New York Times* (November 5, 1899).

—. "Maine Wild Man Sought in Woods." *Boston Daily Globe* (December 03, 1922).

—. "Man Covered with Hair." *Hartford Weekly Times* (August 29, 1895).

—. "Many Things Preyed on Elsie Davis' Mind." *Lewiston Evening Journal* (03 August 1911).

—. "May Search Sunday for Wild Man." *Lewiston Daily Sun* (August 29, 1913).

—. "Mexico Child Found at Gray." *Lewiston Daily Sun* (June 19, 1951).

—. "Miss Crosby, Writer on Hunting, Dies." *Meridian Record* (November 12, 1946).

—. "Missing 411." BigfootForums.com/index.php?/topic/29296-missing-411/

—. "Moji is Safely Chained." *Lewiston Evening Journal* (September 7, 1900).

—. "Mother Bigfoot." *Finding Bigfoot* (Season 3, Episode 2, November 18, 2012).

—. "Motion Media Entertainment." www.facebook.com/pages/Motion-Media-Entertainment/129994827086325?sk=info

—. "Native Americans." www.visitmaine.com/things-to-do/arts-and-culture/native-americans/ (accessed May 21, 2014).

—. "New Hampshire Wood Devils." *Phantoms & Monsters* naturalplane.blogspot.com/2011/09/new-hampshire-wood-devils.html

—. "New York Showmen Report a Scarcity of Good Wild Men." *Biddeford Weekly Journal* (June 14, 1907)

—. "Noted Author Dies at his Norway Home." *Lewiston Daily Sun* (September 23, 1931).

—. "No Trace of 'Wild Man' Seen in Topsham Woods." *Lewiston Daily Sun* (June 6, 1929).

—. "Ocean-Going Wild Man in Maine Woods." *Boston Daily Globe* (August 07, 1919).

—. "October Mountain State Park." www.mass.gov/eea/agencies/dcr/massparks/region-west/october-mountain-state-forest-generic.html

—. "Oral Traditions." *Acadian Culture in Maine* acim.umfk.maine.edu/oral_traditions.html

—. "Peaks Island's Wild Man." *Daily Kennebec Journal* (July 20, 1899).

—. "Poland's Wild Man." *Lewiston Evening Journal* (July 22, 1905).

—. "Police Solve Famous 'Ape Man' Mystery." *Davenport Democrat* (April 04, 1926).

—. "Posse of 60 Men Hunt Wild Man of Durham." *Lewiston Daily Sun* (August 25, 1913).

—. "Puckwudgie." en.wikipedia.org/wiki/Pukwudgie (accessed May 21, 2014).

—. "Queer Animals in Maine." *Boston Daily Globe* (November 05, 1905).

—. "Queer Maine Folks." *Lewiston Evening Journal* (June 25, 1887): 5.

—. "Raccoon Gully Douroucouli in Limelight." *Biddeford Weekly Journal* (February 21, 1919).

—. "Raccoon Gully Duroculi." *Biddeford Journal* (October 5, 1921).

—. "Real 'Indian Devil'." *Lewiston Evening Journal* (July 20, 1900).

—. "Rented Gorilla Costume Has Not Been Returned." *Lewiston Daily Sun*, (July 30, 1973).

—. "Rumor of a Wild Man Seen." *Lewiston Daily Sun* (August 13, 1913).

—. "Run Amuck!" *Lewiston Evening Journal* (July 20, 1900).

—. "Runaway Monkey Likes Maine Air." *Lewiston Daily Sun* (November 03, 1962).

—. "Sabatis' Annual Wild Man." *Lewiston Evening Journal* (July 20, 1899).

—. "Sandford Men See Bear." *Biddeford Journal* (October 7, 1921).

—. "Say it is a Gorilla," *The World* (August 30, 1895).

—. "Searching for the 'Wild Man'." *New York Times* (September 04, 1895).

—. "Search on for Monkey Man." *Lewiston Daily Sun* (April 20, 1973).

—. "Send Organs On to be Analyzed." *Lewiston Evening Journal* (May 5, 1923).

—. "Sheriff Hill Saw Governor Powers." *Lewiston Evening Journal* (January 30, 1897).

—. "Signs Suggest a Return of Timber Wolf to Maine." NYTimes.com (December 22, 1996).

—. "Slipperyskin—Bear, Bigfoot, or Indian?" www.northlandjournal.com/stories/stories18.html

—. "Some Odd Stories." *Daily Advocate* (Ohio) (September 6, 1893).

—. "Some Stuff." *Naugatuck Daily News* (April 21, 1973).

—. "Speaking of the Witch Spring Wild Man." *Bath Independent* (June 25, 1898).

—. "State Chat." *Lewiston Evening Journal* (May 14, 1900).

—. "State Chat." *Lewiston Evening Journal* (August 19, 1905).

—. "Still Searching for Missing Girl." *Lewiston Evening Journal* (August 01, 1911).

—. "Strange Animal Prowling About Raccoon Gully." *Biddeford Journal* (January 31, 1919).

—. "Strange Monster Seen in Maine." *Boston Daily Globe* (August 09, 1895).

—. "That L. J. 'Wild Man'?" *Daily Kennebec Journal* (September 23, 1913).

—. "The Attractions at the Station." *Daily Kennebec Journal* (September 9, 1907).

—. "The Cryptid Zoo: Wildmen." www.newanimal.org/wildmen.htm

—. "The Death of Ki-Ko." *Bath Independent and Enterprise* (February 13, 1909).

—. "The Durham People Still Believe." *Lewiston Evening Journal* (October 2, 1913).

—. "The Game Season." *Bar Harbor Record* (September 21, 1895).

—. "The Indian Devil." *The Hartford Courant* (January 28, 1890).

—. "The Insane Man."*Daily Kennebec Journal* (July 25, 1905).

—. "The Loocivee's Yell." *Lewiston Evening Journal* (October 28, 1901).

—. "The Mysterious Wild Man." *Daily Kennebec Journal* (December 27, 1871).

—. "The 'Never-Never' Nature Yarns that Made Winsted Famous." *Spokesman-Review* (April 16, 1933).

—. "The Story of Sasqua." www.cultreviews.com/interviews/the-story-of-sasqua/ (December 11, 2009).

—. "The Wild Man Has Wandered." *Lewiston Evening Journal* (September 5, 1913).

—. "The Wild Man is Coming to the Front This Fall." *Los Angeles Times* (May 05, 1887).

—. "The Wild Man of Haverhill, Massachusetts." newenglandfolklore.blogspot.com (January 04, 2010).

—. "There had been circuses." *Bangor Daily Whig and Courier* (June 04, 1896).

—. "There is more or less excitement." *Daily Kennebec Journal* (September 15, 1894).

—. "Town's Bugaboo." *Boston Daily Globe* (September 16, 1895).

—. "Tramp Scared Residents." *Biddeford Journal* (October 27, 1920).

—. "Two Grange Exhibits." *Lewiston Evening Journal* (August 29, 1907).

—. "UFO Sighting Reported at Durham; Track Beep Sound." *Lewiston Daily Sun* (April 18, 1966).

—. "UFO Spotted by Campers in Winsted Area." *The Day*, (New London, Connecticut, July 30, 1976).

—. "Was it a bear or Bigfoot?" *Kennebec Journal* (July 4, 1977).

—. "Weird and Uncanny." *Lewiston Evening Journal* (August 19, 1899).

—. "What is it?" *New York Times* (April 26, 1871).

—. "Whistling Wild Boy of the Woods." *Bangor Whig and Courier* (August 04, 1838).

—. "Wild Man?" *Bath Independent* (June 25, 1898).

—. "Wild Man at Large in Minot." *Boston Daily Globe* (July 21, 1896).

—. "Wild Man Caught." *Boston Daily Globe* (July 24, 1905).

—. "'Wild Man' Hunt On in Maine." *Syracuse Telegram* (December 7, 1922).

—. "Wild Man May Be a Gorilla." *New York Times* (August 30, 1895).

—. "Wild Man in Riggsville." *Bath Independent and Enterprise* (May 18, 1907).

—. "Wild Man Seen at South Auburn." *Boston Daily Globe* (August 01, 1896).

—. "Wild Man of Williamstown, VT." newenglandfolklore.blogspot.com/2010/02/wild-man-of-williamstown-vt.html

—. "You Know it was Huskins." *Lewiston Evening Journal* (June 04, 1898).

Arment, Chad. *The Historical Bigfoot*. (Landisville, PA: Coachwhip, 2006).

Barackman, Cliff. "Behind the Scenes of *Finding Bigfoot*: 'Big Rhodey' with Cliff Barackman." www.cryptomundo.com/bigfoot-report/finding-bf-71/

Barackman, Cliff. "Behind the Scenes of *Finding Bigfoot*: 'Mother Bigfoot' with Cliff Barackman." www.cryptomundo.com/Bigfoot-report/finding-bf-116/

Barackman, Cliff. "Vermont Trail Camera Photo Analysis." www.cliffbarackman.com/research/field-investigations/vermont-trail-camera-photo-analysis/

Baron, Rachel, Blair Braverman, and Francis Gassert. *The State of Wood Biomass Energy in Maine*. (Colby Environmental Policy Group 2010). wiki.colby.edu/display/stateofmaine2011/The+State+of+Biomass+Energy (accessed May 11, 2013).

Bartholomew, Robert E., & Brian Regal. "From Wild Man to Monster: the Historical Evolution of Bigfoot in New York State." *Voices* (Volume 35, 2009).

Bartholomew, Robert E., & Paul Bartholomew. *Bigfoot Encounters in New York & New England* (Surrey, BC: Hancock House, 2008).

Bartlett, Kay. "Elusive 'Bigfoot' Is America's Loch Ness Monster," *Youngstown Vindicator* (January 28, 1979): 20.

Belanger, Jeff. *Weird Massachusetts.* (New York: Sterling, 2008).

Berry, Ric. *Bigfoot on the East Coast* (1993).

BFRO.net. "Camper Hears Screams While Camping in Kennebunk, Maine." www.bfro.net/GDB/show_report.asp?id=883

—. "Classic Sighting on Wooded Maine Road." www.bfro.net/GDB/show_report.asp?id=7421

—. "Hunters Find Scat and Have Large Rock Thrown at Them." BFRO Report #8259 (March 13, 2004).

—. "Man Hears Unknown Animal Vocalization on Consecutive Nights While Camping." www.bfro.net/GDB/show_report.asp?id=3229

—. "Mother and Son See Sasquatch Closeup from Road." www.bfro.net/gdb/show_report.asp?id=6643

—. "Report #1192 (Class B)." www.bfro.net/GDB/show_report.asp?id=1192

—. "Series of Sightings Taking Place Over Three Days." BFRO Report #1186 (January 01, 1998).

—. "Sighting by a Motorist." BFRO Report #1189 (March 16, 1997).

—. "The Mystery of the Maine 'Sasquatch' Now Resolved." www.bfro.net/news/porcupine_maine.asp

—. "Two Men in Car See Hairy, Upright Animal on Side of Road Near Turner, Maine." www.bfro.net/GDB/show_report.asp?id=859

—. "Two Men See Huge Hairy Animal Near Poland." www.bfro.net/GDB/show_report.asp?id=2678 (June 11, 2001)

—. "Two Sisters See Dark, Shaggy Haired, Bipedal Creature Standing in Middle of Road." www.bfro.net/GDB/show_report.asp?id=1188

—. "Weird Mountain 'Creature' is Reported by Picnickers." *Berkshire Eagle* (August 23, 1983). www.bfro.net/gdb/show_article.asp?id=414

BigfootEncounters.com. "At Mooselookmeguntic Lake . . . ." www.bigfoot-encounters.com/stories/bemis.htm (September 2004)

—. "Between New Brunswick, Canada and Topsfield, Maine." www.
bigfootencounters.com/sbs/tobey02.htm

—. "Big Footprints found in Aroostook County, Frenchville, Maine."
www.bigfootencounters.com/sbs/Aroostook.htm

—. "Built Like a Horse." *North Adams Daily Transcript* www.bigfoot-
encounters.com/creatures/injun_meadow.htm (August 23, 1895)

—. "Encounter at Prong Pond Nearby Moosehead Lake." www.bigfoot-
encounters.com/stories/piscataquis.htm

—. "Franklin-Oxford County Lines, Maine." www.bigfootencounters.com/
sbs/franklin.htm

—. "Freedom, Knox County." www.bigfootencounters.com/stories/
freedom.htm [Note: Freedom, Maine, is in Waldo County]

—. "Freedom, Waldo County." www.bigfootencounters.com/sbs/freedom-
ME07.htm

—. "Kezar Falls." www.bigfootencounters.com/stories/kezarfallsME.htm

—. "More on the Meddybumbps Howler Area." www.bigfootencounters.com/
sbs/meddybemps2.htm

—. "Oxford County, Maine." www.bigfootencounters.com/sbs/oxford_
countyME06.htm

—. "Piscataquis County." www.bigfootencounters.com/sbs/piscataquis.htm

—. "Piscataquis County, Maine." www.bigfootencounters.com/sbs/
baxter.htm

—. "Sagadahoc County." www.bigfootencounters.com/sbs/sagadahoc.htm

—. "Suddenly There He Was. Less Than 15 Feet in Front of Me." www.
bigfootencounters.com (March 13, 2011).

—. "The Maine Ridge Monster." www.bigfootencounters.com/stories/
maine.htm

—. "The Meddybumps Howler, Washington County, Maine." www.bigfoot-
encounters.com/

—. "Wakefield, South County, Rhode Island." www.bigfootencounters.com/
sbs/wakefield.htm

Bindernagel, John A. *Paranatural—Sasquatch Planet.* (2012).

Bindernagel, John A. *The Discovery of the Sasquatch.* (Courtenay, BC:
Beachcomber Books, 2010).

Blackburn, Lyle. *Lizard Man: The True Story of the Bishopville Monster.*
(Anomalist Books, 2013).

Bourque, Bruce Joseph. *Twelve Thousand Years: American Indians in Maine.* (University of Nebraska, 2001).

Buckley, Ken. "A Sasquatch Search in Bridgton, Maine." *Bangor Daily News* (August 5, 1971).

Buhs, Josuha Blu. "Wildmen on the Cyberfrontier: The Computer Geek as an Iteration in the American Wildman Lore Cycle." *Folklore* (April, 2010): 62.

Burke, Selden. "Romance and Reality of the Wild Men." *The Scrap Book* (November, 1908).

Burnham, Emily. "Animal Planet's 'Finding Bigfoot' not coming to Maine, but welcomes leads on Sasquatch sightings." *Bangor Daily News* (November 18, 2013).

Burns, J. W. "Introducing B.C.'s Hairy Giants." *Maclean's Magazine* (April 1, 1929).

Bush, Grace. "Devil's Footprints' Seen." *Bangor Daily News* (October 11, 1976).

Byrne, Peter. "The Case for the *OMAH*." *Explorers Journal* (June, 1972).

Carkhuff, David. "Maine Ghost Hunters Plan Bigfoot Quest." *Portland Daily Sun* (July 30, 2012).

Carpenter, Scott. "Wolf Pack—Shelley Rockwell-Martin Separates Fact From Fiction." Dogmen—Monsters are Real dogman-monsters-are-real.blogspot.com/2013/08/wolf-pack-shelley-rockwell-martin.html

Chase, George Wingate. *The History of Haverhill, Massachusetts.* (Haverhill, 1861).

ChattahoocheeBigfootRadio. "Chattahoochee Bigfoot Radio—Int. with Mike Merchant." www.blogtalkradio.com/chattahoocheebigfootradio/2013/10/06/chattahoochee-bigfoot-radio—int-with-mike-merchant

Citro, Joseph A. *Passing Strange—True Tales of New England Hauntings and Horrors.* (New York: Houghton Mifflin, 1997).

Citro, Joseph A. *Weird New England.* (Sterling Publishing, 2005): 100, 110-111.

Coatsworth, Elizabeth. *Maine Memories.* (S. Green Press, 1968): 95-96.

Cole, Benjamin C. *It Happened Up in Maine.* (Penobscot Bay Press, 1980).

Coleman, Loren. *Bigfoot! The True Story of Apes in America.* (Paraview Pocket Books, 2003).

Coleman, Loren. "Chain of Ponds, Maine, "Bigfoot" Casts." www.crypto-mundo.com/cryptozoo-news/aug-me-tracks/ (August 21, 2012)

Coleman, Loren. "Injun Devils." www.cryptomundo.com/cryptozoo-news/injun-devils/

Coleman, Loren. "Maine Tree Creature: First Blobsquatch of 2010?" www.cryptomundo.com/cryptozoo-news/1-blobsqu10/

Coleman, Loren. "Pucabob Reviews Monster of the Woods." www.crypto-mundo.com/cryptozoo-news/motw-review/

Coleman, Loren. "The Leeds Loki: 2010's Maine Bigfoot." www.crypto-mundo.com/cryptozoo-news/leeds-loki/

Colin Bord and Janet Bord. *Bigfoot Casebook* (Granada, 1982): 205.

Condit, Jon. "No Stoppin the 'Squatch." www.dreadcentral.com/node/02192

Crowe, Ray. *The Track Record.*

Daegling, David. *Bigfoot Exposed: An Anthropologist Examines America's Enduring Legend.* (AltaMira Press, 2004).

Day, Holman. "Down in Maine," *Hartford Courant*, (October 17, 1900).

Day, Holman. *The Eagle Badge: or the Skokums of the Allagash* (Harper & Brothers, 1908).

Day, Holman. "Old King Spruce." *New England Magazine* 37 (October, 1907).

Day, John S. "Ford May Make N. H. Gains by Handling Hecklers Better Than Reagan." *Bangor Daily News* (February 12, 1976).

Dietz, Lew. *The Allagash.* (Down East Books, 1968).

Dorson, Richard Mercer. *Folktales Told Around the World.* (University of Chicago Press, 1975).

Dudley, John. "The Airline Road." www.mainething.com/alexander/Transportation%20and%20Trade/THE%20AIRLINE%20ROAD.html (2003).

Dunning, Brian. "Killing Bigfoot with Bad Science." *Skeptoid* Podcast (Skeptoid Media, Inc., 03 December, 2006) skeptoid.com/episodes/4011

Easton, Thomas A. *Bigfoot Stalks the Coast of Maine.* (Wildside Press, 2000).

Eberhart, George. *Mysterious Creatures: A Guide to Cryptozoology.* (ABC-CLIO, 2002).

Eisenthal, Bram. *Milford Daily News* [30 October, 2004; last updated 13 November, 2007].

Elder, William. "The Aborigines of Nova Scotia." *The North American Review* CCXXX (January, 1871).

Emblidge, David. *The Appalachian Trail Reader*. (Oxford University Press, 1996).

Erickson, Amy. "Drowning . . . or Murder? Part One of Amy Erickson's Special Report." www.wabi.tv/2009/04/29/drowning-or-murder-part-one-of-amy-ericksons-special-report/

Erickson, Amy. "Drowning or Murder? Part Two of Amy Erickson's Special Report." www.wabi.tv/news/5699/drowning-or-murder-part-two-of-amy-ericksons-special-report

Fairless, Robert and Randy Ashby. "Dr. Jeff Meldrum." 25th Hour Radio Show (December 12, 2012) www.25thradioshow.com/radio-shows/2012-radio-show-interviews/december-2012-interviews/

Faragher, John Mack. *Daniel Boone: The Life and Legend of an American Pioneer*. (Holt, 1992).

Fendler, Donn and Joseph B. Egan. *Lost on a Mountain in Maine*. (Harper Trophy, 1992).

Franklin, James. *The Science of Conjecture: Evidence and Probability before Pascal*. (The Johns Hopkins University Press, 2001).

Frazier, Joseph. "'Natural Man' was Partly a Fraud; But He still Inspired many with his Frontier Spirit." NiagaraFallsReview.ca (February 22, 2008).

French, Scott. "The Man Who Spied Bigfoot Comes Forward." *Concord Monitor* (November 13, 1987). www.bfro.net/gdb/show_article.asp?id=267

Frost, Stanley. "The Air Line to the Big Woods." *Outing*, 78 (July, 1921).

GCBRO. "From the Files of the Gulf Coast Bigfoot Research Organization." www.gcbro.com

Giles, John. *Memoirs of Odd Adventures, Strange Deliverances, Etc. in the Captivity of John Giles, Commander of the Garrison of Saint George River, in the District of Maine*. (1869).

Godfrey, Linda S. *Real Wolfmen: True Encounters in Modern America*. (Tarcher, 2012).

Gray, T. M. *New England Graveside Tales*. (Schiffer Publishing, 2010).

Green, John. *Sasquatch: The Apes Among Us.* (Blaine, WA: Hancock House, 1978).

Green, John. "John Green—Sasquatch Investigator, Chronicler and Pioneer Analyst." www.sasquatchdatabase.com

Griffin, Joseph. *History of the Press of Maine.* (Brunswick, 1872).

Grigg, Bob. *The Winsted Wildman.* www.colebrookhistoricalsociety.org/PDF%20Images/The%20Winsted%20Wildman.pdf

Guarnieri, Catherine. *Twisted History: The Winsted Wild Man.* (January, 2010). www.registercitizen.com

Halpin, Marjorie and Michael Ames. *Manlike Monsters on Trial: Early Records and Modern Evidence.* (University of British Columbia Press, 1980).

Hayward, Ed. "The Bigfoot of Bridgewater." *Boston Herald Library Online* (April 06, 1998).

Hazard, Wendy. "Halloween: Local Legends." *Kennebec Journal* (October 30, 1976).

Hinckley, G. W. *Roughing it with Boys.* (1913).

Hoffenberg, Noah. "Man Spots Bigfoot." *Bennington Banner* (September 26, 2003).

Hornibrook, Isabel. *Camp and Trail: A Story of the Maine Woods.* (Lothrop Publishing Company, 1897).

Hosmer, George L. *An Historical Sketch of the Town of Deer Isle, Maine.* (1886).

Jacobson, G. L., I. J. Fernandez, P. A. Mayewski, and C. V. Schmitt (editors). *Maine's Climate Future: An Initial Assessment.* (Orono, ME: University of Maine, February 2009, revised April, 2009).

Josselyn, John. *Voyages to New England.* (1663).

Landau, Elizabeth. "Why Do We Need to Look for Bigfoot?" CNN.com (June 21, 2010).

LeBlanc, Deanna. "Bigfoot Sightings Bring Vt National Attention." WCAX.com (12 April 2012).

Leland, Charles Godfrey. *The Algonquin Legends of New England: or Myths and Folk Lore of the Micmac, Passamaquoddy, and Penobscot Tribes.* (Houghton, Mifflin and Company, 1884).

LeMoine, James MacPherson. *The Chronicles of the St. Lawrence.* (Dawson Brothers, 1878).

L'Heureux, Juliana. "This is the Time when the Werewolf's Howl Scares Little Children." *Portland Press Herald/Maine Sunday Telegram* [York County Extra] (October 25, 2001).

Lynds, Jen. "One Year Later, Houlton Still Owes for Cleanup Effort of Artist's Property." *Bangor Daily News* (December 03, 2011).

Maine Department of Inland Fisheries and Wildlife. *Prong Pond* www.maine.gov/ifw/fishing/lakesurvey_maps/piscataquis/prong_pond.pdf (1989).

—. "Black Bears." www.maine.gov/ifw/wildlife/species/mammals/bear.html (2013).

—. "Bobcat." www.maine.gov/ifw/education/wildlifepark/wildlife/bobcat.htm (2013).

—. "Canada Lynx." www.maine.gov/ifw/wildlife/species/mammals/canada_lynx.html (2013).

—. "Coyote." www.maine.gov/ifw/education/wildlifepark/wildlife/coyote.htm (2013).

—. "Deer: White-tails in the Maine Woods." www.maine.gov/ifw/wildlife/species/mammals/deer.html (2013).

—. "Eastern cougar." www.maine.gov/ifw/wildlife/species/endangered_species/eastern_cougar/ (2012).

—. "Gray Wolf." www.maine.gov/ifw/wildlife/species/endangered_species/gray_wolf/ (2012).

—. "Gray Wolf." www.maine.gov/ifw/wildlife/endangered/pdfs/graywolf_20_21.pdf (2013).

—. "Lynx." www.maine.gov/ifw/wildlife/management/lynx_avoid.htm (2012).

—. "Moose." www.maine.gov/ifw/wildlife/species/mammals/moose.html (2013).

Maine Tourism Association. "Enjoy Maine's Changing Climates Throughout the Year." www.mainetourism.com/content/4048/Maine_Weather/ (2005-2014).

Marsh, Scarboro. "Cushing's Island." *Pine Tree Magazine* 5:6 (July, 1906).

McGaga, Julius. "Abominable Half-Human Monster Linked to Barefoot Tracks in Winsted Woods." *Hartford Courant* (January 02, 1952).

Meldrum, Jeffrey. *Sasquatch: Legend Meets Science.* (New York: Forge, 2006).

Meldrum, Jeffrey. *Sasquatch Field Guide.* (Paradise Cay Publications, 2013).

Merchant, Michael. "Michael Merchant Interview: Fisherman Encounter Bigfoot (Down East Maine)." www.youtube.com/SnowWalkerPrime

Merchant Michael. "SnowWalkerPrime's Philosophy on Bigfoot." bigfootevidence.blogspot.com/2012/02/snowwalkerprimes-philosophy-on-bigfoot.html (February 20, 2012).

Merchant, Michael. "Watch Witnesses From Ellsworth, Maine Describe Sasquatch Sighting." bigfootevidence.blogspot.com/2012/04/ellsworth-maine-sasquatch-sighting.html

Monahan, Peter Frederick. *The American Wild Man: The Science and Theatricality of Nondescription in the Works of Poe, Jack London, and Djuna Barnes.* (ProQuest, UMI Dissertation Publishing, 2008).

Moran, Mike and Mark Sceurman. *Weird U.S.: Your Travel Guide to America's Local Legends and Best Kept Secrets.* (Sterling, 2009).

Motavalli, Jim. "Naked in the Woods Contest!" www.emagazine.com/daily-news-archive/commentary-the-e-magazine-naked-in-the-woods-contest

Muise, Peter. "Cannibal Giants of the Snowy Northern Forest." newenglandfolklore.blogspot.com/2009/01/cannibal-giants-of-snowy-northern.html

Munn, Charles Clark. *The Hermit.* (Lee and Shepard, 1903).

Murphy, Christopher L. *Sasquatch/Bigfoot Chronicle & Other Unrecognized Relict Hominoids & Wild Men.* www.sasquatchcanada.com/uploads/9/4/5/1/945132/sc_-_earliest_to_1899.pdf (2013).

Muscato, Ross A. "Tales from the Swamp." www.boston.com/news (October 30, 2005).

Nason, R. B. "Jack's Injun Devil." *Outing Magazine* 53 (March, 1909).

North East Sasquatch Researchers Association. "Maine." Encounter Database. www.teamnesra.net

O'Brien, Joseph A. "Girls See UFO Glow; Oddly Shaped Beings." *The Hartford Courant* (September 18, 1967).

ParaBreakdown. "Death by Sasquatch (True Stories from Maine)." www.youtube.com/watch?v=KgpVBsMDa78

Partridge, Emelyn Newcomb. *Glooscap The Great Chief and Other Stories.* (Sturgis & Walton, 1913).

Paulides, David. *Missing 411—Eastern United States.* (Createspace, 2012).

Paulides, David. *Missing 411—Western United States & Canada.* (Createspace, 2011).

Peterson, Arthur. "Mr. Whopper." *Toledo Blade* (August 08, 1949).

Pierce, Josiah. *A History of the town of Gorham, Maine.* (1862).

Pinkham, Steve. *The Mountains of Maine: Intriguing Stories Behind Their Names.* (Down East Books, 2009).

Prince, J. Dyneley. "Some Passamaquoddy Witchcraft Tales." *Proceedings of the American Philosophical Society* 38 (Philadelphia, 1899).

Prince-Hughes, Dawn. *The Archetype of the Ape-Man: The Phenomenological Archaeology of a Relic Hominid Ancestor.* (2000).

Rand, Silas Tertius. *Micmac Indian Legends.* (1894).

Rhodes, Dean. "'Les Diables' Still Rove Allagash Wilds." *Bangor Daily News* (February 18, 1974).

Ricker, Nok-Noi. "Bigfoot in Maine? 10-Foot Tall 'Wild Man' was Killed in 1886, Newspapers Reported." www.bangordailynews.com/2013/10/27/news/state/bigfoot-in-maine-10-foot-tall-wild-man-was-killed-in-1886-newspapers-reported/

RoadsideAmerica.com. "7 Wonders of God creatures." www.roadside-america.com/story/22889 (1996-2014).

Roberts, Leonard. "Curious Legend of the Kentucky Mountains." *Western Folklore* 16 (January, 1957).

Rose, Carol. *Giants, Monsters, and Dragons: An Encyclopedia of Folklore, Legend, and Myth.* (W. W. Norton & Company, 2001).

Rosen, Brenda. *Mythical Creatures Bible: The Definitive Guide to Legendary Beings.* (Sterling Publishing, 2009): 223.

Rumsey, Barbara. "The Baby That Washed Ashore at Hendricks Head." Out of Our Past. www.benrussell.com/HH-historical%20society%20part%20II.htm

Russell, Rick. "Witch's Tale Haunts Bucksport Founder." *Bangor Daily News* (July 28, 2007).

Sarmiento, Esteban. *Paranatural—Sasquatch Planet.* (2012).

Sanderson, Ivan T. "The Universe Nobody Knows." *True* (July 1950).

Schorn, M. Don. *Elder Gods of Antiquity: First Journal of the Ancient Ones.* (Ozark Mountain Publishing, 2008).

Seymour, Tom. *Tom Seymour's Maine: A Maine Anthology*. (iUniverse, 2003).

Shackley, Myra. *Wildmen*. (Thames & Hudson, 1983).

Shaw, Dick. "Mysterious Devil's Footprints in Milo, Maine." *Lewiston Evening Journal* (June 26, 1976).

Skelton, Kathryn. "Bigfoot & Me." *Lewiston Sun Journal* (May 5, 2007).

Skelton, Kathryn. "Coleman Dubs New Sighting 'Leeds Loki'." *Lewiston Sun Journal* (March 09, 2010).

Skelton, Kathryn. "Durham Man's Monster-Hunting TV Show Debuts Thursday." *Lewiston Sun Journal* (September 04, 2014).

Skelton, Kathryn. "Durham's 'Gorilla'." *Lewiston Sun Journal* (August 02, 2008).

Skelton, Kathryn. "Looking for Bigfoot in Maine." *Lewiston Sun Journal* (December 01, 2007).

Skelton, Kathryn. "Weird, In Review." *Lewiston Sun Journal* (December 06, 2008).

Skinner, Charles Montgomery. *American Myths and Legends 1*. (J. P. Lippincott, 1903).

SnowWalkerPrime. "Chilling Allagash Maine Bigfoot Encounter!" www.youtube.com/watch?v=eTzyF4mD_jc

Solunac, Alex, Tom Yamarone, and Lesley Solunac. *Sasquatch Summit: A Tribute to John Green*. Commemorative Program (2011).

Soucy, D. L. *Maine Monster Parade*. (Lulu Enterprises, 2008).

Soucy, D. L. "March Bigfoot Hunt in Maine." www.youtube.com/watch?v=75j69ILjVZk

Soucy, D. L. "The Hunt for Bigfoot." www.youtube.com/watch?v=9MQgxr_0AKY

State of Maine Dept. of Economic & Community Development. "Bigfoot of Maine." www.mainemade.com/members/profile.asp?ID=1302

Stephens, Charles Asbury. *A Busy Year at the Old Squire's*. (C. H. Simonds Company, 1922).

Stephens, Charles Asbury. *Camping Out*. (Henry T. Coates & Co., 1873).

Stephens, Charles Asbury. *The Adventures of Six Young Men in the Wilds of Maine and Canada* (Dean & Son, 1884).

Stephens, Charles Asbury. *The Young Moose Hunters* (Henry L. Shepard & Co., 1883).

Stephens, Charles Asbury. "Was it an Indian Devil?" *Ballou's Monthly Magazine 43* (April, 1876).

Stephens, C. J. *The Supernatural Side of Maine.* (John Wade Pub. 2002).

Strain, Kathy M. *Giants, Cannibals & Monsters.* (Hancock House, 2008).

Sullivan, Jennifer. "Big Foot 'Sightings' Turn Out to be Hoax." *Sun-Journal* (August 14, 1993).

Tarpey, John P. "'Bigfoot' Leaves Impression on 2 Ellington Farmhands." *The Hartford Courant* (November 26, 1982).

Thistle, Scott. "Wolf-like animal trapped, tested." *Sun-Journal* (November 23, 1996).

Thompson, Lucy. *To the American Indian: Reminiscences of a Yurok Woman.* (Eureka, 1916).

Thoreau, Henry David. *The Maine Woods.* (Houghton Miflin, 1884).

Trotti, Hugh H. "Did Fiction Give Birth to Bigfoot?" *Skeptical Inquirer* (1994).

Twain, Mark. *The Republic of Gondor.* (1869).

United States Department of Agriculture. "Trade Promotion Authority: State Specific Information." www.fas.usda.gov/ffpd/economic-overview/overview.html

U.S. Dept. of Commerce, United States Census Bureau. quickfacts.census.gov/qfd/states/23000.html (accessed May 11, 2014).

Vose, Steve. "Finding Maine's Bigfoot." *The Maine* www.themaineblog.com/?p=2560 (May 30, 2012).

Wasson, Barbara. *Tracking the Sasquatch.* (Sisters, Oregon 1994).

Wharton, Eric H., et al. *The Forests of Connecticut.* (Pennsylvania: U.S. Dept. of Agriculture, Forest Service, Northeastern Research Sta., 2004).

Widmann, Richard H. "Trends in Rhode Island Forests: A Half-Century of Change." United States Department of Agriculture. www.dem.ri.gov/programs/bnatres/forest/pdf/ritrends.pdf

Wigginton, Eliot. *Foxfire 4* (New York: Random House, 1975).

Woolheater, Craig. "Spike TV's 10 Million Dollar Bigfoot Bounty Premieres January 10, 2014." www.cryptomundo.com/bigfoot-report/participant-talks-10-million-dollar-bigfoot-bounty/

Zuefle, David Matthew. "Swift, Boone, and Bigfoot: New Evidence for a Literary Connection." *Skeptical Inquirer* (January/February 1997).

IMAGES

INTRODUCTION

Page 11: *Pine Tree Magazine* 5(6): 571. (July 1906).

CHAPTER 1

Page 15: Are they the last cave men? *The Lincoln Star* (July 29, 1934).

CHAPTER 2

Page 22: *Outing* 68(6): 655 (September, 1916). | Page 24: Track depiction by Daniel Green. Page 25: (Top) Creative Commons 2.0-BY, GoToVan. (Bottom) Creative Commons 2.0-BY, Ken Sturm/USFWS. | Page 26: Track depiction by Daniel Green. | Page 27: (Top) Creative Commons 2.0-BY, Ryan Hagerty/USFWS. (Bottom) Creative Commons 2.0-BY, Larry Smith. | Page 28: (Both) Track depiction by Daniel Green. | Page 30: (Top) Creative Commons 2.0-BY, Zac Richter/NPS. (Bottom) Creative Commons 2.0-BY, Connie Bransilver/USFWS. | Page 32: Track depiction by Daniel Green. | Page 33: Track depiction by Daniel Green. | Page 34: (Top) Creative Commons 2.0-BY, USFWS. (Bottom) Creative Commons 2.0-BY, Craig O'Neal. | Page 35: (Both) Track depiction by Daniel Green. | Page 36: (Top) Creative Commons 2.0-BY, USFWS. (Bottom) Creative Commons 2.0-BY, Neal Herbert/Yellowstone NP. | Page 37: Calais (ME) *Advertiser*, April 27,

1949. www.mainething.com/alexander/Transportation%20and%20Trade/THE%20AIRLINE%20ROAD.html | Page 37: Track depiction by Daniel Green.

CHAPTER 3

Page 40: Wikimedia Commons. | Page 43: "The 'never-never' nature yarns that made Winsted famous." *Spokesman-Review* (April 16, 1933). | Page 44: "The 'never-never' nature yarns that made Winsted famous." *Spokesman-Review* (April 16, 1933). | Page 50: "Say it is a Gorilla." *The World* (August 30, 1895). | Page 64: McGaga, Julius. "Abominable half-human monster linked to barefoot tracks in Winsted woods." *Hartford Courant* (January 2, 1952). | Page 72: www.cultreviews.com/wp-content/uploads/2009/12/Sasqua-poster-art-156x300.jpg

CHAPTER 4

Page 96: Leland, Charles Godfrey. *The Algonquin Legends of New England: or, Myths and Folk Lore of the Micmac* (1884).

CHAPTER 5

Page 115: Husband, Timothy. *The Wild Man: Medieval Myth and Symbolism* (1980). | Page 122: *Mansfield Daily Shield* (January 22, 1893). | Page 126: *Newark* (OH) *Daily Advocate* (September 6, 1893): 6. | Page 145: "A successful blueberry trip." *Lewiston Evening Journal* (August 2, 1912). | Page 183: Burke, Selden. "Romance and Reality of the Wild Man." *The Scrap Book* (November, 1908): 908. | Page 186: *Lewiston Evening Journal* (September 7, 1900): 1.

CHAPTER 6

Page 200: Wikimedia Commons. | Page 225: *Kennebec Journal* (October 30, 1976). 17 U.S. Code § 107: Fair use for criticism, scholarship, and commentary.

CHAPTER 7

Page 235: *Lewiston Evening Journal* (August 3, 1911): 5.
| Page 237: *Lewiston Evening Journal* (August 4, 1911).
| Page 240: *Lewiston Evening Journal* (May 05, 1923): 1.
| Page 241: www.wabi.tv/news/5699/drowning-or-murder-part-two-of-amy-ericksons-special-report | Page 243:
Lewiston Evening Journal (May 3, 1923): 7. | Page 244:
Lewiston Evening Journal (May 7, 1923): 14. | Page 246:
Lewiston Daily Sun (June 18, 1951): 1. | Page 247: *Lewiston Daily Sun* (June 19, 1951): 1. | Page 248: *Lewiston Daily Sun* (June 19, 1951). | Page 251: *Lewiston Evening Journal* (July 19, 1954): 1.

CHAPTER 8

Page 255: (Both) www.youtube.com/watch? feature=
player_embedded&v=YQ5EyriQJHU. 17 U.S. Code § 107:
Fair use for criticism, scholarship, and commentary. |
Page 256: www.bfro.net/news/porcupine_ maine.asp.
17 U.S. Code § 107: Fair use for criticism, scholarship, and
commentary. | Page 258: Postcard owned by Daniel
Green.

CHAPTER 10

Page 354: Courtesy Loren Coleman/International Cryptozoology Museum. | Page 356: *The Pittsburgh Press* (July
19, 1914). | Page 358: (Top) dgpaulart; www.zazzle.com/
dgpaulart. | Page 359: www.bigfootofmaine.com | Page 361:
www.roadsideamerica.com/blog/god-and-jerry-cardone/
17 U.S. Code § 107: Fair use for criticism, scholarship, and
commentary. | Page 363: empty-house.sabatos.net/m-
posters.php

CHAPTER 11

Page 369: *Ladies Home Journal* (July 1913): 5. | Page 374:
"A Real Heroine." *Stevens Point Daily Journal* (October
5, 1907). | Page 382: *Lewiston Daily Sun* (September 23,

1931). | Page 391: Farrar, Charles Alden John. *Illustrated Guide Book to Moosehead Lake and Vicinity*. (1879). | Page 397: Stephens, Charles Asbury. *The Young Moose Hunters*. (1883). | Page 402: Tiffany, Howard S. "A Pocket Camping Outfit." *Outing* (May, 1916): 206. | Page 408: Stephens, Charles Asbury. *The Adventures of Six Young Men in the Wilds of Maine and Canada*. (1884). | Page 417: Stephens, Charles Asbury. *The Adventures of Six Young Men in the Wilds of Maine and Canada*. (1884). | Page 422: Munn, Charles Clark. *The Hermit: A Story of the Wilderness*. (1903). | Page 424: "Holman F. Day Dies in California." *Lewiston Evening Journal* (February 21, 1935): 2.

EPILOGUE

Page 442: *Pine Tree Magazine* 7(1): 3. (February 1907).

FOOTNOTES

Introduction

[1] Sanderson, Ivan T. "The Universe Nobody Knows," *True* (July 1950): 83.

[2] "The presence of two American panthers in this vicinity is causing much concern. For some time Davis Pierce, landlord of the West Forks Hotel, has insisted that a pair of these dangerous beasts was in the vicinity. He claims to have heard their blood-curdling calls, and that his cattle have been chased out of the woods by them."—"Real Indian Devil." *Lewiston Evening Journal* (July 20, 1900): 8.

[3] Franklin, James. *The Science of Conjecture: Evidence and Probability before Pascal.* (The Johns Hopkins University Press, 2001).

[4] "Current News and Notes." *Lewiston Evening Journal* (December 04, 1879): 1.

[5] "State Chat." *Lewiston Evening Journal* (May 14, 1900): 4.

[6] "State Chat." *Lewiston Evening Journal* (August 19, 1905): 4.

CHAPTER 1

[1] Citro, Joseph A. *Passing Strange—True Tales of New England Hauntings and Horrors* (New York: Houghton Mifflin, 1997). Citro's Introduction discussed the choice of the book's title which was inspired by his adopted aunt. Citro writes in his Introduction,

> There is much I remember about this independent old New Englander, including some of her colorful language: she called a bicycle a wheel and a porch a piazza, and she could name every wild flower and bird.
>
> But there is a certain phrase she used that has lodged itself forever in my writer's memory. If something was strange, she would say so: "That's strange."
>
> But if an event surpassed even that, if something was exceptionally uncanny, she'd say, "That's passing strange."

[2] LeBlanc, Deanna. "Bigfoot Sightings Bring Vt National Attention." www.wcax.com/story/17402034 /bigfoot-sightings-bring-vt-national-attention (12 April 2012).

[3] Meldrum, Jeffrey. *Sasquatch: Legend Meets Science.* (New York: Forge, 2006).

[4] Meldrum, Jeffrey. *Sasquatch Field Guide.* (Paradise Cay Publications, 2013).

[5] Bindernagel, John A. *Paranatural—Sasquatch Planet* (2012). www.dailymotion.com/video/ x11a93w_paranatural-sasquatch-planet_tv? search_algo=2#.UexPKW3YH5k

[6] "Whistling Wild Boy of the Woods." *Bangor Whig and Courier* (August 4, 1838).

[7] Animal Planet's *Finding Bigfoot* has aired four seasons as of this writing, producing over forty

one-hour episodes. This series was actually pre-dated by the Outdoor Life Network's *Mysterious Encounters* hosted by Bigfoot researcher Autumn Williams. *Mysterious Encounters* was a documentary television series focused on investigating Bigfoot sightings airing 13 episodes in 2003-2004.

8 Sarmiento, Esteban. *Paranatural—Sasquatch Planet* (2012). www.dailymotion.com/video/x11a93w_paranatural-sasquatch-planet_tv?search_algo=2#.UexPKW3YH5k

9 Landau, Elizabeth. "Why Do We Need to Look for Bigfoot?" www.cnn.com/2010/HEALTH/06/21/bigfoot.psychology.monsters/ (June 21, 2010).

10 Bindernagel, John A. *The Discovery of the Sasquatch.* (Beachcomber Books, 2010).

CHAPTER 2

1 Thoreau, Henry David. *The Maine Woods.* (1864).

2 Maine's 2013 population estimate: 1,328,302. Source: U.S. Department of Commerce, United States Census Bureau. quickfacts.census.gov/qfd/states/23000.html

3 *Ibid* 1 :9.

4 Vose, Steve. "Finding Maine's Bigfoot." *The Maine* www.themaineblog.com/?p=2560 (May 30, 2012).

5 Baron, Rachel, Blair Braverman, and Francis Gassert. *The State of Wood Biomass Energy in Maine.* (Colby Environmental Policy Group 2010). wiki.colby.edu/display/stateofmaine2011/The+State+of+Biomass+Energy

6 "Enjoy Maine's Changing Climates Throughout the Year." Maine Tourism Association. www.maine-tourism.com/content/4048/Maine_Weather/

[7] Jacobson, G. L., I. J. Fernandez, P. A. Mayewski, and C. V. Schmitt (eds). *Maine's Climate Future: An Initial Assessment.* (Orono, ME: University of Maine, February 2009, Revised April, 2009).

[8] Seymour, Tom. *Tom Seymour's Maine: A Maine Anthology.* (iUniverse, 2003).

[9] "Black Bears." Maine Department of Inland Fisheries & Wildlife. www.maine.gov/ifw/wildlife/species/mammals/bear.html (2013.)

[10] Byrne, Peter. "The Case for the *OMAH.*" *Explorers Journal* (June, 1972).

[11] "Maine's Giant Moose." *New York Times* (November 5, 1899).

[12] www.maine.gov/ifw/wildlife/species/mammals/moose.html

[13] www.maine.gov/ifw/wildlife/species/mammals/deer.html

[14] www.maine.gov/ifw/wildlife/species/endangered_species/eastern_cougar/

[15] www.maine.gov/ifw/wildlife/species/mammals/canada_lynx.html

[16] www.maine.gov/ifw/wildlife/management/lynx_avoid.htm

[17] www.maine.gov/ifw/education/wildlifepark/wildlife/bobcat.htm

[18] "General and Personal." *Philadelphia Record* (January 23, 1890): 4.

[19] "Killed Animal a Wolf." *Bangor Daily News* (March 29, 1994): 1.

[20] Adams, Glenn. "Wolves may have returned to Maine." *Lawrence Journal-World* (October 10, 1993): 3C.

[21] Thistle, Scott. "Wolf-like animal trapped, tested." *Sun-Journal* (November 23, 1996): 3B.

[22] www.maine.gov/ifw/wildlife/species/endangered_species/gray_wolf/

[23] "Signs Suggest a Return of Timber Wolf to Maine."
New York Times (December 22, 1996).

[24] www.maine.gov/ifw/wildlife/endangered/pdfs/
graywolf_20_21.pdf

[25] www.maine.gov/ifw/education/wildlifepark/wild-
life/coyote.htm

[26] "Deer and Caribou in Maine." *New York Times*
(January 29, 1888).

[27] "Miss Crosby, Writer on Hunting, Dies." *Meridian Record* (November 12, 1946): 6.

[28] "Caribou Return to Maine Woods." *New York
Times* (March 24, 1912).

[29] "Bull Caribou Spied in Moxie Lake Region."
Lewiston Daily Sun (January 7, 1930).

CHAPTER 3

[1] www.bfro.net/GDB/show_report.asp? id=1192

[2] "Lee's Wild Man." *Boston Daily Globe* (November
06, 1896): 5.

[3] newenglandfolklore.blogspot.com/2010/02/wild-
man-of-williamstown-vt.html

[4] Solunac, Alex, Tom Yamarone, and Lesley Solunac.
Sasquatch Summit—A Tribute to John Green.
Commemorative Program. (2011).

[5] Wharton, Eric H., et al. *The Forests of Connecticut*
(Pennsylvania: U.S. Department of Agriculture,
Forest Service, Northeastern Research Station,
2004).

[6] *Ibid.*

[7] Winsted is known for other unusual events besides
Louis Stone's stories and the Wild Man. The town
experienced a UFO flap beginning in February,
1967 and on-and-off again UFO sightings over
the next ten years. Two example stories follow.

Joseph O'Brien wrote a story titled, "Girls See UFO Glow; Oddly Shaped Beings," published in the September 17, 1967 edition of *The Hartford Courant*. The story began, "September UFO sightings are getting thicker than August blueberries in northwestern Connecticut. The latest is an elliptical glow and oddly shaped beings that cowered by its light spotted near a Winsted barn Friday night."

The event was witnessed by two fourteen-year-old girls looking out their second story bedroom window of their Winsted home. O'Brien wrote of the girl's experience, "We saw a bright light over that tall tree in the distance," said one, "It was bright, it looked as big as a Volkswagen."

The girls also witnessed several humanoid figures of unusual aspect. Obrien wrote, "The figures beside the mailbox were dark and "looked like the same person all over." There were no distinguishing lines between top or bottom but they seemed to be straight from toes, if any, to head."

[O'Brien, Joseph A. "Girls See UFO Glow; Oddly Shaped Beings." *The Hartford Courant*, (September 18, 1967):7.]

A 1976 AP story carried by New London's *The Day* wrote of fourteen campers watching a saucer-like object hovering over Blueberry Mountain for about twenty seconds before the object moved away rapidly. The account concludes with comments made from a dispatcher.

"It was silver colored, egg shaped and saucer-liked. It was bigger than a car and had a red revolving light and made a noise like feedback on a PA system." Said the dispatcher after consulting the report.

"After it took off rapidly in one direction, the campers took off rapidly in another direction. During the sighting they were awed but of course when they realized what had happened, they were quite shaken."

["UFO Spotted by Campers in Winsted Area." *The Day*, (New London, Connecticut, July 30, 1976).]

8 "Some Stuff," *Naugatuck Daily News* (April 21, 1973).

9 Peterson, Arthur. "Mr. Whopper." *Toledo Blade* (August 08, 1949): 11.

10 "Every Day Was April Fool's for Connecticut Publisher." *Record-Journal* (March 30, 1986).

11 "Some Stuff." *Naugatuck Daily News* (April 21, 1973): 5.

12 Grigg, Bob. "The Winsted Wildman." www.colebrookhistoricalsociety.org/PDF%20Images/The%20Winsted%20Wildman.pdf

13 "Built Like a Horse." *North Adams Daily Transcript* (August 23, 1895); www.bigfootencounters.com/creatures/injun_meadow.htm

14 Murphy, Christopher L. *Sasquatch/Bigfoot Chronicle & Other Unrecognized Relict Hominoids & Wild Men.* www.sasquatchcanada.com/uploads/9/4/5/1/945132/sc_-_earliest_to_ 1899.pdf (2013).

15 *Ibid* 12.

16 "Man Covered with Hair." *Hartford Weekly Times* (August 29, 1895).

17 "'Wild Man' May Be a Gorilla." *New York Times* (August 30, 1895).

18 "Say it is a Gorilla." *The World* (August 30, 1895).

19 "Drummer" is an older term for travelling salesman first used in the 1880s.

[20] "A Hairy Giant." *Lewiston Evening Journal* (August 30, 1895): 1.

[21] "Searching for the 'Wild Man'." *New York Times* (September 4, 1895).

[22] "Town's Bugaboo." *Boston Daily Globe* (September 16, 1895): 10.

[23] McGaga, Julius. "Abominable Half-Human Monster Linked to Barefoot Tracks in Winsted Woods." *Hartford Courant* (January 2, 1952).

[24] Guarnieri, Catherine. "Twisted History: The Winsted Wild Man." www.registercitizen.com (January, 2010).

[25] *Ibid.*

[26] Tarpey, John P. "'Bigfoot' Leaves Impression on 2 Ellington Farmhands." *The Hartford Courant* (November 26, 1982).

[27] Chase, George Wingate. *The History of Haverhill, Massachusetts*. (Haverhill, 1861): 496.

[28] "The Wild Man of Haverhill, Massachusetts." newenglandfolklore.blogspot.com (January 24, 2010).

[29] "A Wild Man of the Mountains." *New York Times* (18 October, 1879).

[30] Belanger, Jeff. *Weird Massachusetts*. (Sterling, 2008): 88.

[31] www.thetranscript.com/news/ci_22966543

[32] www.tauntongazette.com/x163303980/Hockomock-Swamp-can-be-a-devil-of-a-time-especially-around-Halloween?zc_p=1

[33] Belanger, Jeff. *Weird Massachusetts*. (Sterling, 2008): 96.

[34] Muscato, Ross A. "Tales from the Swamp." www.boston.com/news (October 30, 2005).

[35] Hayward, Ed. "The Bigfoot of Bridgewater; Is it a Man-Beast or Hockomock Crock?" *Boston Herald Library Online* (April 6, 1998).

36 "October Mountain State Forest." www.mass.gov/
    eea/agencies/dcr/massparks/region-west/
    october-mountain-state-forest-generic.html

37 "Weird Mountain 'Creature' is Reported by Pic-
    nickers." *The Berkshire Eagle* (Tuesday, August
    23, 1983), reported by BFRO: www.bfro.net/
    gdb/show_article.asp?id=414

38 "The Story of Sasqua." www.cultreviews.com/in-
    terviews/the-story-of-sasqua/

39 Maine has the distinction of the only State within
    the lower 48 that borders only one other State:
    New Hampshire. All other States within the con-
    tiguous 48 border two or more States.

40 www.fas.usda.gov/ffpd/economic-overview/over-
    view.html

41 naturalplane.blogspot.com/2011/09/new-hamp-
    shire-wood-devils.html

42 Berry, Rick. *Bigfoot on the East Coast.* (1993):
    150.

43 French, Scott. "The Man Who Spied Bigfoot Comes
    Forward." *Concord Monitor* (November 13,
    1987); accessed as BFRO Media Article #267;
    www.bfro.net/gdb/show_article.asp?id=267

44 "Hiking: Myth Versus Reality of Sasquatch in
    Ossipee Range." www.mid-americaBigfoot.com/
    forums/viewtopic.php?f=47&t=6732

45 "Lowell Man Flees Hollis." *The Nashua Telegraph*
    (May 10, 1977).

46 "Behind the Scenes of *Finding Bigfoot*: 'Big
    Rhodey' with Cliff Barackman." www.crypto-
    mundo.com/Bigfoot-report/finding-bf-71/

47 Widmann, Richard H. "Trends in Rhode Island
    Forests: A Half-Century of Change" United
    States Department of Agriculture. www.dem.
    ri.gov/programs/bnatres/forest/pdf/rit-
    rends.pdf

48 "Wakefield, South County, Rhode Island." www.bigfootencounters.com/sbs/wakefield.htm

49 "Mother and Son See Sasquatch Closeup from Road." www.bfro.net/gdb/show_report.asp?id=6643

50 "Behind the Scenes of *Finding Bigfoot*: 'Big Rhodey' with Cliff Barackman." www.cryptomundo.com/Bigfoot-report/finding-bf-71/

51 "Behind the Scenes of *Finding Bigfoot*: 'Big Rhodey' with Cliff Barackman" www.cryptomundo.com/Bigfoot-report/finding-bf-71/

52 Whitehall, New York is a geographical area of study of researcher Paul Bartholomew.

53 "Bigfoot Sightings Bring VT National Attention." www.wcax.com/story/17402034/Bigfoot-sightings-bring-vt-national-attention

54 "Slipperyskin—Bear, Bigfoot, or Indian?" www.northlandjournal.com/stories/stories18.html

55 Citro, Joseph A. *Weird New England.* (2005): 100.

56 Hoffenberg, Noah. "Man Spots Bigfoot." *Bennington Banner* (September 26, 2003).

57 *Ibid* 55: 110-111.

58 "Mother Bigfoot." *Finding Bigfoot.* Season 3, Episode 2 (November 18, 2012).

59 "Behind the Scenes of *Finding Bigfoot*: 'Mother Bigfoot' with Cliff Barackman." www.cryptomundo.com/Bigfoot-report/finding-bf-116/

60 "Vermont Trail Camera Photo Analysis." www.cliffbarackman.com/research/field-investigations/vermont-trail-camera-photo-analysis/

CHAPTER 4

1 "Native Americans." www.visitmaine.com/things-to-do/arts-and-culture/native-americans/

2 Hosmer, George L. *An Historical Sketch of the Town of Deer Isle, Maine.* (1886).

3 Strain, Kathy M. *Giants. Cannibals & Monsters.* (Hancock House, 2008).

4 Schorn, M. Don. *Elder Gods of Antiquity: First Journal of the Ancient Ones.* (Ozark Mountain Publishing, 2008).

5 Halpin, Marjorie and Michael Ames. *Manlike Monsters on Trial: Early Records and Modern Evidence.* (University of British Columbia Press, 1980): 143-146.

6 Coleman, Loren. *Bigfoot! The True Story of Apes in America.* (Paraview Pocket Books, 2003): 27-29.

7 *Ibid* 3.

8 Bourque, Bruce Joseph. *Twelve Thousand Years: American Indians in Maine.* (University of Nebraska, 2001): 19.

9 *Ibid.*

10 Giles, John. *Memoirs of Odd Adventures, Strange Deliverances, Etc. in the Captivity of John Giles, Commander of the Garrison of Saint George River, in the District of Maine.* (1869): 35-36.

11 *Ibid* 6.

12 *Ibid* 5.

13 *Ibid* 5.

14 *Ibid* 5.

15 Rose, Carol. *Giants, Monsters, and Dragons: An Encyclopedia of Folklore, Legend, and Myth.* (W. W. Norton & Company, Inc., 2001): 24.

16 *Ibid* 3.

17 *Ibid* 6.

18 Elder, William. "The Aborigines of Nova Scotia." *The North American Review*, CCXXX (January, 1871).

19 *Ibid* 6.

[20] Partridge, Emelyn Newcomb. *Glooscap The Great Chief and Other Stories*. (Sturgis & Walton, 1913).

[21] "Cannibal Giants of the Snowy Northern Forest." newenglandfolklore.blogspot.com/2009/01/cannibal-giants-of-snowy-northern.html

[22] *Ibid*.

[23] Prince, J. Dyneley. "Some Passamaquoddy Witchcraft Tales." *Proceedings of the American Philosophical Society* 38 (1899): 185.

[24] Prince-Hughes, Dawn. *The Archetype of the Ape-Man: The Phenomenological Archaeology of a Relic Hominid Ancestor*. (2000): 64-66.

[25] Frost, Stanley. "The Air Line to the Big Woods." *Outing* 78 (July, 1921): 156. Frost's story focuses on a Lieutenant Graham, a pilot from Massachusetts, and the benefits from the growing use of airplanes in the wilderness for "mapping, exploration, timber cruising, fire patrol, and transportation of officials". Frost recounts the following anecdote regarding Lac Windigo, approximately 300 miles west of the Maine border in Quebec Province:

The Indians at first labeled the plane a "Kitche Chshee" or "big duck," but afterwards, when the mechanism was explained as well as possible, changed the name to "flying buggy." Lieutenant Graham has forgotten the Indian for this term.

Friendly Indians took occasion to warn Graham that it would be very unwise for him to get in the neighborhood of Lac Windigo or Grand Lac because of the danger that the natives in that neighborhood would take the plane for a reincarnation of a particularly malicious Indian devil with which they had been at war. This devil, called a "windigo," was supposed to inhabit the lake of that name and to have devoured two braves who went into the neighborhood and vanished.

Three or four years ago, a party of white men hunting in the neighborhood killed a moose which had a very badly crumpled horn and were amazed to see their Indian guides strip, paint, and do a war dance around the carcass. They learned that the Indians believed that this moose, because of his curious horn, was an incarnation of the "windigo" and that his death would bring them good fortune.

[26] "Queer Animals in Maine." *Boston Daily Globe* (November 5, 1905): 31.

[27] *Ibid.*

[28] "Are the Sea Serpent and Maine Windigo Doomed?" *Boston Daily Globe* (January 11, 1920): E1.

[29] Skinner, Charles Montgomery. *American Myths and Legends 1.* (J. P. Lippincott, 1903): 38.

[30] Rosen, Brenda. *Mythical Creatures Bible: The Definitive Guide to Legendary Beings.* (Sterling Publishing, 2009): 223.

[31] Leland, Charles Godfrey. *The Algonquin Legends of New England: or Myths and Folk Lore of the Micmac, Passamaquoddy, and Penobscot Tribes.* (Houghton, Mifflin and Company, 1884): 235-236. "The Chenoo is not only a cannibal, but a ghoul. He preys on nameless horrors. In this case, 'having yielded to the power of kindness, he has made up his mind to partake of the food and hospitality of his hosts,' 'to change his life; to adapt his system to the new regimen, he must thoroughly clear it of the old.'—Rand Manuscript.

[Rand, Silas Tertius. *Micmac Indian Legends* (1894): 192] This is a very naïve and curious Indian conception of moral reformation. It appears to be a very ancient Eskimo tale, recast in modern time by some zealous recent Christian convert."

[32] *Ibid* 32: 233-244.

[33] Rand, Silas Tertius. *Micmac Indian Legends.* (1894): 192.

[34] *Ibid* 32: 251-254.

[35] *Ibid* 32: 246-251.

[36] An elegant womanly form above, ending in form of a fish below, i.e., a mermaid.

[37] LeMoine, James MacPherson. *The Chronicles of the St. Lawrence.* (Dawson Brothers, 1878): 133-136.

[38] en.wikipedia.org/wiki/Pukwudgie

[39] Author C. J. Stephens has a small section on Bigfoot in Maine in his book, *The Supernatural Side of Maine* (John Wade Pub. 2002), which includes the following note on smaller beings:

There is a rush of stories about hunters who see small human-like beings. (This brings to mind Washington Irving's "Rip Van Winkle" in the Catskills; this character met several little bowlers and slept for 20 years.) There are Iroquois who insist that little people are present today—in Maine and all of New England—seldom seen because there is less hunting along deserted ridges of mountains and foothills. [Stephens, C. J. *The Supernatural Side of Maine* (2002): 93]

[40] "Puckwudgie," en.wikipedia.org/wiki/Pukwudgie

[41] Gray, T. M. *New England Graveside Tales.* (Schiffer Publishing, 2010): 10.

[42] Coatsworth, Elizabeth. *Maine Memories.* (S. Green Press, 1968): 95-96.

[43] Eberhart, George. *Mysterious Creatures: A Guide to Cryptozoology.* (ABC-CLIO, 2002):325.

CHAPTER 5

[1] "The Cryptid Zoo: Wildmen." www.newanimal.org/wildmen.htm

[2] "What is it?" *New York Times* (April 26, 1871).

[3] Burns, J.W. "Introducing B.C.'s Hairy Giants." *MacLean's Magazine* (April 1, 1929).

[4] Shackley, Myra. *Wildmen.* (Thames & Hudson Ltd., 1983): 16.

[5] Bartholomew, Robert E., and Brian Regal. "From Wild Man to Monster: the Historical Evolution of Bigfoot in New York State." *Voices* (Volume 35, 2009).

[6] Buhs, Josuha Blu. "Wildmen on the Cyberfrontier: The Computer Geek as an Iteration in the American Wildman Lore Cycle." *Folklore* (April, 2010): 62.

[7] Monahan, Peter Frederick. *The American Wild Man: The Science and Theatricality of Nondescription in the Works of Poe, Jack London, and Djuna Barnes.* (ProQuest, UMI Dissertation Publ., 2008): 40.

[8] *Ibid*: 2.

[9] Twain, Mark. *The Republic of Gondor, The Wild Man Interviewed.* (1869).

[10] Allusions to suffering and disappointment infuse many hermit and Wild Men stories of a century ago. More-civilized society's longing to understand the choices, passions, and motivations of such men fueled newspaper, magazine, and popular accounts of these recluses with voids in knowledge filled in with ideas like cannibalism and exotic origins.

The Otis Cannibal—Wrecked When A Boy And Brought Up As A Man-Eater—Pittsield, Mass., Feb. 7.—The "Otis Cannibal," as he is known throughout Berkshire County, has figured for a day or two in the Criminal Court as defendant in a case for robbing an old woman's pork barrel, and to-day was given a month in jail for the theft. "Yer'd better have stuck to man meat and let the pork alone," said an old granger to the prisoner as he passed out of the court-room in charge of an officer. The prisoner evidently did not enjoy the banter, and replied: "I wouldn't want ter chew your tough old carkiss." The cannibal's real name is Edward Hazard, and many people believe that he has really eaten human flesh. For several years he has been one of the attractions at neighboring cattle shows, the little tent that concealed his not over-attractive person bearing the card, "Only 5 cents to see the oldest cannibal in Berkshire County." While Western Massachusetts has had a good many queer characters, no one had previously supposed that the county which boasts a college and such Summer resorts as Stockbridge, Lenox, and Williamstown did really possess a collection of cannibals of assorted ages.

Till within a year or so Hazard has led a semi-savage existence in a hut on one of the Otis hills. He fished and hunted, and if reports are true, more often ate his fish and game raw than cooked, expressing a decided taste for it in that state. In his hut he kept a "boudish," a hideous idol, which he worshiped, performing strange ceremonies before it in the dead of night. This he finally burned, saying that it was impossible to have it bring him good luck unless he could offer human sacrifices before it. He continued, however, to worship any freaks of nature which he found in the woods or fields, such as strange rock formations or gnarled tree branches.

According to his story he was thrown on one of the South Sea islands when a boy, the ship on which he was serving as cabin boy being wrecked. The island was occupied by cannibals who made short work of Hazard's companions who escaped to land with him. Two were offered up to the big "boudish," or idol, of the tribe, after being tortured terribly with fire and by other means. The three others were fattened after they had apparently been taken into the tribe, and then slaughtered for a banquet which the King gave in honor of friends who came to see him from a neighboring island. Hazard's youth saved him for the time being, and he won the good-will of the King by his ingenuity in the use of various tools taken from the wrecked ship. He was given a wife, taught to eat human flesh, and raised a family. He also learned to worship the idols of the islanders, and says he cannot entirely lay aside the "religious habits" he formed there. After living with the savages a dozen years, he escaped to a passing vessel, paddling out to the ship on the pretense of decoying the crew to the island. With all his eccentricities he is apparently a harmless sort of a man, and the yearning for roast baby which he occasionally expresses is laughed at by the people who live about him.—*New York Times* (February 8, 1884).

[11] Buhs, Josuha Blu. "Wildmen on the Cyberfrontier: The Computer Geek as an Iteration in the American Wildman Lore Cycle." *Folklore* (April, 2010): 64. This is also an *Ibid*.

[12] Pierce, Josiah. *A History of the town of Gorham, Maine* (1862): 236-237.

[13] "The Mysterious Wild Man." *Daily Kennebec Journal* (December 27, 1871): 3.

[14] "A Wild Man's Prisoner." *Mansfield Daily Shield* (January 22, 1893).

[15] *Ibid.*

[16] "Some Odd Stories." *Daily Advocate* (Ohio) (September 6, 1893): 6.

[17] "Hunting a Wild Man." *Lewiston Evening Journal* (November 21, 1893).

[18] "There is More or Less Excitement." *Daily Kennebec Journal* (September 15, 1894): 1.

[19] Burke, Selden. "Romance and Reality of the Wild Men." *The Scrap Book* (November, 1908): 912.

[20] "The Game Season." *Bar Harbor Record* (September 21, 1895): 5.

[21] "A Man Thought to be the Wild Man." *Bar Harbor Record* (December 04, 1895): 13.

[22] "There Had Been Circuses." *Bangor Daily Whig and Courier* (June 04, 1896): 1.

[23] "50 Years Ago Today." *Lewiston Evening Journal* (July 9, 1946).

[24] "Chased by a Wild Man." *Montreal Daily Herald* (June 25, 1888).

[25] "Wild Man at Large in Minot." *Boston Daily Globe* (July 21, 1896): 5.

[26] "Wild Man Seen at South Auburn." *Boston Daily Globe* (August 01, 1896): 9.

[27] "A Wild Man." *Bangor Daily Whig and Courier* (August 17, 1896): 3.

[28] "A Wild Man." *Lewiston Evening Journal* (November 30, 1896).

[29] "Again the Wild Man." *Lewiston Evening Journal* (January 27, 1897): 7.

30 "Sheriff Hill Saw Governor Powers." *Lewiston Evening Journal* (January 30, 1897).

31 "History from our Files." *Portland Press Herald* (November 17, 1947): 4.

32 "Stories of the Streets and Town." *Lewiston Evening Journal* (June 4, 1898).

33 "Bad Man with Red Hair." *Bath Independent* (June 18, 1885): 2.

34 "Speaking of the Witch Spring Wild Man." *Bath Independent* (June 25, 1898): 1.

35 "Wild Man?" *Bath Independent* (June 25, 1898): 3.

36 "Aroostook County Woman Believes Danville Wild Man is Her Better Half." *Daily Kennebec Journal* (August 08, 1898): 3.

37 "Sabatis' Annual Wild Man." *Lewiston Evening Journal* (July 20, 1899).

38 "Peaks Island's Wild Man." *Daily Kennebec Journal* (July 20, 1899): 1.

39 "Weird and Uncanny." *Lewiston Evening Journal* (August 19, 1899).

40 Gray, T. M. *New England Graveside Tales.* (Schiffer Publishing, 2010): 13.

41 "Run Amuck!" *Lewiston Evening Journal* (July 20, 1900).

42 Day, Holman "Down in Maine." *Hartford Courant,* (October 17, 1900).

43 Ibid.

44 "Poland's Wild Man." *Lewiston Evening Journal* (July 22, 1905).

45 Mulespinner refers to a machinist working in the textile industry.

46 "Wild Man Caught." *Boston Daily Globe* (July 24, 1905).

47 "The Insane Man."*Daily Kennebec Journal* (July 25, 1905): 2.

[48] "Gorham is the Latest Town," *Daily Kennebec Journal* (August 15, 1905): 6.

[49] "Wild Man in Riggsville," *Bath Independent and Enterprise* (May 18, 1907): 6.

[50] "Is not Insane," *Lewiston Evening Journal* (November 20, 1907).

[51] "Acted Like a Wild Man," *Biddeford Weekly Journal* (July 19, 1912).

[52] "Rumor of a Wild Man Seen," *Lewiston Daily Sun* (August 13, 1913).

[53] "Durham Men on Wild Man Hunt," *Lewiston Evening Journal* (August 20, 1913): 10.

[54] "Durham Wild Man Not Yet Captured," *Lewiston Daily Sun* (August 21, 1913): 6.

[55] "All-Day Hunt for Wild Man," *Lewiston Evening Journal* (August 21, 1913): 3.

[56] "Durham Wild Man on Rampage," *Lewiston Evening Journal* (August 22, 1913): 1, 11

[57] "Durham Wild Man May Be Artist John Knowles," *Lewiston Daily Sun* (August 22, 1913): 6.

[58] "Hunt for Wild Man," *Lewiston Daily Sun* (August 23, 1913): 6.

[59] "Posse of 60 Men Hunt Wild Man of Durham," *Lewiston Daily Sun* (August 25, 1913): 8.

[60] "Fifty Searchers Out," *Lewiston Evening Journal* (August 25, 1913): 5.

[61] "A Wild Man," *The Toronto World* (August 25, 1913): 6.

[62] "Fired at Wild Man," *Lewiston Daily Sun* (August 27, 1913).

[63] "May Search Sunday for Wild Man," *Lewiston Daily Sun* (August 29, 1913): 6.

[64] "The Wild Man Has Wandered," *Lewiston Evening Journal* (September 5, 1913).

[65] The thought that a Wild Man guise could be leveraged to scare away unwanted attention brings to

mind another case, carried by the *Kennebec Journal*, taking place outside of Maine but relatively nearby in Stonington, Connecticut. An "ape man" was seen several times near the Miner farmstead soon after Muriel Miner, 19 and her sister Mildred, 16 inherited interests in the property after the passing of their father Horace D. Miner. The girls lived alone in the family farmhouse, and a brother helps take care of some of the 2,000 acre farm. The girls described the ape man as a man dressed in skins, appearing only at a distance and fled the scene when approached by one of the sisters. Other farmers in the township claimed to have seen the creature. The Miners ascribe its activities as a plot to drive them away from the farm in order that it may be purchased at a low price.

Aped "Ape Man" to Frighten Farm Owners—Prospective Purchaser of Connecticut Property Adopted Unusual Methods—Providence, R.I., April 3—An "ape man" reported to be frightening residents of North Stonington, Conn., is merely a prospective purchaser of a far, wearing a fur coat and endeavoring by his antics to frighten away two orphan sisters who inhabit the farm he seeks. This was the report made Friday by George Denison, game warden after a day's search for the "ape man" on the Horace D. Minor farm.

"If that fellow goes out there again with the skin of an animal over his head," said Denison, "they ae going to put the lead to him. I wouldn't want to try it if I were he."

Denison said he had found no trace of the man Friday.

The Miner girls Muriel aged 19 and Mildred, aged 16 inherited the farm from their father who died recently. Denison said they are armed and ready to shoot and other residents of the neighborhood also had their shot guns at hand and will brook no "ape man" real estate tactics in the vicinity in the future.

The man creature has succeeded in frightening Frank Miller, aged farm hand from the place in spite of the stoical attitude of his youthful employers.

"You won't catch me going back," said Miller.

The Miner sisters, however, insist that they are not afraid, and will retain and till the acres of the Miner homestead.

The "ape man" has not been seen for several days and is felt to have been frightened by threats of shooting. He had been frequently seen bounding about on boulders in the field and once or twice on haystacks. By these antics, it is believed, he sought to frighten away the Miner girls and so be able to buy the property cheaply. [*Daily News* [Frederick, MD] (April 03, 1926).]

The plot has led some to suggest that the affair could have been inspired by *The Cat and the Canary* theatrical play (the film would open the following year in September, 1927).

The story received a denouement in the April 4, 1926, edition of the *Davenport Democrat*:

Police Solve Famous "Ape Man" Mystery—Was Found to be April Fool Joke Perpetrated by Two Young Ladies. (AP) New London, Conn., April 3— North Stonington's now famous "ape man" has been captured. It was found to be nothing but a dummy. The originator of the scheme has confessed.

This was the result of a state police investigation announced today by Sergeant Clifford Gorgas of the Groton barracks.

The ape-man dummy was found near the home of the Miner Sisters, Muriel and Mildred, and was constructed, Miss Muriel Miner admitted, to scare away visitors not wanted. Directed to Haystack Rock by Miss Miner, the state policeman found the dummy wrapped in a sheet with a large April Fool sign hanging from it. It was then, the state police report stated, that Miss Muriel Miner confessed to the hoax.

Thus ends a mystery which has caused considerable uneasiness in the North Stonington vicinity for more than a week and a mystery that had led to the formation of at least one posse and several expeditions in search of the "ape man."

[66] "25 Years Ago Today." *Lewiston Evening Journal* (September 12, 1938).

[67] "Durham Wild Man Now at Pownal." *Lewiston Daily Sun* (September 15, 1913): 8.

[68] "Durham Wild Man Left Clothes." *Lewiston Evening Journal* (September 22, 1913): 5.

[69] "That L. J. 'Wild Man'?" *Daily Kennebec Journal* (September 23, 1913): 7.

[70] "The Durham People Still Believe." *Lewiston Evening Journal* (October 2, 1913): 12.

[71] *Ibid*: 2.

[72] "Because his Mother." *Lewiston Evening Journal* (March 20, 1915).

[73] "An Eerie Character." *Lewiston Evening Journal* (August 4, 1914).

74 "Ocean-Going Wild Man in Maine Woods." *Boston Daily Globe* (August 07, 1919): 7.

75 "Tramp Scared Residents." *Biddeford Journal* (October 27, 1920): 3.

76 "Maine Wild Man Sought in Woods." *Boston Daily Globe* (December 03, 1922): 32.

77 "'Wild Man' Hunt On in Maine." *Syracuse Telegram* (December 7, 1922).

78 The Pownal State School opened in 1907 as The Maine School for the Feeble-Minded. The institution closed its doors in 1996 by which time it was known as Pineland Center. At its largest size during the 1950s, some 1,700 residents and several hundred staff occupied the some 50 buildings of the institution's campus. Through its almost ninety years, the place offered care, treatment, and housing for orphans, persons with mental retardation, youth with psychiatric needs, and disorderly children.

79 "No Trace of 'Wild Man' Seen in Topsham Woods." *Lewiston Daily Sun* (June 6, 1929).

80 "New York Showmen Report a Scarcity of Good Wild Men." *Biddeford Weekly Journal* (June 14, 1907)

81 Stephens, Charles Asbury. *A Busy Year at the Old Squire's.* (C. H. Simonds Company, 1922): 205-210.

82 "It is All There." *Lewiston Evening Journal* (September 07, 1900): 2.

83 *Ibid.*

84 *Ibid.*

85 *Ibid.*

86 "Moji is Safely Chained." *Lewiston Evening Journal* (September 7, 1900).

87 "The Attractions at the Station." *Daily Kennebec Journal* (September 9, 1907): 3.

[88] "The Death of Ki-Ko." *Bath Independent and Enterprise* (February 13, 1909): 5.

[89] "Two Grange Exhibits." *Lewiston Evening Journal* (August 29, 1907).

[90] "A Hermit in Maine." *New York Times* (August 05, 1883).

[91] "Queer Maine Folks." *Lewiston Evening Journal* (June 25, 1887): 5.

[92] Day, Holman. "Down in Maine." *Hartford Courant* (October 17, 1900).

[93] *Ibid.*

CHAPTER 6

[1] Blackburn, Lyle. *Lizard Man: The True Story of the Bishopville Monster.* (Anomalist Books, 2013).

[2] Josselyn, John. *Voyages to New England.* (1663).

[3] "Runaway Monkey Likes Maine Air." *Lewiston Daily Sun* (November 03, 1962).

[4] "Strange Animal Prowling About Raccoon Gully." *Biddeford Journal* (January 31, 1919).

[5] "Raccoon Gully Douroucouli in Limelight." *Biddeford Weekly Journal* (February 21, 1919): 8.

[6] "Closing of Glodthwaite's Much Regretted." *Biddeford Journal* (July 27, 1920).

[7] "Farmer Snow's Sweet Corn." *Biddeford Journal* (July 31, 1920): 5

[8] "Sandford Men See Bear." *Biddeford Journal* (October 03, 1921): 4.

[9] "It Was Not a Bear." *Biddeford Journal* (October 04, 1921): 1.

[10] "Raccoon Gully Duroculi." *Biddeford Journal* (October 05, 1921): 5.

[11] "Looking for the Duroculi." *Biddeford Journal* (October 6, 1921): 1.

[12] Hinckley, G. W. *Roughing it with Boys.* (1913) pp. 176-178.

[13] "The Loocivee's Yell." *Lewiston Evening Journal* (October 28, 1901): 6.

[14] "Leaves from the Experience of an Old Hunter." *Lewiston Evening Journal* (January 26, 1884): 5.

[15] "Down-East Backwoodsman." *The New York Times* (August 7, 1892).

[16] Fendler, Donn, and Joseph B. Egan. *Lost on a Mountain in Maine.* (Harper Trophy, 1992): 70-71.

[17] "A Hunter's Encounter with the Devil." *Lewiston Evening Journal* (October 17, 1876).

[18] The Internet Movie Database gives cast credits for *Mysterious Monsters*, listing Dr. Lawrence Bradley as portraying himself.

[19] Cole, Benjamin C. *It Happened Up in Maine.* (Penobscot Bay Press, 1980): 51.

[20] Rhodes, Dean. "'Les Diables' Still Rove Allagash Wilds." *Bangor Daily News* (February 18, 1974): 17.

[21] "Oral Traditions." *Acadian Culture in Maine.* acim.umfk.maine.edu/oral_traditions.html

[22] *Ibid.*

[23] *Ibid.*

[24] L'Heureux, Juliana. "This is the Time when the Werewolf's Howl Scares Little Children." *Portland Press Herald/Maine Sunday Telegram* (October 25, 2001) [York County Extra]: 2E.

[25] Godfrey, Linda S. *Real Wolfmen: True Encounters in Modern America.* (Tarcher, 2012): 128.

[26] *Ibid.*, p. 129.

[27] *Ibid.*, p. 130.

[28] "Wolf Pack—Shelley Rockwell-Martin Separates Fact From Fiction." *Dogmen—Monsters are Real.* dogman-monsters-are-real.blogspot.com/2013/08/wolf-pack-shelley-rockwell-martin.html

[29] Meldrum, Jeffrey. *Sasquatch: Legend Meets Science.* (Forge, 2006): 221.

[30] Wasson, Barbara. *Tracking the Sasquatch.* (Sisters, Oregon: Barbara Wasson, 1994).

[31] "Footprints of the Devil." *Bath Independent* (July 13, 1901): 7.

[32] "Footprints on Maine Shores." *Lewiston Evening Journal* (December 04, 1897).

[33] Shaw, Dick. "Mysterious Devil's Footprints in Milo, Maine." *Lewiston Evening Journal* (June 26, 1976): 8A.

[34] *Ibid.*

[35] Skelton, Kathryn. "Bigfoot & Me." *Lewiston Sun Journal* (May 05, 2007).

[36] "Footprints on Maine Shores." *Lewiston Evening Journal* (December 04, 1897).

[37] Moran, Mike and Sceurman, Mark. *Weird U.S.: Your travel Guide to America's Local Legends and Best Kept Secrets.* (Sterling, 2009): 57.

[38] Hazard, Wendy. "Halloween: Local Legends." *Kennebec Journal* (October 30, 1976).

[39] *Ibid* 34.

[40] *Ibid* 30.

[41] Bush, Grace. "Devil's Footprints' Seen." *Bangor Daily News* (October 11, 1976).

[42] *Ibid* 30.

[43] *Ibid* 38.

[44] "Devil's Tracks." *Reading Eagle* (May 11, 1902): 12.

[45] *Ibid* 30.

[46] Russell, Rick. "Witch's Tale Haunts Bucksport Founder." *Bangor Daily News* (July 28, 2007): C1.

CHAPTER 7

[1] Paulides, David. *Missing 411—Western United States & Canada.* (Createspace, 2011): xvi.

2 "Missing 411." www.bigfootforums.com/index.php?/topic/29296-missing-411/

3 "Heard Boy Crying." *Lewiston Evening Journal* (August 05, 1911).

4 Paulides, David. *Missing 411—The Devil's in the Details*. (Creatspace, 2014): 124.

5 "Lost Child Found at Masardis." *Lewiston Evening Journal* (August 12, 1897): 1.

6 "Many Things Preyed on Elsie Davis' Mind." *Lewiston Evening Journal* (August 03, 1911): 5.

7 *Ibid*.

8 "Still Searching for Missing Girl." *Lewiston Evening Journal* (August 01, 1911).

9 *Ibid*.

10 "Her Memory is a Blank." *Lewiston Evening Journal* (August 04, 1911).

11 *Ibid*. 8.

12 *Ibid*. 4.

13 *Ibid*. 4.

14 *Ibid*. 4.

15 *Ibid*. 4.

16 *Ibid*. 8.

17 *Ibid*. 8.

18 *Ibid*. 10.

19 *Ibid*. 4.

20 Erickson, Amy. "Warden Jarred Herrick, Drowning . . . or Murder? Part One of Amy Erickson's Special Report." www.wabi.tv/2009/04/29/drowning-or-murder-part-one-of-amy-ericksons-special-report/ (April 29, 2009)

21 "Send Organs on to be Analyzed." *Lewiston Evening Journal* (May 05, 1923): 1.

22 *Ibid*. 18.

23 Erickson, Amy. "Drowning or Murder? Part Two of Amy Erickson's Special Report." www.wabi.tv/news/5699/drowning-or-murder-part-two-of-amy-ericksons-special-report (April 30, 2009)

[24] "Hunt Boy, 3, in Woodland at East Gray." *Lewiston Daily Sun* (June 18, 1951): 1, 16.

[25] "Mexico Child Found at Gray." *Lewiston Daily Sun* (June 19, 1951): 1, 11.

[26] "Find Missing Thorndike Boy Safe." *Lewiston Evening Journal* (July 19, 1954): 1,7.

[27] "2-Year-Old Girl Missing in Woods." *Meriden Journal* (July 14, 1956).

[28] "Lost 24 Hours, Girl, 2, Found." *Pittsburgh Press*, Sunday (July 15, 1956): 4.

CHAPTER 8

[1] Coleman, Loren. "Maine Tree Creature: First Blobsquatch of 2010?" www.cryptomundo.com/cryptozoo-news/1-blobsqu10/

[2] BFRO. "The Mystery of the Maine 'Sasquatch' Now Resolved." www.bfro.net/news/porcupine_maine.asp

[3] Griffin, Joseph. *History of the Press of Maine.* (Brunswick, 1872): 193.

[4] Arment, Chad. *The Historical Bigfoot.* (Coachwhip Publications, 2006): 170-171.

[5] "Finding Bigfoot in Maine!" www.topix.com/forum/city/lincoln-me/TNUT3RGL25KBGANHT (November 25, 2007).

[6] Ricker, Nok-Noi. "Bigfoot in Maine? 10-Foot Tall 'Wild Man' was Killed in 1886, Newspapers Reported." www.bangordailynews.com/2013/10/27/news/state/bigfoot-in-maine-10-foot-tall-wild-man-was-killed-in-1886-newspapers-reported/

[7] "The Wild Man is Coming to the Front This Fall." *Los Angeles Times* (May 05, 1887): 11.

[8] "Strange Monster Seen in Maine." *Boston Daily Globe* (August 09, 1895): 9.

9 "A Terror in the Woods." *San Francisco Call* (November 27, 1895).

10 "The Meddybumps Howler, Washington County, Maine." www.bigfootencounters.com

11 *Ibid.*

12 Soucy, D. L. *Maine Monster Parade.* (Lulu Enterprises, 2008): 73.

13 "John Green—Sasquatch Investigator, Chronicler and Pioneer Analyst." www.sasquatchdatabase.com

14 "Suddenly There He Was. Less Than 15 Feet in Front of Me." www.bigfootencounters.com (March 13, 2011).

15 "I Believe in Bigfoot or Sasquatch." www. experienceproject.com/stories/Believe-In-Bigfoot-Or-Sasquatch/1488085

16 "Maine IBS Sighting Reports." www.mid-america-bigfoot.com/forums/viewtopic.php?t=4070 &p=13483

17 "Hunters Find Scat and Have Large Rock Thrown at Them." BFRO Report #8259. www.bfro.net (March 13, 2004).

18 Pinkham, Steve. *The Mountains of Maine: Intriguing Stories Behind Their Names.* (Down East Books, 2009).

19 ParaBreakdown. "Death by Sasquatch (True Stories from Maine)." www.youtube.com/watch?v= KgpVBsMDa78

20 Buckley, Ken. "A Sasquatch Search in Bridgton, Maine." *Bangor Daily News* (August 5, 1971).

21 *Ibid.* 12.

22 "Sighting by a Motorist." BFRO Report #1189. www.bfro.net (March 16, 1997).

23 "UFO Sighting Reported at Durham; Track Beep Sound." *Lewiston Daily Sun* (April 18, 1966).

24 *Ibid.*

25 "Search on for Monkey Man." *Lewiston Daily Sun* (April 20, 1973): 8.

[26] *Ibid.*

[27] "Rented Gorilla Costume has not been Returned." *Lewiston Daily Sun*, (July 30, 1973): 28.

[28] "Series of Sightings Taking Place over Three Days." BFRO Report #1186. www.bfro.net (January 1, 1998).

[29] *Ibid.*

[30] Skelton, Kathryn. "Durham's 'Gorilla'." Weird, Wicked Weird. *Lewiston Sun Journal* (August 02, 2008).

[31] "Anyone Lose an Ape at Durham?" *Lewiston Daily Sun* (July 27, 1973): 2.

[32] *Ibid.*

[33] *Ibid.*

[34] *Ibid 22.*

[35] "Beast Reported Prowling in Durham Area." *Bangor Daily News* (July 27, 1973): 19.

[36] "Gorilla is Back." *Lewiston Evening Journal* (August 13, 1973): 22.

[37] *Ibid 30.*

[38] "Gorilla Spectators Cause Durham Property Damage." *Lewiston Daily Sun* (August 15, 1973): 28.

[39] *Ibid.*

[40] *Ibid 22.*

[41] Day, John S. "Ford may Make N. H. Gains by Handling Hecklers Better than Reagan." *Bangor Daily News* (February 12, 1976): 1.

[42] "Camper Hears Screams while Camping in Kennebunk, Maine." www.bfro.net/GDB/show_report.asp?id=883 [Report #833 (Class B), submitted December 07, 2000.]

[43] Bord, Colin and Janet Bord. *Bigfoot Casebook.* (Granada, 1982): 205.

[44] Green, John. *Sasquatch: The Apes Among Us.* (Hancock House, 1978): 229.

45 www.sasquatchdatabase.com/%28S%283n5boq 55bpsho245dlsmgfyo%29%29/SearchResult .aspx?i_state_prov=Maine

46 *Ibid* 33: 207.

47 *Ibid* 12.

48 "From the Files of the Gulf Coast Bigfoot Research Organization." www.gcbro.com/MEsomerset 0002.html

49 "Was it a bear or Bigfoot?" *Kennebec Journal* (July 4, 1977): 10.

50 "Kezar Falls." www.bigfootencounters.com/stories/ kezarfallsME.htm

51 "From the Files of the Gulf Coast Bigfoot Research Organization." www.gcbro.com/MEoxford0002.html

52 "1977 Incident Unreported." www.bigfootforums.com/ index.php/topic/2165-1977-incident-unreported/

53 *Ibid* 12. www.sasquatchdatabase.com/%28S%28 emdq4545vrytx3qodlsljb55%29%29/Search Result.aspx?i_state_prov=Maine

54 "Piscataquis County." www.bigfootencounters.com/ sbs/piscataquis.htm

55 "Maine Bigfoot: Maine Ridge Monster, Meddy- bumps Howler or Pomoola." (December 31, 2011). www.examiner.com/new-age-in-bangor/maine- Bigfoot-maine-ridge-monster-meddybemps- howler-or-pomoola

56 *Ibid* 12.

57 "Maine IBS Sighting Reports." MABRC Forums. www.mid-americaBigfoot.com/forums/view- topic.php?t=167&t=4070

58 Crowe, Ray. *The Track Record #96*: 12.

59 "Two Men in Car See Hairy, Upright Animal on Side of Road Near Turner, Maine." www.bfro.net/ GDB/show_report.asp?id=859

60 "From the Files of the Gulf Coast Bigfoot Research Organization." www.gcbro.com/MEpenob001.html

[61] The International Bigfoot Society (IBS) originates from Ray Crowe's Western Bigfoot Society and the creation of the *Track Record* newsletter. As a database, the IBS sought to collect sightings and reports worldwide.

[62] "Maine IBS Sighting Reports." www.mid-america-bigfoot.com/forums/viewtopic.php?f=167&t=4070

[63] "The Maine Ridge Monster." www.bigfootencounters.com/stories/maine.htm

[64] *Ibid* 45.

[65] Merchant, Michael. "Michael Merchant Interview: Fisherman Encounter Bigfoot (Down East Maine)." youtube.com/SnowWalkerPrime

[66] Dietz, Lew. *The Allagash*. (Down East Books, 1968): 25.

[67] SnowWalkerPrime. "Chilling Allagash Maine Bigfoot Encounter!" www.youtube.com/watch?v=eTzyF4mD_jc

[68] "More on the Meddybumps Howler Area." *Bigfoot-Encounters.com*

[69] "From the Files of the Gulf Coast Bigfoot Research Organization." www.gcbro.com/MEandroo01.html

[70] "Franklin-Oxford County Lines, Maine." www.bigfootencounters.com/sbs/franklin.htm

[71] "John Green—Sasquatch Investigator, Chronicler and Pioneer Analyst." www.sasquatchdatabase.com/%28S%28oeiuj2nom0eebh45ve43kxj5%29%29/SearchResult.aspx?i_state_pov=Maine

[72] Sullivan, Jennifer. "Big Foot 'Sightings' Turn Out to be Hoax." *Sun-Journal* (August 14, 1993).

[73] "Maine." *North East Sasquatch Researchers Association: Encounter Database*. www.teamnesra.net/drupal/?q=taxonomy/term/32

[74] Maine Department of Inland Fisheries and Wildlife. *Prong Pond*. www.maine.gov/ifw/fishing/lakesurvey_maps/piscataquis/prong_pond.pdf

(1989)

75 "Encounter at Prong Pond Nearby Moosehead Lake." www.bigfootencounters.com/stories/piscataquis.htm

76 "Two Sisters See Dark, Shaggy Haired, Bipedal Creature Standing in Middle of Road." www.bfro.net/GDB/show_report.asp?id=1188

77 "From the Files of the Gulf Coast Bigfoot Research Organization." www.gcbro.com/MEhanco001.html

78 "Classic Sighting on Wooded Maine Road." www.bfro.net/GDB/show_report.asp?id=7421

79 "Two Men See Huge Hairy Animal Near Poland." (June 11, 2001) www.bfro.net/GDB/show_report.asp?id=2678

80 "From the Files of the Gulf Coast Bigfoot Research Organization." www.gcbro.com/MEkenno001.htm

81 "Sagadahoc County." www.bigfootencounters.com/sbs/sagadahoc.htm

82 "Big Footprints found in Aroostook County, Frenchville, Maine." www.bigfootencounters.com/sbs/Aroostook.htm

83 "From the Files of the Gulf Coast Bigfoot Research Organization." www.gcbro.com/MEsomer001.html

84 "Man Hears Unknown Animal Vocalization on Consecutive Nights while Camping." www.bfro.net/GDB/show_report.asp?id=3229

85 "Between New Brunswick, Canada and Topsfield, Maine." www.bigfootencounters.com/sbs/tobe-y02.htm

86 "Piscataquis County, Maine." www.bigfoot-encounters.com/sbs/baxter.htm

87 "At Mooselookmeguntic Lake, Bemis, Maine." www.bigfootencounters.com/stories/bemis.htm

88 "From the Files of the Gulf Coast Bigfoot Research Organization." www.gcbro.com/MEkennebec0003.html

[89] "Oxford County, Maine." www.bigfootencounters.com/sbs/oxford_countyME06.htm

[90] "Jackman, Maine Encounter: Did Greg Hear a Bear of Bigfoot." bigfootevidence.blogspot.com/2013/09/jackman-maine-encounter-did-greg-hear.html

[91] Robert E. Bartholomew & Paul Bartholomew. *Bigfoot Encounters in New York & New England.* (Hancock House Publishers, 2008): 138.

[92] "From the Files of the Gulf Coast Bigfoot Research Organization." www.gcbro.com/MEknox0001.html

[93] Merchant, Michael. "Watch Witnesses From Ellsworth, Maine Describe Sasquatch Sighting." bigfootevidence.blogspot.com/2012/04/ellsworth-maine-sasquatch-sighting.html

[94] "Freedom, Waldo County." www.bigfootencounters.com/sbs/freedomME07.htm

[95] "From the Files of the Gulf Coast Bigfoot Research Organization." www.gcbro.com/MEyork0001.html

[96] "From the Files of the Gulf Coast Bigfoot Research Organization." www.gcbro.com/MEpiscataquis0001.html

[97] "Freedom, Knox County." www.bigfootencounters.com/stories/freedom.htm

[98] "From the Files of the Gulf Coast Bigfoot Research Organization." www.gcbro.com/MEaroostook0001.html

[99] "From the Files of the Gulf Coast Bigfoot Research Organization." www.gcbro.com/MEsomerset0003.html

[100] Coleman, Loren. "The Leeds Loki: 2010's Maine Bigfoot." www.cryptomundo.com/cryptozoo-news/leeds-loki

[101] Skelton, Kathryn. "Coleman Dubs New Sighting 'Leeds Loki'." *Lewiston Sun Journal* (March 09, 2010).

[102] *Ibid* 68.

CHAPTER 9

[1] Skelton, Kathryn. "Looking for Bigfoot in Maine." *Sun Journal* (December 01, 2007): A1, A10.

[2] *Ibid.*

[3] The Georgia Bigfoot Hoax was one of the biggest Bigfoot stories of 2008 and centered on the claim by several men of a dead Bigfoot body in their possession.

[4] Skelton, Kathryn. "Weird, in Review." *Sun Journal* (December 06, 2008).

[5] Merchant Michael. "SnowWalkerPrime's Philosophy on Bigfoot." bigfootevidence.blogspot.com/2012/02/snowwalkerprimes-philosophy-on-bigfoot.html (February 20, 2012).

[6] *Ibid 5.*

[7] *Ibid 5.*

[8] The Hovey photograph is a color photo purported to show the broad, hairy back of a Bigfoot. Several commentators maintained it was a photo of an actor in the monster suit for the movie, "Clawed."

[9] *Ibid 5.*

[10] *Ibid 5.*

[11] *10 Million Dollar Bigfoot Bounty* premiered on Spike on January 10, 2014.

[12] Woolheater, Craig. "Spike TV's 10 Million Dollar Bigfoot Bounty Premieres January 10, 2014." www.cryptomundo.com/bigfoot-report/participant-talks-10-million-dollar-bigfoot-bounty/

[13] "Chattahoochee Bigfoot Radio—Int. with Mike Merchant." www.blogtalkradio.com/chattahoocheebigfootradio/2013/10/06/chattahoochee-bigfoot-radio—int-with-mike-merchant

[14] Coleman, Loren. "Chain of Ponds, Maine, 'Bigfoot' Casts." www.cryptomundo.com/cryptozoo-news/aug-me-tracks (August 21, 2012).

[15] *Ibid.*

[16] Bravo, Damian. "Breaking News! Possible Bigfoot Tracks Found in Same Area of the 1973 Durham Ape Sightings in Maine." bigfootevidence. blogspot.com/2012/12/breaking-news-possible-bigfoot-tracks.html

[17] Skelton, Kathryn. "Durham Man's Monster-Hunting TV Show Debuts Thursday." *Sun Journal* www.sunjournal.com/news/lewiston-auburn-maine/2014/09/04/weird-wicked-weird/1573663# (September 04, 2014).

[18] Dysart's, a favorite area restaurant, is located at Exit 180 off I-95 in Bangor, Maine and is frequented by truck drivers and area families alike.

[19] Carkhuff, David. "Maine Ghost Hunters Plan Bigfoot Quest." *Portland Daily Sun* (July 30, 2012).

[20] Ibid. 4.

[21] Soucy, D. L. "The Hunt for Bigfoot." www.youtube.com/watch?v=9MQgxr_oAKY

[22] Soucy, D. L. "March Bigfoot Hunt in Maine." www.youtube.com/watch?v=75j69ILjVZk

CHAPTER 10

[1] "Maine Best Attractions 2010." www.yankee-magazine.com/article/travel/best-me-2010/attractions-me-2.

[2] "Bigfoot Still Lurks in Crookston." *Grand Forks Herald* (January 11, 1999).

[3] Frazier, Joseph. "'Natural Man' Was Partly a Fraud; But He still Inspired Many with his Frontier Spirit." NiagaraFallsReview.ca (February 22, 2008).

[4] Motavalli, Jim. "Naked in the Woods Contest!" *Commentary: The E Magazine.* www.emagazine.com/daily-news-archive/commentary-the-e-magazine-naked-in-the-woods-contest

[5] *Ibid.*

[6] "Bigfoot of Maine." State of Maine, Department of Economic & Community Development. www. mainemade.com/members/profile.asp?ID=1302

[7] Easton, Thomas A. *Bigfoot Stalks the Coast of Maine.* (Wildside Press, 2000): 97.

[8] "7 Wonders of God Creatures." www.roadside-america.com/story/22889

[9] Lynds, Jen. "One Year Later, Houlton Still Owes for Cleanup Effort of Artist's Property." *Bangor Daily News* (December 03, 2011).

[10] Condit, Jon. "No Stoppin the 'Squatch." www. dreadcentral.com/node/02192 (June 01, 2007).

[11] "Motion Media Entertainment." www.facebook. com/pages/Motion-Media-Entertainment/ 129994827086325?sk=info

[12] Coleman, Loren. "Pucabob Reviews Monster of the Woods." www.cryptomundo.com/cryptozoo-news/motw-review/

[13] "Finding Bigfoot Coming to Maine." maine. craigslist.org/rnr/4187198740.html (November 12, 2013).

[14] Burnham, Emily. "Animal Planet's 'Finding Bigfoot' not coming to Maine, but welcomes leads on Sasquatch sightings." *Bangor Daily News* cultureshock.bangordailynews.com/2013/11/18/ small-screen/animal-planets-finding-bigfoot-not-coming-to-maine-but-welcomes-leads-on-sasquatch-sightings/ (November 18, 2013).

CHAPTER 11

[1] In his book, *Daniel Boone: The Life and Legend of an American Pioneer* (Holt, 1992), John Mack Faragher writes of Boone's choice of reading

materials when out in the wilderness: "He frequently carried along a copy of the Bible, or a book of history, which he loved, or Gulliver's Travels, his favorite book, to read by the light of the Campfire."

2 Trotti, Hugh H. "Did Fiction Give Birth to Bigfoot?" *Skeptical Inquirer* (1994).

3 Roberts, Leonard. "Curious Legend of the Kentucky Mountains," *Western Folklore Volume 16* (January, 1957): 48-51. "Four or five versions of this curious and strange legend come into my collection over a period of about six years (1948 to 1954) from an isolated region of the Kentucky Mountains. At first I did not know what to make of it but, having also collected a few versions of the 'Bear's Son' story minus the half-bear, half-man introduction, I guessed that this was that introduction now broken away and told separately.

"It now appears to be a distinct legend since Dr. Archer Taylor refers me to the long search for American versions by Mr. Rudolph Altrocchi and now that I reflect on this item I realize that it is not unique to Kentucky mountain folklore.

"During my youth in these mountains it was not unusual to hear a rumor of some half-wild man, naked and hairy, being found in the woods, living close to animal state.

"This kind of Romulus-Remus legend seems to stick in the minds of the folk. But how this particular legend made its way into eastern Kentucky is a mystery to me.

"The following version was taken down in pencil in 1950 from the lips of lee Maggard, who lived in a small cabin on the south slope of the Pine Mountain range near the small lumber town of Putney, Harlan County, Kentucky. He had heard it on Maggard's Branch, Leslie County, KY.

"The Yeahoh

"'Once there was man out huntin', he got lost and after a while he begin to get hungry. He came to a big hole in the ground and he thought he would venture down into it. He went down in there and he found that the old Yeahoh lived in there. There was deer meat hangin' up and other food piled around the walls. The man was afraid at first, but Yeahoh didn't bother him and he went toward that meat to get him some. The Yeahoh walked over and looked at the knife and said, "Yeahoh, Yeahoh," a time or two. He cut it off a piece of the meat and he started eatin' it.

"'Well the man stepped over to the middle of the pit and took out his flint and built him up a fire. And the Yeahoh watched him and looked at the fire and at the flint and said, "Yeahoh, Yeahoh" again. The man put his meat on a stick and br'iled him a piece and started eatin' it. The Yeahoh watched him and acted like it wanted a piece. The man cut it off a piece of the br'iled meat and reahed it over, and the Yeahoh commenced to eatin' it up and smackin' its lips saying, "Yeahoh, Yeahoh."

"'Well the man lived there with it a long time and they got along all right. After so long they was a young'un born to 'em, and it was half-man and half-Yeahoh. And the Yeahoh took such a liking to the man it wouldn't let him leave. He got to wanting to get away and go back home. One day he slipped off and the Yeahoh follered him and made him go back. Went on that way for a good while, but he picked him a good time and slipped away. This time he got to the shore where they was a ship ready to sail.

"'He got oon this ship and he looked and saw the Yeahoh comin' with the young'un. It screamed and hollered for him to come back and when it saw he wasn't goinn' to come, why, it tore the baby in two and held it out one-half to him and said, "Yeahoh, Yeahoh."He sailed off and left it standing there.'

"The version that Dr. Taylor refers to in my book *South from Hell fer Sartin* is called 'The Origin of Man.' Another version was given to me by this teller's grandson. It has the same title and contents, except that the Yeahoh has six children and tears them all in two and throws them after the embarked man. Another text, similar to the one given above, was accidentally erased from my tapes.

"The following text was recorded from Joe Couch, Appalaicia, Virginia, in 1954. He had heard it from his people while he lived in Perry County, Kentucky.

"The Hairy Woman

"'One time I's prowling in the wilderness, wandering about, kindly got lost and so weak and hungry I couldn't go. When it began to get cool, I found a big cave and crawled back in there to get warm. Crawled back in and come upon a leaf bed and I dozed off to sleep. I heard an awful racket coming into that cave, and something ome in and crawled right over me and laid down like a big old bear. It was a hairy thing and when it laid down it went chomp, chomp, chewing on something. I thought to myself, "I'll see what it is and find out what it is eating."

"'I reached over and a hairy like woman was there eating chestnuts, had about a half bushel there. I got me a big handful of the, and went to chewing on them too. Well, in a few minutes she handed me over another big handful, and I eat chestnuts until I was kindly full and wasn't hungry any more. D'rectly she got up and took off and out of sight.

"'Well, I stayed on there till next morning and she come in with a young deer. Brought it in and with her big long fingernails she ripped its hide and skinned it, and then she sliced the good lean meat and handed me a bite to eat. I kindly slipped it behind me, afraid to eat it raw and afraid not to eat it being she gave it to me. She'd cut off big pieces of deer meat and eat it raw. Well, I laid back and the other pieces she give over as she eat her'n. She was going' to see I didn't starve.

"'When she got gone again I built up a little fire and br'iled my meat. After being hungry for two or three days, it was good cooked—yes, buddy. She come in while I had my fire built br'iling my meat, and she run right into that fire. She couldn't understand because it kindly burnt her a little. She jumped back and looked at me like she was going to run through me. I said, "Uh-oh, I'm going to get in trouble now."

"'Well, it was cold and bad out, so I just stayed another night with her. She was a woman but was right hairy all over. After several days I learnt her how to br'ile meat and that fire would burn her. She got shy of the fire and got so she like br'iled meat and wouldn't eat it raw any

more. We went on through the winter that way.
She would go out and carry in a deer and bear.
So I lived there about two year, and when we had
a little kid, one side of it was hairy and the other
side was slick.

"' I took a notion I would leave there and go
back home. I begin to build me a boat to go away
across the lake in. One time after I had left, I took
a notion I would slip back and see what she was
doing. I went out to the edge of the cliff and
looked down into the mountain, and it looked
like two or three dozen of hairy people coming
up the hill. They were all pressing her and she
would push them back. They wanted to come on
up and come in. I was scared to death, afraid
they's going to kill me. She made them go back
and wouldn't let them come up and interfere.

"'Well, I took a notion to leave one day when
my boat was ready. I told her one day I was going to leave. She follered me down to my boat
and watched me get ready to go away. She was
crying, wanting me to stay. I said, "No, I'm tired
of the jungles. I'm going back to civilization
again, going back."

"'When she knowed she wasn't going to keep
me there, she just grabbed the little young'un and
tore it right open with her nails. Throwed me the
hairy part and she kept the slick side. That's the
end of that story.'" [Retrieved from: http://
www.bfro.net/legends/algonkian.htm; accessed
April 20, 2014].

4 Zueffle, David Matthew. "Swift, Boone, and Big-
    foot: New Evidence for a Literary Connection,"
    Skeptical Inquirer (January/February 1997).

5 Wigginton, Eliot. *Foxfire 4*. (New York: Random
    House, 1975): 483.

[6] Dorson, Richard Mercer. *Folktales Told Around the World*. (University of Chicago Press, 1975).

[7] Green, John. *Sasquatch: The Apes Among Us*. (Hancock House, 1978): 228.

[8] *Ibid.*

[9] Emblidge, David. *The Appalachian Trail Reader*. (Oxford University Press, 1996): 322.

[10] Stephens, Charles Asbury. *A Busy Year at the Old Squire's*. (C. H. Simonds Company, 1922).

[11] Thoreau, Henry David. *The Maine Woods*. (Houghton, Mifflin and Company, 1884).

[12] "The Indian Devil." *The Hartford Courant* (January 28, 1890).

[13] Hornibrook, Isabel. *Camp and Trail: A Story of the Maine Woods*. (Lothrop Publishing, 1897).

[14] *Ibid.*

[15] "Injun Devil." *Boston Daily Globe* (March 12, 1898).

[16] "Real 'Indian Devil'." *Lewiston Saturday Journal* (July 20, 1900).

[17] "An 'Injun Devil' Yarn." *Daily Kennebec Journal* (July 23, 1907).

[18] "'Injun Devil' Was a Big Maltese Cat." *Lewiston Evening Journal* (January 10, 1912).

[19] *Ibid.*

[20] *Ibid.*

[21] Seymour, Tom. *Tom Seymour's Maine: A Maine Anthology*. (iUniverse, 2003).

[22] Thompson, Lucy. *To the American Indian: Reminiscences of a Yurok Woman*. (Eureka, 1916).

[23] *Ibid.*

[24] *Ibid.*

[25] *Ibid.*

[26] *Ibid.*

[27] "Ghostly Figure haunts Indians of North Wilds." *Berkeley Daily Gazette* (November 4, 1935).

28 Coleman, Loren. "Injun Devils." www.crypto-mundo.com/cryptozoo-news/injun-devils/

29 Nason, R. B. "Jack's Injun Devil." *Outing Magazine 53* (March, 1909): 760-761.

30 "Noted Author Dies at his Norway Home." *Lewiston Daily Sun* (September 23, 1931): 5.

31 Arment, Chad. *The Historical Bigfoot.* (Coachwhip Publications, 2006).

32 Green, John. *Sasquatch—The Apes Among Us.* (Hancock House, 1978): 228.

33 Stephens, Charles Asbury. "Was it an Indian Devil?" *Ballou's Monthly Magazine 43* (April, 1876).

34 Stephens, Charles Asbury. *The Young Moose Hunters.* (Henry L. Shepard & Co., 1883): 22.

35 *Ibid.* pp. 185-186.

36 *Ibid.* pp. 208-210.

37 *Ibid.* pp. 264-271.

38 Stephens, Charles Asbury. *Camping Out.* (Henry T. Coates & Co., 1873): 127-138.

39 Stephens, Charles Asbury. *The Adventures of Six Young Men in the Wilds of Maine and Canada.* (Dean & Son, 1884): 200-209.

40 Rumsey, Barbara. "The Baby That Washed Ashore at Hendricks Head." *Out of Our Past.* www.ben-russell.com/HH-historical%20society%20part%20II.htm

41 Munn, Charles Clark. *The Hermit.* (Lee and Shepard, 1903): 18-24.

42 Day, Holman. *The Eagle Badge: or the Skokums of the Allagash.* (Harper & Brothers, 1908): 96.

43 "Holman F. Day Dies in California." *Lewiston Evening Journal* (February 21, 1935): 2.

44 Day, Holman. "Old King Spruce." *New England Magazine 37* (October, 1907).

[45] "Hobgoblins of the Wild North Woods." *Boston Evening Transcript* (December 31, 1902).

CHAPTER 12

[1] Eisenthal, Bram. *Milford Daily News* (posted 30 October, 2004; last updated 13 November, 2007).

[2] Bartlett, Kay. "Elusive 'Bigfoot' Is America's Loch Ness Monster." *Youngstown Vindicator* (January 28, 1979): 20.

[3] Fairless, Robert and Randy Ashby. "Dr. Jeff Meldrum." www.25thradioshow.com/radio-shows/2012-radio-show-interviews/december-2012-interviews/ (December 12, 2012)

[4] Dunning, Brian. "Killing Bigfoot with Bad Science." *Skeptoid* Podcast. www.skeptoid.com/episodes/4011 (December 3, 2006).

[5] Daegling, David. *Bigfoot Exposed: An Anthropologist Examines America's Enduring Legend.* (AltaMira Press, 2004): 4, 6.

EPILOGUE

[1] Barta, Roger *The Artificial Savage* (1997): 287.

COACHWHIP PUBLICATIONS

COACHWHIPBOOKS.COM

The
Historical
Bigfoot

RANCHERS BEGIN 'WILD MAN' HUNT ALONG ROCKIES

Prospectors Say Wild Man Rules Arctic Kingdom

TERRIBLE SASQUATCH ABROAD IN THE LAND

Chad Arment

Cavemen Roam the Rockies?

VARMINTS
CHAD ARMENT

MYSTERY CARNIVORES
OF NORTH AMERICA

COACHWHIP PUBLICATIONS

COACHWHIPBOOKS.COM

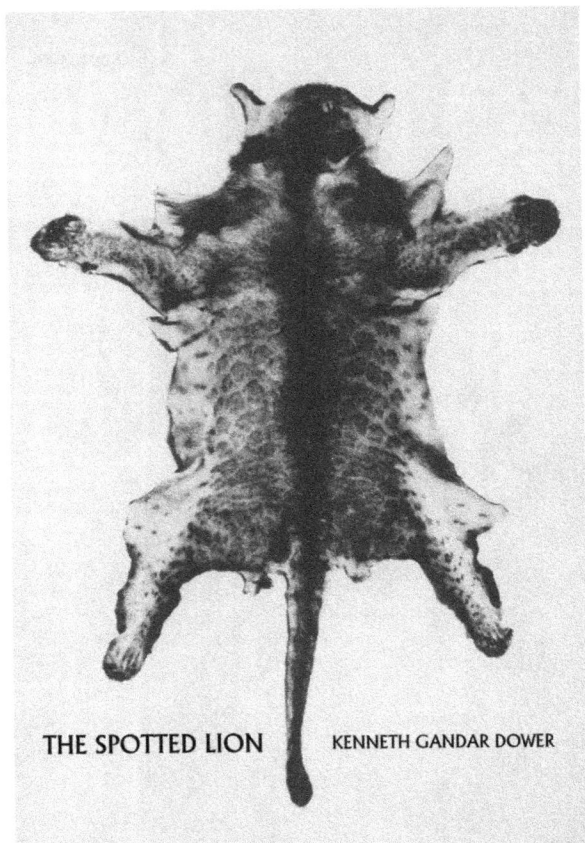

THE SPOTTED LION   KENNETH GANDAR DOWER

COACHWHIP PUBLICATIONS

ALSO AVAILABLE

BIOFORTEAN NOTES

VOLUME 4

BIPES RUMORS IN THE WESTERN UNITED STATES
Chad Arment

CRYPTOFICTION: A RENAISSANCE
Matt Bille

MINIATURE HORSES OF THE GRAND CANYON: POSTCARDS . . .
Chad Arment

POLISH EXPLORATIONS ACROSS ASIAN BORDERS
Tomasz Pietrzak

NOTES ON THE INTRODUCTION OF THE FROG TO IRELAND
Richard Muirhead

SEEKING NEW TURTLES IN NORTH AMERICA
Chad Arment

EXORCISING THE PHANTOM KANGAROO
Chad Arment

BIOFORTEAN MISCELLANY
Chad Arment

2015

COACHWHIP PUBLICATIONS

COACHWHIPBOOKS.COM

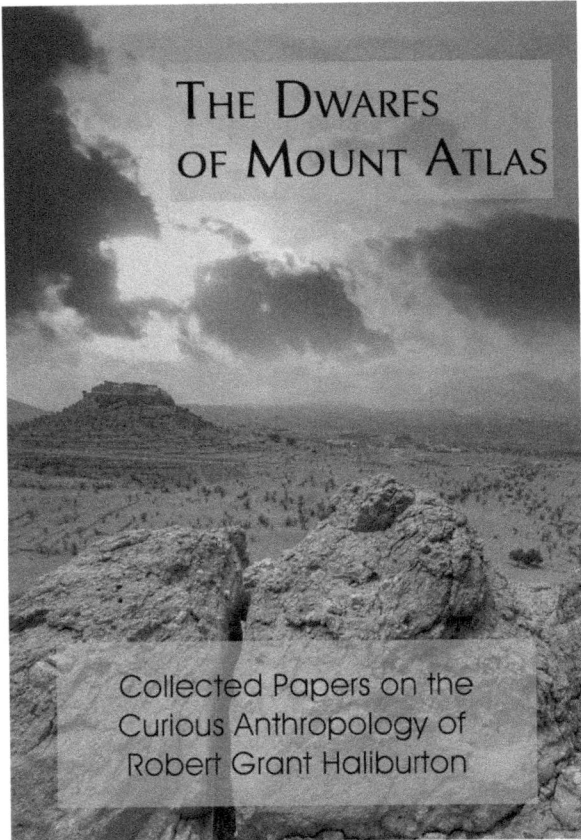

THE DWARFS
OF MOUNT ATLAS

Collected Papers on the
Curious Anthropology of
Robert Grant Haliburton

COACHWHIP PUBLICATIONS

ALSO AVAILABLE

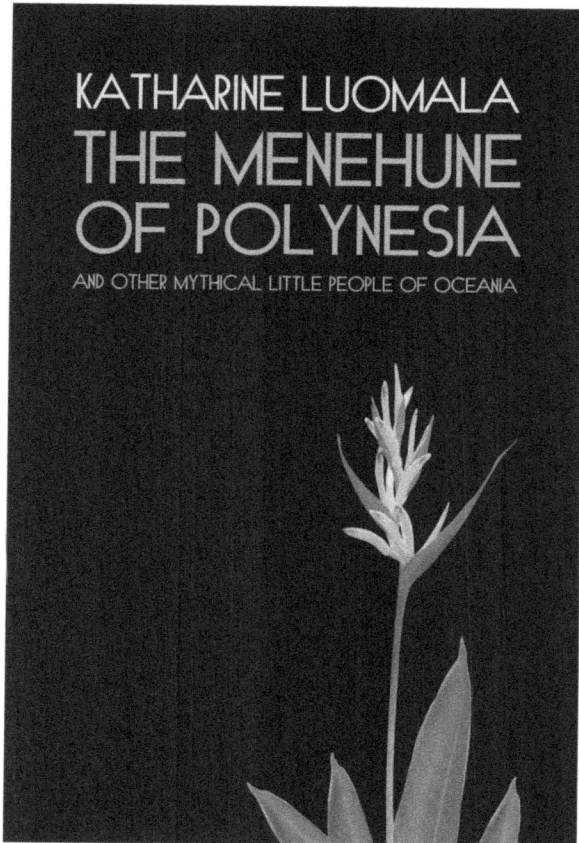

KATHARINE LUOMALA
THE MENEHUNE
OF POLYNESIA
AND OTHER MYTHICAL LITTLE PEOPLE OF OCEANIA